Light and Optics

Principles and Practices

Light and Optics
Principles and Practices

Abdul Al-Azzawi

CRC Press
Taylor & Francis Group
Boca Raton London New York

CRC Press is an imprint of the
Taylor & Francis Group, an **informa** business

CRC Press
Taylor & Francis Group
6000 Broken Sound Parkway NW, Suite 300
Boca Raton, FL 33487-2742

First issued in paperback 2019

© 2007 by Taylor & Francis Group, LLC
CRC Press is an imprint of Taylor & Francis Group, an Informa business

No claim to original U.S. Government works

ISBN-13: 978-0-8493-8313-7 (hbk)
ISBN-13: 978-0-367-38956-7 (pbk)

Visit the Taylor & Francis Web site at
http://www.taylorandfrancis.com

and the CRC Press Web site at
http://www.crcpress.com

Preface

We live in a world bathed in light. Light is one of the most familiar and essential things in our lives. For many thousands of years, the Sun was our only source of light. Eventually, the ability to create fire, and its by-product, light, led to a profound change in the way humans managed their time. Today, there are many options for creating light. Our understanding of light has spawned many applications of light. Light can be used in fiber communications; early applications included ship-to-ship communications using Morse code. Infrared remote controls for televisions demonstrated free-space optical communications, using many of the same principles. Optical fiber has revolutionized the way we interact with the world. Light is used to treat those with seasonal disorders. Lasers are now used in medical applications, such as re-shaping our corneas, cauterizing blood vessels, and removing tattoos. Lasers are also used in industrial applications, such as cutting metal, welding, and sensing. New imaging technology permits the creation of flat-panel displays, night vision devices, and autonomous product inspection systems. With so many applications of light, the need for Photonics technology and innovation will most certainly grow in the future, as new applications emerge to light.

A unique approach is taken in this book to present light and optics and their applications. This book covers the basic theoretical principles and industrial applications of light and optics, suitable for students, professionals, and professors. Each chapter is presented in two parts: theoretical and practical. The theoretical part has adequate material to cover the whole aspect of the subject. In the experimental part, students will apply the learned theoretical concepts in simple and advanced experimental works. In this way, students will learn and gain practical hands-on experience in the light and optics subjects. This will assist the students to apply theoretical knowledge to real-world applications. The step-by-step approach and technical illustrations in this book will guide students through each experiment. The experimental work has more than one case in most of the chapters, and sometimes have sub-cases.

This book is written with simple language, and gives adequate information and instruction to enable students to achieve maximum comprehension. An effort has been made to use the international system of units (SI) throughout the book. The organization of the chapters is designed to provide a solid foundation for today's light and optics students, and to upgrade their knowledge. Universal tools, devices, and equipment, which are used throughout the experiments, are available in any Photonics, Physics, and material labs. This book abounds in theoretical and practical aids, and is an effective teaching tool, helpful to both professors and students. Simple and advanced subjects are presented by an expert author, and some new subjects appear for the first time in this book.

Care has been taken to label parts clearly, and to use colours in diagrams wherever it will aid understanding. Some figures are drawn in three-dimensions, where applicable, for easy understanding of the concepts. Colour pictures are used to clearly show parts in a device, system, and experimental set-up.

The book is structured in seventeen chapters. The book includes the following chapters:
- Chapters 1 through 7 cover light, light and shadow, thermal radiation, light production, light intensity, light and color, and the laws of light.
- Chapters 8 through 16 cover plane mirrors, spherical mirrors, lenses, prisms, beamsplitters, light passing through optical components, optical instruments for viewing applications, polarization of light, and optical materials.
- Chapter 17 covers laboratory safety.

The book includes 375 figures, 46 tables, and 55 experimental cases. The book was developed with generous input from members of the photonics industry, research scientists, and members from academia.

Abdul Al-Azzawi
Algonquin College
Canada

Acknowledgments

This book would not have been possible without the enthusiasm and teamwork of my colleagues and family support. In particular, the author would like to thank Mietek Slocinski for his support, time and energy in working long hours to set-up the labs, taking pictures, and fruitful discussion during the years to complete the book.

The author would like to thank Steve Finnegan, Kathy Deugo and Nicole McGahey for their support and solving the difficulties.

The author would like to thank his daughter Abeer and son Abaida for their help in reviewing chapters and making drawings and figures.

The author also extends his thanks to Eng. Monica Havelock for her contribution in working long hours in reviewing and editing the materials, and support.

The author wishes to express his gratitude to colleagues Prof. Devon Galway and Prof. Rao Kollipara for their comments and feedback in reviewing some materials in this book.

The author wishes to thank Gergely Horvath for hard work in reviewing and proofreading most chapters in this book. The author likes to thank Madeleine Camm, Andrew Lynch and Nicolas Lea for reviewing a few chapters in this book.

The author would like to extend his thanks to Peter Casey for his reviewing and proposing the materials in this book.

The author would like to extend his thanks and appreciation to Dr Charley Bamber for writing and reviewing Chapter 16: Optical Materials.

Author

Abdul Al-Azzawi, PhD, graduated from the University of Strathclyde in Glasgow, Scotland, UK. He has worked in the photonics manufacturing industry, research (NRC/Canmet), and teaching at Algonquin College, Ontario, Canada. While employed at NRC, he participated in studying energy saving in a residential building and developing the green building assessment programme. As a photonics engineer, he designed new production lines, modified products, developed manufacturing process, and designed new jigs.

At Algonquin College, he has taught mechanical and photonics courses in the mechanical and photonics engineering programmes. He was a member of the founding team of the Photonics Engineering Programmes. He has published three books and many papers, and he has participated in many workshops and conferences around the world. He is the author of the book, *Fibre Optics—Principles and Practices*.

He is the coordinator of the photonics engineering programme at Algonquin College, Ottawa, Ontario, Canada. His special area of interest is optic and optical fibre devices, fibre optic lighting, and fibre optic sensors. He is a member of the professional photonics societies in Canada. He is the recipient of the NISOD Excellence award from the University of Texas at Austin in 2005.

Table of Contents

Chapter 7

Chapter 9

Chapter 10

Chapter 11

Chapter 12

Chapter 13

Chapter 15

Chapter 17

1 The Nature of Light

1.1 INTRODUCTION

Throughout history, humankind has been fascinated with the properties and behaviour of light. Light from the sun served as a catalyst in the formation of life on Earth. Solar and lunar light provided humanity with the celestial timepieces required to measure time.

We live in a world bathed in light. Light is one of the most familiar things in our lives. We see things with eyes that sense the intensity (brightness) and wavelength (colour) of light. We experience light in a variety of other ways as well. For example, we sense radiant heat when our skin is near a warm object. This is due to our skin's reaction to infrared radiation.

We learn almost all of what we know about the world around us from the interaction of materials with electromagnetic waves. Often, the word light is used a little more broadly to include electromagnetic waves, such as in the ultraviolet and infrared waves that are just outside the visible range.

Much of what we know about light has been discovered during the past five centuries. Initially, light was understood to be a particle. Light is now widely understood to be one part of a much larger electromagnetic spectrum. Photons, the smallest resolvable quanta of light, were initially described using particle theory; however, a wave model has since been widely adopted. In the context of the wave model, photons are energy packets moving through space and time.

1.2 THE EVOLUTION OF LIGHT THEORY

In the 17th century, light was considered to be stream of particles that were emitted by a light source. These particles stimulate the sense of sight when entering the eye. The English physicist and mathematician Isaac Newton (1642–1727) was the inventor of the particle theory of light. Newton regarded rays of light as streams of very small particles emitted from a source of light and travelling in straight lines. Newton was able to provide a simple explanation for some known experimental facts concerning the nature of light, specifically, the laws of reflection and refraction.

Most scientists accepted Newton's particle theory of light. However, during Newton's lifetime, another theory was proposed by the Dutch physicist and astronomer Christian Huygens (1629–1695). In 1678, Huygens presented his theory, in which light might be some sort of wave motion. His experiment demonstrated that when the two beams of light intersected, they emerged unchanged, just as in the case of two water or sound waves. Huygens was able to adopt a wave theory of light to derive the laws of reflection and refraction, and to explain double refraction in calcite. This wave theory did not receive immediate acceptance from the scientific community for several reasons. The only waves known at that time were sound and water. It was known that these waves travelled through some sort of medium. On the other hand, light could travel to us from the sun through the vacuum of space. It was agreed that if light was some form of wave motion, the waves would be able to bend around obstacles and corners. This bending is easily observed with both water and sound waves. In this case, it would be easy to see the light around corners. It is now known that light does actually bend around the edges of objects; this phenomenon is known as the diffraction of light, which will be discussed later in this book.

In 1660, experimental evidence for the diffraction of light was discovered by Francesco Grimaldi (1618–1663). Most scientists still rejected the wave theory and continued to adhere to Newton's particle theory for more than a century.

In 1801, the first clear demonstration of the wave theory of light was provided by the English physician, Thomas Young (1773–1829). He performed a significant experiment, which showed that light exhibits interference behaviour. There are two types of interference: constructive and destructive. When two waves are moving in the same direction, the vertical displacement (amplitude) of the combined waveform is greater than that of either wave; this situation is referred to as constructive interference. Conversely, if one wave has a negative displacement, the two waves work to cancel each other when they overlap, and the amplitude of the combined waveform is smaller than that of either wave. This is referred to as destructive interference. Light interference behaviour will be explained later in this book. Most scientists accepted the wave theory of light and more theoretical and experimental work was conducted to further explore it.

In 1821, the French physicist Augustin Fresnel (1788–1827) published the results and analysis of a number of detailed experiments, which dealt with interference of polarized light and diffraction phenomena. He obtained circularly polarized light by means of a special glass prism now known as a Fresnel rhomb. For each of the two components of the polarized light, Fresnel developed the Fresnel Equations, which give the amplitude of light reflected and transmitted at a plane interface separating two optical media.

In 1850, Jean Foucault (1791–1868) provided further evidence of the inadequacy of the particle theory by showing that the speed of light in liquids is less than that in air. According to the particle model of light, the speed of light would be higher in a glass and liquid than in air. Further experimental and theoretical developments during the 19th century led to the general acceptance of the wave theory of light.

In 1873, the most important development concerning the theory of light was the work of a Scottish physicist, James C. Maxwell (1831–1879). Maxwell asserted that light was a form of high-frequency electromagnetic wave. Working in the field of electricity and magnetism, Maxwell created known principles in his set of four Maxwell Equations. These equations predict the speed of an electromagnetic wave in the ether; this turned out to be the true measured speed of light. His theory predicted that these waves should have a speed of about 3×10^8 m/s. Within experimental error, his predicted value is nearly equal to the speed of light measured by sophisticated instruments today. From then on, light was viewed as a particular region of the electromagnetic spectrum of radiation.

In 1887, Heinrich Hertz (1857–1894), a German physicist and pioneering investigator of electromagnetic waves, provided experimental confirmation of Maxwell's theory by producing and detecting electromagnetic waves. Hertz also defined the frequency. Furthermore, Hertz and other scientists and investigators showed that these waves exhibited reflection, refraction, and all the other characteristic properties of waves.

Although the classical theory of electricity and magnetism was able to explain most known properties of light, some subsequent experiments could not be explained by assuming that light is a wave. The most striking discovery of the experiments is the photoelectric effect, which was discovered by Hertz. The photoelectric effect is the ejection of electrons from a metal when its surface is exposed to light. As one example of the difficulties that arose, experiments showed that the kinetic energy of an ejected electron is independent of the light intensity. This was in contradiction to the wave theory, which held that a more intense beam of light should add more energy to the electron. In 1905, an experiment demonstrating of this phenomenon was proposed by Albert Einstein (1879–1955), a German-Swiss physicist. In 1900, Einstein's theory used the concept of the quantum theory developed by Max Planck (1858–1947), a German theoretical physicist. The quantization model assumes that the energy of a light wave is present in bundles of energy called photons. Therefore, the energy is said to be quantized. According to Einstein's theory, the energy of a photon is proportional to the frequency of the electromagnetic wave. The energy of a photon can be defined by Equation (1.1):

$$E = n(hf) \quad n = 0,1,2,3,\ldots \tag{1.1}$$

where

n is a positive integer number
f is the frequency of light (Hz)
h is a constant known as Planck's constant and has a value of 6.6261×10^{-34} (J s).

This idea suggests that light can behave as discrete quanta or "particles" of energy, rather than as waves. Equation (1.1) is thus the mathematical "connection" between the wave nature of light (a wave of frequency f) and the particle nature (photons each with an energy E). Given light of a certain frequency (or wavelength), this equation can be used to calculate the amount of energy in each photon, or vice versa. It is important to note that this theory retains some features of both the wave theory and the particle theory of light.

Einstein used the quantum concept to explain the photoelectric effect, for which the classical physical description was inadequate. Certain metallic materials are photosensitive. That is, when light strikes their surface, electrons can be emitted. The radiant energy supplies the work necessary to free the electrons from the material's surface. The electron interacts with one photon of light as if the electron had been struck by a particle. Yet the photon has wave-like characteristics as implied by the fact that light exhibits interference phenomena of light.

In 1913, the Danish physicist Niels Bohr (1885–1962) incorporated the quantum behaviour of radiation in his explanation of the emission and absorption processes of the hydrogen atom; this provided a physical basis for understanding the hydrogen spectrum. In 1922, the photon model of light came again to the rescue for the American scientist Arthur Compton (1892–1962), who explained the scattering of x-rays from electrons as particle collisions between photons and electrons in which both energy and momentum were conserved.

All such victories for the photon or particle model of light indicated that light could be treated as a particular kind of matter, possessing both energy and momentum. It was the French scientist Luis de Broglie (1892–1987) who saw the other side of the picture. In 1924, Broglie published his theory that subatomic particles exhibit wave properties. He suggested that a particle with momentum M had an associated wavelength λ of

$$M = \frac{h}{\lambda} \tag{1.2}$$

where h is Planck's constant.

Broglie suggested that because electromagnetic waves show particle characteristics, particles should, in some cases, also exhibit wave properties. This prediction was verified experimentally within a few years by the American physicists Clinton Davisson (1881–1958) and Lester Germer (1896–1971) and the British physicist George Thomson (1892–1975). They showed that a beam of electrons scattered by a crystal produces a diffraction pattern characteristic of a wave. The wave concept of a particle led the Austrian physicist Erwin Schrodinger (1887–1961) to develop a so-called wave equation to describe the wave properties of a particle and, more specifically, the wave behaviour of the electron in the hydrogen atom.

Thus the wave-particle duality came full circle. Light behaved like waves in its propagation and in the phenomena of interference and diffraction. Light could, however, also behave as particles in its interaction with matter, as in the photoelectric effect. On the other hand, electrons usually behaved like particles, as observed in the point-like scintillations of a phosphor exposed to a beam of electrons. In other situations, electrons were found to behave like waves, as in the diffraction produced by an electron microscope. Photons and electrons that behaved both as particles and as waves seemed at first an impossible contradiction, since particles and waves are very different entities indeed. Gradually it became clear, to a large extent through the reflections of Niels Bohr and especially in his principle of complementarily, that photons and electrons were neither waves nor particles, but something more complex than either.

In attempting to explain physical phenomena, it is natural use physical models, such as waves and particles. However, it turns out that the behaviour of a photon or an electron is not fully explained by either model. In certain situations, wavelike behaviour may predominate; in other situations, particle-like behaviours may stand out. It is noticeable that there is no singular physical model that adequately handles all cases.

Quantum mechanics, or wave mechanics, as it is often called, deals with all particles, which are localized in space, and so describes both light and matter. Combined with relativity equations, momentum M, wavelength λ, and speed v for both material particles and photons are given by the same general equations:

$$M = \frac{\sqrt{E^2 - m^2 c^4}}{c} \tag{1.3}$$

$$\lambda = \frac{h}{M} = \frac{hc}{\sqrt{E^2 - m^2 c^4}} \tag{1.4}$$

$$v = \frac{Mc^2}{E} = c\sqrt{1 - \frac{m^2 c^4}{E^2}} \tag{1.5}$$

where

m is the rest-mass,
E is the total energy,
mc^2 is the sum of rest-mass energy
v is the speed of light in an optical medium, and
kinetic energy provides the work done to accelerate the particle from rest to its measured speed.

The relativistic mass is given by γm, where γ is the ratio $1/\sqrt{1-(v/c)^2}$. The proper expression for kinetic energy is no longer simply $(1/2)mv^2$, but $mv^2 (\gamma - 1)$. The relativistic expression for kinetic energy approaches $(1/2)mv^2$ for $v \ll c$.

A crucial difference between particles such as electrons and neutrons and particles like photons is that the photons have zero rest mass. Equation (1.3) through Equation (1.5) then take the simpler form for photons:

$$M = \frac{E}{c} \tag{1.6}$$

$$\lambda = \frac{h}{M} = \frac{hc}{E} \tag{1.7}$$

$$v = \frac{Mc^2}{E} = c \tag{1.8}$$

Thus, while nonzero rest-mass particles like electrons have a limiting speed of c, Equation (1.8) shows that zero rest-mass particles like photons must travel with the constant speed c in a vacuum. The energy of a photon is not a function of its speed but of its frequency, as expressed in Equation (1.1) or in Equation (1.6) and Equation (1.7) taken together. Notice that since a photon has a zero rest mass, there is no distinction between its total energy and its kinetic energy.

In summary, from the above view of these developments, one must regard light as having a dual nature. That is, in some cases light acts like a wave and in others it acts like a particle.

1.3 MEASUREMENTS OF THE SPEED OF LIGHT

Light travels at a high speed of about 3.00×10^8 m/s. Early experimental attempts to measure its speed were unsuccessful. It was believed that light travelled with infinite speed. Many attempts were made to obtain an exact value of the speed of light. The following shows the experimental attempts and theoretical methods that were tried in the last few centuries.

1.3.1 GALILEO'S ATTEMPTS

The first scientific attempt to measure the speed of light was made by the Italian physicist Galileo Galilei (1564–1642), but without success. He used two shuttered lanterns for his experiment. Galileo stationed himself on one hilltop with one lantern and an assistant on another hilltop with a similar lantern. The distance between the two hilltops was measured. Galileo would first open his lantern for an instant, sending a short flash of light to the assistant. As soon as the assistant saw this light he opened his own lantern, sending a flash back to Galileo, who measured the total time elapsed. After numerous repetitions of this experiment at greater distances between observers, Galileo came to the conclusion that they could not open their lanterns fast enough and that light probably travels with an infinite speed.

1.3.2 ROEMER'S METHOD

In 1675, the first successful estimate of the speed of light was made by the Danish astronomer Ole Roemer (1644–1710). His method involved astronomical observations of one of the moons of Jupiter, called Io. At that time, four of Jupiter's 14 moons had been discovered, and the periods of their orbits were known. Io, the innermost moon, has a period of about 42.5 h. This was measured

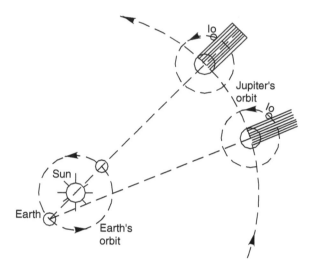

FIGURE 1.1 Schematic diagram of Roemer's method.

by observing the eclipse of Io as it passed behind Jupiter, as shown in Figure 1.1. The period of Jupiter is about 12 years, so as the Earth moves through 180° around the sun, Jupiter revolves through only 15°.

Using the orbital motion of Io as a clock, one would expect a constant period in its orbit over long time intervals. However, Roemer observed a systematic variation in Io's period during each Earth orbit. He found that the periods were larger than average when the Earth receded from Jupiter and smaller than average when approaching Jupiter. However, when Roemer checked to see if the second eclipse did occur at the predicted time, he found that if the Earth was receding from Jupiter, the eclipse was late. In fact, if the interval between observations was three months, the delay was approximately 600 s. Roemer attributed this variation in period to the fact that the distance between the Earth and Jupiter was changing between the observations. In three months (one quarter of the period of the Earth orbit), the light from Jupiter has to travel an additional distance equal to the radius of the Earth's orbit.

Roemer noted that when the Earth is closest to Jupiter, Io's reappearance was 11 min early, and when further from Jupiter, it was 11 min late. Roemer concluded that this total discrepancy of about 22 min is equal to the time required for the light from Jupiter's moon to cross the diameter of the Earth's orbit, as shown in Figure 1.1. Such measurements required the development of a reasonably sophisticated timekeeping device, so that a discrepancy of 22 min could be observed over a 6-month interval. The diameter of the Earth's orbit around the sun (in Roemer's day) was thought to be $2.90 10^{11}$ m. This gives a speed of light c:

$$c = \frac{d}{t} \approx \frac{2.90 \times 10^{11} \text{m}}{22 \text{ min} \times \frac{60 \text{ s}}{\text{min}}} \approx 2.20 \times 10^8 \text{ m/s} \tag{1.9}$$

where

d is the distance (m),
t is time (s).

In 1676, Roemer announced a value for the speed of light of 2.25×10^8 m/s. This experiment is important historically because it demonstrated that the speed of light does have a finite numerical value.

1.3.3 FIZEAU'S METHOD

In 1849, the first successful laboratory measurement of the speed of light was performed by the French scientist Armand Fizeau (1819–1896). The basic elements of his experimental arrangement are shown in Figure 1.2. The basic idea is to measure the total time it takes light to travel from an intense source to a distant mirror and back. To measure the transit time, Fizeau used a rotating toothed wheel. Light passing through one notch travels to a mirror a considerable distance away, is reflected back, and then, if the rotational speed of the toothed wheel is adjusted properly, passed through the next notch in the wheel. By measuring the rotational speed of the wheel and the distance from the wheel to the mirror, Fizeau was able to obtain a value of 3.13×10^8 m/s for the speed of light using Equation (1.10) and Equation (1.11).

$$t = \frac{\theta}{\varpi} \tag{1.10}$$

$$c = 2\frac{d}{t} \tag{1.11}$$

where

θ is the number of revolutions (rev)
ω is the angular speed of the wheel (rev/s)
d is the distance between the light source and the mirror (m)
t is the transit time for one round trip (s).

In 1862, the French scientist Jean Foucault (1819–1868), who worked with Fizeau, greatly improved the accuracy of the experiment by substituting a rotating mirror for the wheel. He was able to obtain a value of 2.977×10^8 m/s for the speed of light in air.

Figure 1.2 demonstrates Fizeau's method, in which the wheel has 450 notches and rotates with a speed of 35 rev/s. A mirror is located at a distance of 9500 m from the notched wheel. Light passing through one notch travels to the mirror and back just in time to pass through the next notch. To calculate the speed of light, the distance travelled $2d$ is divided by the time Δt. To find the time, the wheel rotates from one notch to the next during this time. The time is calculated by the wheel rotation through an angle $\Delta\theta = (1/450)$ rev. Knowing the rotational speed ω of the wheel, it is possible to find the time using the relation $\Delta\theta = \omega\Delta t$. Therefore:

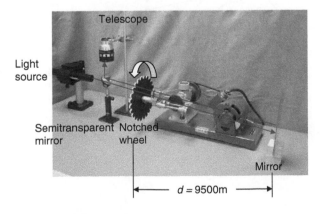

FIGURE 1.2 Fizeau's method to measure the speed of light.

The time required for the wheel to rotate from one notch to the next:

$$\Delta t = \frac{\Delta \theta}{\omega} = \frac{(1/450 \text{ rev})}{35 \text{ rev/s}} = 6.3 \times 10^{-5} \text{ s}$$

Calculate the speed of light by dividing the distance into the time:

$$c = \frac{2d}{\Delta t} = \frac{2(9500 \text{ m})}{6.3 \times 10^{-5} \text{ s}} \approx 3.0 \times 10^8 \text{ m/s}$$

1.3.4 MICHELSON'S MEASUREMENTS

In the years that followed these earliest experiments, several investigators improved upon Fizeau's apparatus and methods of observation, and obtained more accurate values for the speed of light. The German American physicist Albert A. Michelson (1852–1931) was celebrated for the invention and development of the interferometer, an optical instrument, now named the Michelson stellar interferometer in his honour. In 1877, Michelson used his own improvement of Foucault's rotating mirror method, replacing the toothed wheel by a small eight-sided mirror, as shown in Figure 1.3. If the angular speed of the rotating eight-sided mirror is adjusted correctly, light reflected from one side travels to the fixed mirror, reflects, and can be detected after reflecting from another side that has rotated into place at just the right time. The minimum angular speed must be such that one side of the mirror rotates one-eighth of a revolution during the time it takes for the light to make the round trip between the mirrors. For one of his experiments, Michelson placed mirrors on Mt. San Antonio and Mt. Wilson in California, a distance of 35 km apart. From the value of the minimum angular speed in such experiments, he obtained the value of the speed of light of $c = (2.997 \ 96 \pm 0.00004) \times 10^8$ m/s in 1926.

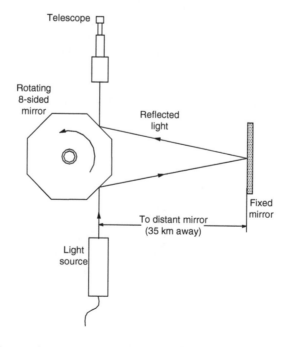

FIGURE 1.3 Michelson's experiment to measure the speed of light.

1.3.5 Maxwell's Electromagnetic Waves Method

In 1865, Maxwell's theoretical description of electromagnetic waves allowed him to obtain a simple expression for c in terms known as physical quantities. Figure 1.4 shows the propagation of an electromagnetic wave in free space. The wave speed should be a function of the properties of the optical medium through which it propagates. The propagation velocity of an electromagnetic wave is given in Equation (1.12). Maxwell determined theoretically that electromagnetic waves propagate through a vacuum at a speed given by:

$$c = \frac{1}{\sqrt{\varepsilon_0 \mu_0}} \tag{1.12}$$

where the (electric) permittivity of free space is $\varepsilon_0 = 8.85 \times 10^{-12}$ (C^2/N m^2) and the (magnetic) permeability of free space is $\mu_0 = 4\pi \times 10^{-7}$ (T m/A).

$$c = \frac{1}{\sqrt{(8.85 \times 10^{-12} \ C^2/Nm^2) \times (4\pi \times 10^{-7} \ Tm/A)}} = 3.00 \times 10^8 \text{ m/s}$$

The experimental and theoretical values for c agree. Maxwell's success in predicting c provided a basis for inferring that light behaves as a wave consisting of oscillating electric and magnetic fields.

Today the speed of light has been determined with such high accuracy that it is used to define the metre. Now the speed of light in air is defined as:

$$c = 299, \ 796, \ 458 \text{ m/s}$$

Although a value of the speed of light in air of $c = 3.00 \times 10^8$ m/s is adequate for most calculations, the speed of light in a vacuum is $c = 299, \ 792, \ 458$ m/s. The speed of light has a different value for each medium, such as water, oil, or glass.

Each material has a refractive index. The refractive index or the index of refraction (n) is defined as the ratio between the speed of light in vacuum and the speed of light in a medium (v). For example, the index of refraction for air, water, and crown glass are 1.0003, 1.33, and 1.52, respectively. The refractive index of materials is discussed in detail in this book.

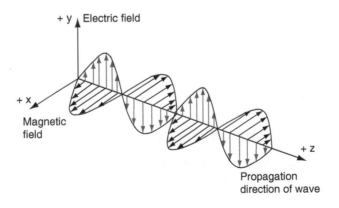

FIGURE 1.4 Propagation of an electromagnetic wave in free space.

1.4 LIGHT SOURCES

Figure 1.5 briefly describes the classification of light sources, which have been built recently and are used by industry for many applications.

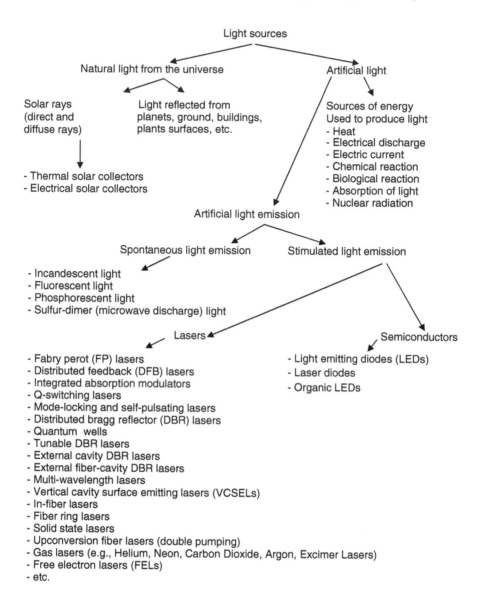

FIGURE 1.5 Classifications of light sources.

1.5 THE ELECTROMAGNETIC SPECTRUM

A few decades after the publication of Maxwell's theory, Hertz showed that electromagnetic waves could indeed be produced by an oscillating magnetic spark. In further experiments, he showed that electromagnetic waves undergo reflection, refraction, diffraction, and interference. Electromagnetic waves behave exactly like light, except that their wavelengths are much greater. The work of Hertz and others laid the foundations for the use of electromagnetic waves in radio communication.

Electromagnetic waves can exist at virtually any frequency (f) or wavelength (λ). The frequency and wavelength are related to the speed (v) of the wave, defined by:

$$v = f\lambda \tag{1.13}$$

For electromagnetic waves travelling through a vacuum or, to a good approximation, through air, the speed is $v = c$, so $c = f\lambda$. Equation (1.13) shows that the frequency and wavelength are inversely related by the travelling wave relationship $\lambda = c/f$. Therefore, the greater the frequency, the shorter the wavelength.

Figure 1.6 shows that electromagnetic waves exist with an enormous range of frequencies, from values less than 10^4 Hz to greater than 10^{22} Hz. It also shows the whole range of electromagnetic

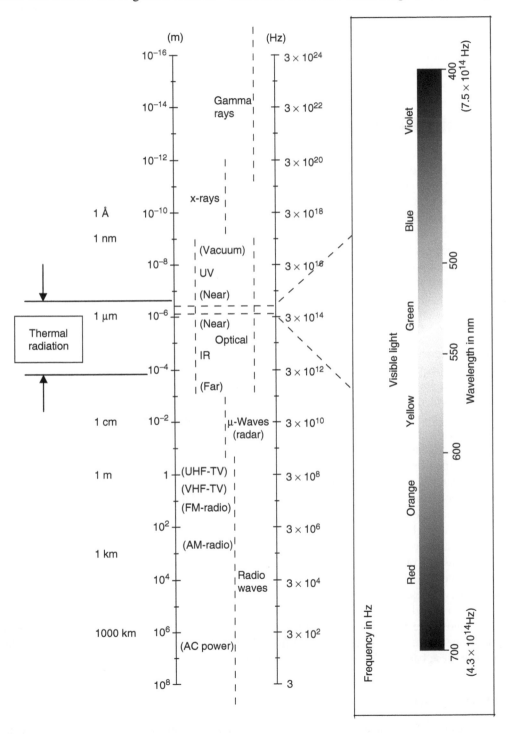

FIGURE 1.6 Electromagnetic wave spectrum.

waves, in order of increasing frequency or decreasing wavelength. Since all these waves travel through a vacuum at the same speed of light $c = 3.00 \times 10^8$ m/s, Equation (1.13) can be used to find the correspondingly wide range of wavelengths. Each range of wavelengths is referred to as a band. A small group of wavelengths within a band is called a channel.

The ordered series of electromagnetic wave frequencies or wavelengths shown in Figure 1.6 is called the electromagnetic spectrum. Historically, regions of the spectrum have been given names, such as visible waves, radio waves, and infrared waves. Although the boundary between adjacent regions is shown as a sharp line in Figure 1.6, the boundary is not so well defined in practice, and the regions often overlap.

1.6 TYPES OF ELECTROMAGNETIC WAVES

In 1867, when Maxwell published the first extensive account of his Electromagnetic Theory, the frequency band was only known to extend from the infrared, across the visible light, to the ultraviolet. Although this region is of major concern in optics, it is a small range of the electromagnetic spectrum. This section lists the main categories into which the electromagnetic spectrum is usually classified, by ranges of frequencies or wavelengths. Table 1.1 lists the frequency and wavelength ranges for the general types of electromagnetic waves.

Somewhere, not far from the centre of the spectrum, there is visible light. Notice that the visible part of the electromagnetic spectrum, so important to life on Earth, is actually the smallest of the frequency bands named above.

1.6.1 RADIOFREQUENCY WAVES

The lowest-frequency electromagnetic waves of particular importance are radio and television waves, in the frequency range of roughly 10^6 to 10^9 Hz. Waves in this frequency range are produced in a variety of ways. The low-frequency (LF), medium-frequency (MF), and high-frequency (HF) bands are reflected by the upper layers of the ionosphere. The short-wave radio frequencies in the HF band are reflected off the E and F layers of the ionosphere at a considerable altitude and can travel great distances.

The straight-line radio waves are normally transmitted around the curvature of the Earth. This is accomplished by reflection off ionic layers in the upper atmosphere. Energetic particles from the sun ionize gas molecules, giving rise to several ion layers. Certain layers reflect radio waves below a specific frequency. By "bouncing" radio waves off these layers, it is easy to send radio transmissions beyond the horizon, to any region of the Earth. Such reflection of radio waves requires the ionic layers to have uniform density. When, from time to time, a solar disturbance produces a shower of energetic particles that upsets this uniformity, a communication blackout can occur, as the radio waves are scattered in many directions rather than reflected in straight lines. To avoid such disruptions, global communications have, in the past, relied largely on transoceanic cables. Now, communications satellites can provide line-of-sight transmission to any point on the globe.

The ionosphere is essentially transparent to the VHF, UHF, and SHF portions of the electromagnetic spectrum. Because transmission from the radiating antenna to the receiver must be line-of-sight in these bands, television and FM radio signals have a limited range.

1.6.2 MICROWAVES

Microwaves with frequencies from 10^9 to about 10^{12} Hz are produced by special vacuum electron tubes called klystrons and magnetrons. Nowadays microwaves are used in communications,

TABLE 1.1
Classifications of Electromagnetic Waves

Types of Waves		Approximate Frequency Range (f) (Hz)	Approximate Wavelength Range (λ) (m)	Source
Power Waves		60	5×10^6	Electric Currents
Radio Waves	AM	$(0.53 \times 10^6) - (1.70 \times 10^6)$	$570 - 186$	Electric Circuits
	FM	$(88.00 \times 10^6) - (108.00 \times 10^6)$	$3.40 - 2.80$	
	TV	$(54.00 \times 10^6) - (890.00 \times 10^6)$	$5.60 - 0.34$	
Microwaves		$10^9 - 10^{11}$	$10^{-1} - 10^{-3}$	Special Vacuum Tubes
Infrared Radiation		$10^{11} - 10^{14}$	$10^{-3} - 10^{-7}$	Warm and Hot Bodies
Visible Light		$(4.00 \times 10^{14}) - (7.00 \times 10^{14})$	10^{-7}	Sun and Lamps Light
Ultraviolet Radiation		$10^{14} - 10^{17}$	$10^{-7} - 10^{-10}$	Very Hot Bodies and Special Lamps
X-rays		$10^{17} - 10^{19}$	$10^{-10} - 10^{-12}$	High-Speed Electron Collisions and Atomic Processes
Gamma Rays		Above 10^{19}	Below 10^{-12}	Nuclear Reactions, Processes in Particle Accelerators, and Natural radioactivity

inter-station television, microwave ovens, and radar applications. They are used in studying the origin of the universe, opening garage doors, guiding planes, and viewing the surface of the planet. They are also quite useful for studying physical optics with experimental arrangements that are scaled up to convenient dimensions.

1.6.3 Infrared Waves

The infrared region of the electromagnetic spectrum lies just beneath red light of the visible light range. Infrared waves have frequencies from about 10^{12} to 4.3×10^{14} Hz. They were first detected by the renowned astronomer Sir William Herschel (1738–1822) in 1800. The infrared radiation, or IR, is often subdivided into four regions: the near IR, near the visible (780–3000 nm); the intermediate IR (3000–6000 nm); the far IR (6000–15000 nm); and the extreme IR (15000 nm–1.0 mm).

Infrared rays are often generated by the rotations and vibrations of molecules. In turn, when infrared rays are absorbed by an object its molecules rotate and vibrate more vigorously, resulting in an increase in the object's temperature. The molecules of any object at a temperature above absolute zero ($-273°C$) will radiate IR, even if only weakly. On the other hand, infrared is emitted in a continuous spectrum from hot bodies, such as electric heaters, glowing coals, and ordinary house radiators. Roughly half the electromagnetic energy from the sun is IR, and the common light bulb actually radiates far more IR than light. The human body radiates IR quite weakly, starting at around 3000 nm. IR is also associated with maintaining the Earth's warmth or average temperature through the greenhouse effect. Incoming visible light (which passes relatively easily through the atmosphere) is absorbed by the Earth's surface and re-radiated as infrared (longer-wavelength) radiation, which is trapped by greenhouse gases, such as carbon dioxide and water vapour.

IR energy is generally measured with a device that responds to the heat generated upon absorption of IR by a blackened surface. There are many types of IR detectors used in different fields, for example, thermocouple, pneumatic, pyroelectric, and bolometer detectors. These detectors rely on temperature-dependent variations in induced voltage, gas volume, permanent electric polarization and resistance, respectively.

The detector can be coupled to a cathode ray tube by way of a scanning system to produce an instantaneous television-like IR picture, known as a thermograph. There are IR spy satellites that look out for rocket launchings, IR resource satellites that look out for crop diseases, and IR astronomical satellites that look out into space. There are "heat-seeking" missiles guided by IR, and IR lasers and telescopes looking out into universe.

Most remote controls operate on a beam of infrared light with a wavelength of about 1000 nm. This infrared light is so close to the visible spectrum and so low in intensity that it cannot be felt as heat.

1.6.4 Visible Light

The visible light region occupies only a very small portion of the total electromagnetic spectrum. The visible spectrum covers a range of less than 4% of the width of the electromagnetic spectrum, as shown in Figure 1.6. It runs from 4×10^{14} to about 7×10^{14} Hz, or a wavelength range of 780–400 nm, respectively. Only the radiation in this region can activate the receptors in our eyes. Visible light emitted or reflected from the objects around us provides visual information about our world. Visible light can be emitted by many sources, such as an incandescent bulb or a hot metal filament. The resulting broad emission spectrum is referred to as thermal radiation, and it is one of the major sources of light. Light is also generated by passing an electric discharge through a gas-filled tube. The atoms of the gas will become excited and emit a stream of photons as visible light. This occurs in an ordinary florescent tube or a gas laser device, such as a Helium Neon laser tube.

Newton was the first to recognize that white light is actually a mixture of all the colours of the visible spectrum. Approximate frequency and wavelength ranges for the various colours of white light are given in Table 1.2. When white light passes through a prism, the white light disperses into its various spectral components from the red to violet colour, like a rainbow. The sun emits white light, as do incandescent light bulbs. Most sources of illumination are white light sources.

TABLE 1.2
Frequency and Wavelength Ranges for the Various Colours in White Light

Colour	Wavelength (λ) (nm)	Frequency (f) (Hz)
Red	780 – 622	$(384 – 482) \times 10^{12}$
Orange	622 – 597	$(482 – 503) \times 10^{12}$
Yellow	597 – 577	$(503 – 520) \times 10^{12}$
Green	577 – 492	$(520 – 610) \times 10^{12}$
Blue	492 – 455	$(610 – 659) \times 10^{12}$
Violet	455 – 390	$(659 – 769) \times 10^{12}$

When visible light falls on an object, that object absorbs and reflects wavelengths in the visible spectrum. A white object, such as the white page forming the background to this writing, reflects all wavelengths, and our eyes see white. A black object, such as the letters printed on this page, absorbs all wavelengths, and no light is reflected.

Colours are seen because of special absorption and reflection of light by an object. The green grass and leaves are perceived as green because the chlorophyll in green plant matter strongly absorbs the wavelengths surrounding green and reflects wavelengths in the green region of the visible spectrum. A red traffic light is red when light passes through it because it passes wavelengths in the red part of the visible light spectrum and absorbs light in the remaining parts of the visible spectrum. If our eyes intercept predominantly red wavelengths coming from an object, the eye perceives that object as having a red colour. Colour is the result of a filtered white light spectrum, in cases where the illumination light is white.

1.6.5 ULTRAVIOLET LIGHT

Ultraviolet radiation is the portion of the electromagnetic spectrum that lies between x-rays and visible light. Ultraviolet radiation was discovered by Johann Wilhelm Ritter (1776–1810). It is not seen by the human eye, but some regions of UV act in much the same way as visible light. There is no sharp dividing line between UV and x-rays, or between UV and visible light. The approximate wavelength range of UV is between 4 and 400 nm. It is not often realized that UV radiation covers such a wide range of wavelengths; the ratio of the longest-to-shortest of these wavelengths is nearly 100 to 1. By comparison, the entire range of visible light seen by humans extends from approximately 380 to 760 nm, which is only a 1 to 2 ratio in wavelength. The portions of the UV spectra are commonly referred to as near UV (from approximately 300–400 nm) and far UV (from approximately 185–300 nm). Medical literature divides regions into UVA long wave (315–380 nm), UVB midrange (280–315 nm), and UVC shortwave (<280 nm).

Ultraviolet radiation is produced by special lamps and very hot bodies. The sun emits large amounts of UV radiation. Fortunately, most of the UV radiation received by the Earth is absorbed in the ozone (O_3) layer in the upper atmosphere at an altitude of about 40 to 50 km. Sunglasses are now labeled to indicate the UV protection standards they meet in shielding the eyes from this potentially harmful radiation.

Artificial sources of UV radiation include incandescent and arc lamps. Lamps used for the generation of UV light are usually enclosed arcs containing mercury. Other sources include argon, krypton, and xenon gases, as well as zinc and cadmium vapour.

UV may be produced directly by these arcs or indirectly by fluorescence of phosphors deposited in thin layers on the lamp envelope. These phosphors absorb UV and re-emit this energy at longer UV wavelengths or as visible light as follows. High frequency UV photons collide with atoms and part of the photon's energy is transferred to the atoms by boosting electrons to higher energy states. Upon de-excitation, as electrons fall back to lower energy states, energy is released as photons of light. Since only a portion of the incoming photon's energy was transferred to an electron, these emitted photons have less energy than the incoming UV photons, so their wavelengths are longer than the excitation photons.

In some phosphoric materials, the fluorescence lingers and disappears, slowing after the UV source is removed. Here, the electron returns slowly to its original state, and this delayed fluorescence is called phosphorescence.

1.6.6 X-Rays

X-rays were discovered in 1895 by the German physicist Wilhelm C. Röntgen (1845–1923) when he noted the glow of a piece of fluorescent paper caused by some mysterious radiation coming from a cathode ray tube. Because of the apparent mystery involved, these were named x-radiation or x-rays. Extending in frequency from roughly 2.4×10^{16} to 5×10^{19} Hz, they have extremely short wavelengths; most are smaller than an atom. The photon energies of x-rays (100 eV to 0.2 MeV) are large enough so that x-ray quanta can interact with particles of matter one at a time in a clearly granular fashion, almost like bullets of energy. One of the most practical mechanisms for producing x-rays is the rapid deceleration of high-speed charged particles. The resulting broad frequency braking radiation arises when a beam of energetic electrons is fired at a material target, such as a copper plate. Collisions with the Cu nuclei produce deflections of the beam electrons, which in turn radiate x-ray photons.

Typically, the x-rays used in medicine are generated by the rapid deceleration of high-speed electrons projected against a metal target. These energetic rays, which are only weakly absorbed by the skin and soft tissues, pass through our bodies rather freely, except when they encounter bones, teeth, or other relatively dense material. However, at low intensities, x-rays can be used with relative safety to view the internal structure of the human body and other opaque objects. X-rays can pass through materials that are opaque to other types of radiation. The denser the material is, the greater its absorption of x-rays, and the less intense the transmitted radiation will be. For example, as x-rays pass through the human body, many more x-rays are absorbed by bone than by tissue, as shown in Figure 1.7. If the transmitted radiation is directed onto a photographic plate or film, the exposed areas show variations in intensity and thus form a picture of internal structures. This property makes x-rays most valuable for medical diagnosis, research, and treatment. Still, x-rays can cause damage to human tissue, and it is desirable to reduce unnecessary exposure to these rays as much as possible.

X-ray processes take place in colour television picture tubes, which use high voltages and electron beams. When the high-speed electrons are stopped as they hit the screen, they can emit x-rays into the environment. Fortunately, all modern televisions have the shielding necessary to protect viewers from exposure to this radiation. Examples of optical instruments using x-rays include x-ray microscopes, picosecond x-ray streak cameras, x-ray diffraction grating, interferometers, and x-ray holography.

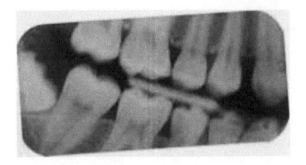

FIGURE 1.7 X-ray photograph of teeth.

1.6.7 GAMMA RAYS

Electromagnetic waves with frequencies above 10^{20} Hz are often referred to as gamma (γ) rays. They are the highest energy (10^4 eV to about 10^{19} eV), shortest wavelength electromagnetic waves. This high-frequency radiation is produced in nuclear reactions, in particle accelerators, and also in certain types of nuclear radioactivity. Gamma rays are emitted by particles undergoing transitions within the atomic nucleus. A single gamma ray carries so much energy that it can be detected with little difficulty. At the same time, its wavelength is so small that it is difficult to observe any wavelike properties. Gamma rays are also highly penetrating and destructive to living cells. It is for this reason that they are used to treat cancer cells and more recently to kill microorganisms in food processing.

1.7 PROPERTIES OF LIGHT

Light travels in a straight line through a uniform optical medium. The eye depends on back-projecting rays entering the eye that direct to the origin of the light rays. Eyes can locate objects, as long as the light from them has travelled to the eye in a straight line. However, the path or nature of light can be altered by optical materials, which can absorb, reflect, or transmit light, or do a combination of all three. Details of light properties are presented throughout this book.

1.7.1 ABSORPTION

Light falling on an optical material may be absorbed, transmitted, or reflected. Light that is neither reflected nor transmitted by an optical surface is absorbed and usually converted into heat energy.

The absorbed light is responsible for the heating effects that occur when light strikes a surface. The more light that is reflected or transmitted, the smaller the absorption and the less heating that takes place. This is most evident in the heating of common objects by sunlight. This depends on the colour of the optical material; a red material reflects red light and absorbs much of the rest of the other colours. The colour of a material is that colour which is reflected rather than absorbed. Black or dark-coloured painted objects or dyed fabrics have extremely low reflectance and transmit little or no light. Consequently, they absorb most of the radiant energy that falls on them. This energy is converted to heat, which explains why the dark materials tend to heat up much more rapidly than light-coloured ones when in the presence of bright sunlight. On the other hand, white and other colours that have high reflectance over large areas of the visible spectrum tend to reflect most of the incident radiation. Because little energy is absorbed by such colours, only a small amount of energy is converted into heat.

Greenhouses also utilize the same effect to capitalize on the energy from the sun. In cool periods of the year, visible light enters through the glass and is absorbed by the dark soils and vegetation.

1.7.2 TRANSMISSION

In addition to the colour that depends on the detailed nature of the source, light may take on a particular colour as it passes through a material that selectively absorbs some wavelengths. For example, a piece of red glass is a filter that selectively passes red light but absorbs shorter wavelengths. Similarly, a blue filter transmits blue but not green, yellow, or red. Thus, colour may be produced by selective absorption.

1.7.3 REFLECTION

Reflection is the most familiar property of light. Reflection enables us to see objects around us. Light from a source, such as the sun or a lamp, travels in a straight line until it strikes a material, at which point it may be either absorbed, transmitted, or reflected, or a combination of two or three processes.

Reflection occurs when waves encounter a boundary that does not absorb the light's waves and bounces the waves off the surface. The incoming light wave is referred to as an incident wave and the wave that is bounced from the material's surface is called the reflected wave. Surfaces that reflect the most light appear white, silver, or mirror-finished. A highly polished, smooth, flat silver surface acts as a mirror, reflecting a perfect image of the objects around it.

Colour is also produced by reflection as well as transmission. The perceived colour of most objects is due to the selective reflection of light. Thus, a red object reflects red light but absorbs green and blue. A red apple appears a vivid red when illuminated with red light, but it looks dark under blue light because it does not reflect blue light well.

The absorption, transmission, and reflection properties of light depend on the following major factors, which are discussed in detail throughout the book:

1. optical characteristics of the optical materials,
2. intensity of a light source,
3. wavelength of a light source,
4. angle of incidence of a light beam on an optical component surface,
5. mechanical characteristics of an optical component surface,
6. environmental and operating temperatures of optical components,
7. colours of the optical components,
8. thickness and density of the optical components, and
9. dopant and codopant percentage in the optical components.

1.7.4 REFRACTION

Light that is transmitted through an optical medium will usually deviate somewhat from the straight path it was previously following. This phenomenon is familiar with transparent optical materials, such as glass, plastic, water, and lenses. Objects seen through them appear larger, smaller, or distorted. Place half of a stick's length into water and the stick appears to be bent at the surface. Refraction is an important characteristic of lenses, and allows them to focus a beam of light onto a single point. Refraction occurs as light passes from one optical material to another when there is a difference in the index of refraction between the two optical materials.

1.7.5 INTERFERENCE

Interference is the net effect of the combination of two or more wave types moving on intersecting or coincident paths. The effect is the addition of the amplitudes of the individual waves at each point affected by more than one wave. There are two types of interference. Assume two waves have the same frequency and amplitude. If two of the waves are in phase with each other, the two waves are reinforced, producing constructive interference. If the two waves are out of phase, the result is destructive interference, producing complete cancellation. One of the best examples of interference is demonstrated by the light reflected from a film of oil floating on water or a soap bubble, or a computer compact disk (CD). The disk reflects a variety of rainbow colours when illuminated by natural or artificial white light sources.

1.7.6 DIFFRACTION

Diffraction occurs when light waves pass by a corner or through an opening or slit. The opening size must be smaller than the light's wavelength. Diffraction is a specialized case of light scattering, in which an object with regularly repeating features (such as a diffraction grating) produces an orderly diffraction of light, seen as a diffraction pattern. In the real world, most objects are very complex in shape and thus are considered to be composed of many individual diffraction features that can collectively produce a random scattering of light.

1.7.7 POLARIZATION

Natural sunlight and most forms of artificial illumination transmit light waves whose electric fields vibrate in all perpendicular planes with respect to the direction of propagation; this is called un-polarized light. The light has two components: the electric field vibrating in the vertical plane and the magnetic field vibrating in the horizontal plane. The two fields combine to form the shape of the electromagnetic wave. When the electric fields of light waves are restricted to a single plane by filtration, the light is polarized with respect to the direction of propagation so that all waves vibrate in the same plane.

FURTHER READING

Beiser, Arthur, *Physics*, 5th ed., Addison-Wesley Publishing Company, Reading, MA, 1991.

Born, M. and Wolf, E., *Principles of Optics: Electromagnetic Theory of Propagation, Interference, and Diffraction of Light*, 7th ed., Cambridge University Press, Cambridge, England, 1999.

Bouwkamp, C. J., Diffraction theory, *Rep. Prog. Phys.*, 17, 35–100, 1949.

Bromwich, T. J. I' A., Diffraction of waves by a wedge Proc. London Math. Soc. 14, 450–468, 1916.

Camperi-Ginestet, C., Young W, *Kim*, Micro-opto-mechanical devices and systems using epitaxial lift off. JPL, Proceedings of the Workshop on Microtechnologies and Applications to Space Systems., 305–316, 1993.

Cox, Arthur, *Photographic Optics*, 15th ed., Focal Press, London, New York, 1974.

Cutnell, John D. and Johnson, Kenneth W., *Physics*, 5th ed., Wiley, New York, 2001.

Cutnell, John D. and Johnson, Kenneth W., *Student Study Guide: Physics*, 5th ed., Wiley, New York, 2001.

Douglas, C., *Giancoli: Physics*, 5th ed., Prentice Hall, 1998.

Evans, R. D., *The Atomic Nucleus*, McGraw-Hill, New York, 1955.

Ewen, Dale, *Physics for Career Education*, 4th ed., Prentice Hall, Englewood Cliffs, NJ, 1996.

Ghatak, Ajoy K., *An Introduction to Modern Optics*, McGraw-Hill Book Company, New York, 1972.

Griot, Melles, *The Practical Application of Light*, Melles Griot Catalog, San Jose, CA, 2001.

Halliday, Resnick, and Walker, *Fundamental of Physics*, 6th ed., Wiley, New York, 1997.

Heath, Robert W., *Fundamentals of Physics*, D.C. Heath Canada Ltd, Toronto, 1979.

Hewitt, Paul G., *Conceptual Physics*, 8th ed., Addison-Wesley, Inc., Reading, MA, 1998.

Jones, Edwin and Richard, Childers, *Contemporary College Physics*, McGraw-Hill Higher Education, New York, NY, 2001.

Lehrman, Robert L., *Physics-The Easy Way*, 3rd ed., Barron's Educational Series, Inc., New York, 1998.

McDermott, Lillian C., *Introduction to Physics*. Preliminary Edition, Prentice Hall, Inc., Englewood Cliffs, NJ, 1988.

McDermott, Lillian C., *Tutorials in Introductory Physics*. Preliminary Edition, Prentice Hall, Inc., Englewood Cliffs, NJ, 1998.

Nichols, Daniel H., *Physics for Technology with Applications in Industrial Control Electronics*, Prentice Hall, Englewood Cliffs, NJ, 2002.

Nolan, Peter J., *Fundamentals of College Physics*, Wm. C. Brown Publishers, Boston, 1993.

Robinson, Paul, *Laboratory Manual to Accompany Conceptual Physics*, 8th ed., Addison-Wesley, Inc., Reading, MA, 1998.

Romine, Gregory S., *Applied Physics Concepts into Practice*, Prentice Hall, Inc., Englewood Cliffs, NJ, 2001.

Salah, B. E. A. and Teich, M. C., *Fundamentals of Photonics*, John Wiley and Sons, New York, 1991.

Sears, Francis W., *University Physics - Part II*, 6th ed., Addison-Wesley Publishing Company, Reading, MA, 1998.

Tippens, Paul E., *Physics*, 6th ed., Glencoe McGraw-Hill, Westerville, OH, U.S.A., 2001.

Urone, Paul Peter, *College Physics*, Brooks/Cole publishing Company, Pacific Grove, CA, 1998.

Walker, James S., *Physics*, Prentice Hall, Englewood Cliffs, NJ, 2002.

Warren, Mashuri L., *Introduction to Physics*, W.H. Freeman and Company, San Francisco, CA, 1979.

White, Harvey E., *Modern College Physics*, 6th ed., Van Nostrand Reinhold Company, New York, 1972.

Williams, John E., *Teacher's Edition Modern Physics*, Holt, Rinehart and Winston, Inc., New York, 1968.

Wilson, Jerry D., *Physics: A Practical and Conceptual Approach Saunders Golden Sunburst Series*, Saunders College Publishing, London, 1989.

Wilson, Jerry D. and Buffa, Anthony J., *College Physics*, 5th ed., Prentice Hall, Inc., Englewood Cliffs, NJ, 2000.

Yeh, C., *Applied Photonics*, Academic Press, New York, 1994.

2 Light and Shadows

2.1 INTRODUCTION

To cast a shadow, we need light from a concentrated source, such as the Sun during the day or the Moon at night. A shadow is also cast by light that comes from a point source, which gives parallel light rays. Consider a flat white screen located a short distance away from the light source and an object placed between the screen and the light source. Part of the light rays are blocked by the object making a dark area, and part of the rays that reach the screen make that part of the screen bright. The dark area is called shadow. The shadow's location can be figured out by drawing straight lines from the point source to the edge of the object and continuing them to the screen. The resulting shadow resembles the object and we can easily recognize the shape of the object in two dimensions.

Perhaps the most spectacular astronomical events that one can observe without a telescope are solar and lunar eclipses, which were considered omens of great fortune or complete disaster in ancient times. We now know that the occurrence of an eclipse is a consequence of the shadows cast by the orbits of the Earth and Moon with respect to the Sun.

X-ray pictures are simply shadows in x-ray light made visible by a fluorescent screen or photographic film.

This chapter presents the principles, types, and applications of shadows. Solar and lunar eclipses are also presented in detail. Two experimental studies of light shadow formation using different optical objects are included in this chapter.

2.2 SHADOWS

The electromagnetic wave of light emitted in all directions from point source S of light is shown in Figure 2.1. Light propagation can be represented by a series of spherical wavefronts moving away from the point light source at the speed of light. The point source of light can be defined as a source whose dimensions are small in comparison with the distances that light travels. Notice that the spherical wavefronts become essentially plane wavefronts in all directions at a large distance from

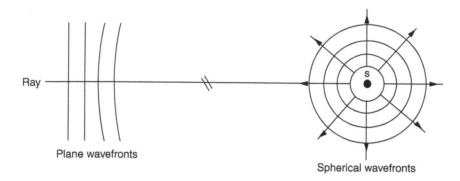

FIGURE 2.1 Point source of light emits rays in all directions.

the light source. More details on this subject will be presented later in this book. An imaginary straight line drawn perpendicular to the wavefront in the direction of the moving wavefronts is called a ray, as shown in Figure 2.1. There are an infinite number of rays starting from the point light source propagating in all directions.

Any dark object absorbs some wavelengths of light, as discussed in the properties of light in Chapter 1. A black object absorbs nearly all of the visible light wavelengths when they are incident on it. Light that is not absorbed when incident upon an object is either transmitted or reflected. If all light incident upon an object is transmitted, the object is called transmissive. If all light incident upon an object is reflected or absorbed, the object is called opaque. Since light cannot pass through an opaque object, a shadow will be produced in the space behind the object. The shadow formed by a point source S of light is shown in Figure 2.2. Since light is propagated in straight lines, rays drawn from the point source past the edges of the opaque object form a sharp shadow. The shape of the shadow is proportional to the shape of the object. The shape of the shadow depends on the opaque object location relative to the point source. The region in which no light has entered is called the umbra.

When the source of light is an extended type, the shadow will consist of two portions, as shown in Figure 2.3. The inner portion receives no light from the extended source; this portion is the umbra, as explained above. The outer portion is called the penumbra. An observer within the penumbra would see a portion of the source but not all of the source. An observer located outside both regions would see the source completely. Shadows are produced when objects are exposed to a light source or natural light. A light shadow occurs in the universe when solar and lunar eclipses occur by a similar formation of shadows, as explained above. One consequence of the Moon's orbit around the Earth is that the Moon can shadow the Sun's light as viewed from the

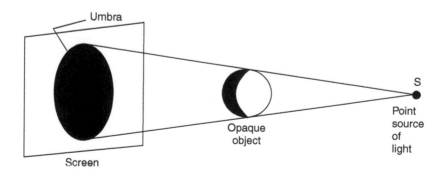

FIGURE 2.2 Shadow formed by an opaque object placed between a screen and a point light source.

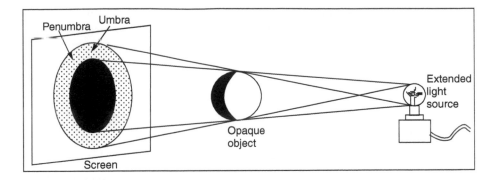

FIGURE 2.3 Shadow formed by an extended source of light.

Earth, or when the Moon passes through the shadow cast by the Earth. The former is called a solar eclipse, and the later is called a lunar eclipse.

2.3 SOLAR ECLIPSE

A solar eclipse occurs when the Moon passes between the Sun and the Earth, blocking the light from the Sun. This causes a shadow to be cast on a small area of the Earth. Due to the relative sizes of the Moon and the Sun and their distances from Earth, at times they appear to be the same size in the sky. When the Moon exactly covers the Sun, a total solar eclipse occurs. Three types of solar eclipses (as observed from any particular point on the Earth) can be defined as:

1. *Total Solar Eclipses* occur when the umbra of the Moon's shadow touches a region on the surface of the Earth. This occurs when the Moon exactly covers the Sun. The alignment of the Earth, Moon, and Sun have to be exact for a total eclipse to occur.
2. *Partial Solar Eclipses* occur when the penumbra of the Moon's shadow passes over a region of the Earth's surface. If the alignment is not exact, a partial eclipse may occur. This is when the Moon only partially overlaps the Sun and blocks only part of the Sun from view.
3. *Annular Solar Eclipses* occur when a region on the Earth's surface is in line with the umbra, but the distance is such that the tip of the umbra does not reach the Earth's surface, as shown in Figure 2.4. An annular eclipse occurs when the apparent size of the Moon is smaller than that of the Sun, and the Moon does not fully block the Sun from view.

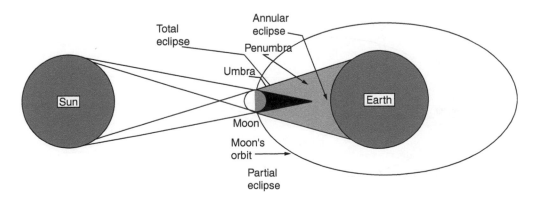

FIGURE 2.4 Geometry of annular solar eclipse.

This occurs because the Moon's orbit is elliptical, causing the Moon's distance from the Earth to vary. When the Moon is farther away from the Earth, it appears to be smaller. Therefore, there are times when the Moon appears to be smaller than the Sun. If an eclipse occurs at this time, it is called an annular eclipse.

2.4 LUNAR ECLIPSE

A lunar eclipse occurs when the Moon passes through the Earth's shadow, as shown in Figure 2.5. Because the Earth is much larger than the Moon, usually the entire Moon is eclipsed. Since the full phase can be seen from anywhere on the night side of the Earth, a lunar eclipse can be seen by more people than a solar eclipse. Similar to solar eclipses, lunar eclipses can be partial or total, depending on whether the light of the Sun is partially or completely blocked from reaching the Moon. Since the Moon is moving through the Earth's shadow, and the size of the Earth is much greater than the size of the Moon, a lunar eclipse lasts for about 3.5 h (as opposed to a solar eclipse, which lasts for about 7.5 min).

Although the Moon is always in a new phase during a solar eclipse, a solar eclipse does not occur every time the Moon is in the new phase. This is because the orbit of the Moon is tilted relative to the Earth's orbit around the Sun. Though the Moon tilt is only 5°, it is enough to allow the alignment of the Earth, Moon, and Sun to occur only about once every six months. This holds true for lunar eclipses as well. In fact, lunar and solar eclipses generally occur together; that is, if the alignment is correct for a lunar eclipse during the full phase of the Moon, it will also be correct for a solar eclipse during the next new phase of the Moon. There are three basic types of lunar eclipses that occur in the universe, as shown in Figure 2.6.

1. *Penumbral Lunar Eclipses* occur when the Moon passes through the Earth's penumbral shadow. These events are only of academic interest, since they are subtle and quite difficult to observe.
2. *Partial Lunar Eclipses* occur when a portion of the Moon passes through the Earth's umbral shadow.
3. *Total Lunar Eclipses* occur when the entire Moon passes through the Earth's umbral shadow. These events are quite striking because of the vibrant range of colors the Moon can display during the total phase.

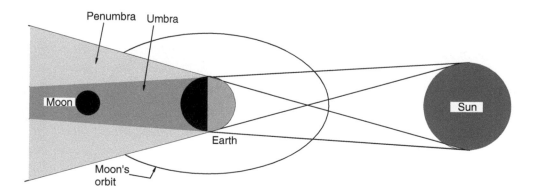

FIGURE 2.5 Geometry of a lunar eclipse—full moon.

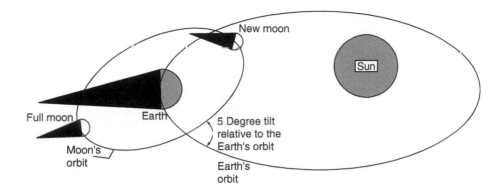

FIGURE 2.6 Types of lunar eclipse.

2.5 APPLICATIONS OF SHADOWS

The following are some of the applications of shadows:

1. The x-ray images of teeth and bones are shadows, made by a kind of light that goes through most things, but is scattered more by dense materials, such as metals.
2. A sundial is a way to measure the position of the Sun (and thus determine the time) by observing its shadow.
3. The first measurement of the size of the Earth was based on the length of shadows exactly at noon in different places. For the lengths of the shadows to be very different, you have to go north a significant fraction of the radius of the Earth. Observing the shadow lengths allows the size of the Earth to be calculated.
4. During an eclipse of the Sun, we are briefly in the shadow of the Moon; during an eclipse of the Moon, it is passing through the Earth's shadow. The duration of the eclipses was one of the first hints to early astronomers that the Moon is smaller than the Earth (but not much smaller), and that the Moon's distance from Earth is about 100 Earth-diameters, very much less than the distance to the Sun.
5. There is another kind of astronomical shadow that played an important role in the universe. About every 100 years, the planet Venus passes directly in front of the Sun, so that it appears as a small dark region on the Sun's disk. From different places on the surface of the Earth, different views are seen, and the path of Venus across the Sun's disk is slightly different. This provided the first means by which astronomers could measure the distance from the Earth to the Sun. The reason that the British Explorer Captain James Cook (1728–1780) sailed to the South Pacific (accidentally discovering New Zealand on the way) was to make the transit of Venus that occurred in 1769.
6. The results of Thomas Young's (1773–1829) experiment predicted the location of bright and dark fringes generated by the diffraction of light when light passes through an opening. The diffraction of light waves is of extreme importance in optical instruments because it sets the ultimate limit on the possible magnification. The interference of light waves is the principle of the American physicist A. Michelson (1852–1931), who invented the interferometer, which is used to make accurate length measurements.

2.6 EXPERIMENTAL WORK

This experiment studies the formation of a light shadow using an opaque object. This experimental work can be divided into two cases:

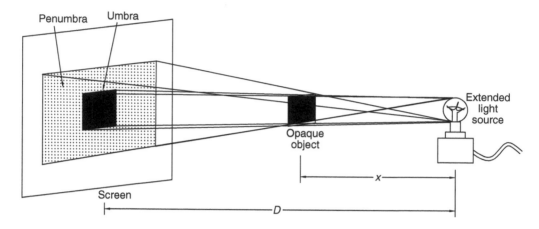

FIGURE 2.7 Shadow formed by an object located between a light source and a screen.

a. The student will practise forming shadows from a light source using an opaque object. The student will measure the sizes of the umbra and penumbra areas that form for at least five different distances from the light source. Then the size of the umbra and penumbra will be drawn as a function of a distance. A comparison among light shadows will be carried out. This case is illustrated in Figure 2.7.

b. The student will practise forming shadows from a light source using three different optical objects. The objects are an opaque object, which blocks light; a translucent object, which allows some light to pass through, but the light is scattered; and a transparent object, which allows light to pass through (relatively) undisturbed. The student will measure the size of the umbra and penumbra that are formed for at least five different distances from the light source, for each object. The student will also measure the light intensity at the umbra and penumbra areas for each object. Then the size of the umbra and penumbra and the light intensity will be drawn as a function of the distance for each object. A comparison among light shadows and light intensity will be carried out. This case is illustrated in Figure 2.8.

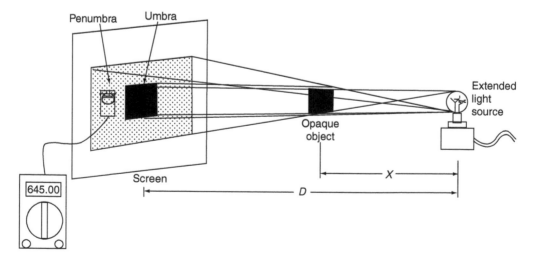

FIGURE 2.8 Shadow formation and light intensity measurements.

2.6.1 TECHNIQUE AND APPARATUS

Appendix A presents the details of the devices, components, tools, and parts.

1. Opaque object and object holder, as shown in Figure 2.9
2. Unfrosted bulb in a light source, as shown in Figure 2.9
3. Opaque, translucent, and transparent objects and object holder
4. Hardware assembly (clamps, posts, positioners, etc.)
5. Screen
6. Light meter
7. Ruler

2.6.2 PROCEDURE

Follow the laboratory procedures and instructions given by the professor and/or instructor.

2.6.3 SAFETY PROCEDURES

Follow all safety procedures and regulations regarding the use of optical instruments and measurement devices, and light source devices.

2.6.4 APPARATUS SET-UP

Case (a): Formation of Shadows

1. Figure 2.9 shows the experimental apparatus set-up.
2. Prepare an opaque optical object on the table.
3. Mount a screen on the table or wall.
4. Mount a light source at a distance D from the screen.
5. Mount an opaque object at a distance x_1 from the light source facing the screen.
6. Connect the light source to the power outlet.
7. Align the light source and the opaque object to face each other.
8. Measure the distance D, the size of the object, and the distance x_1. Fill out Table 2.1.
9. Switch on the power to the light source.
10. Turn off the lights in the lab before taking measurements.

FIGURE 2.9 Formation of shadows apparatus set-up.

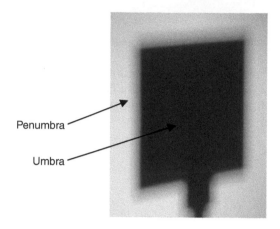

FIGURE 2.10 Umbra and penumbra of a shadow.

11. Measure the size of the umbra and penumbra, as shown in Figure 2.10. Fill out Table 2.1.
12. Switch off the power to the light source.
13. Turn on the lights in the lab.
14. Repeat steps 5 to 13 for four different distances between the object and light source.
15. Illustrate the location of the light source, object, and screen in a diagram.

Case (b): Measuring Light Intensity of Shadows Formed by Optical Materials

Repeat the procedure of Case (a) using three optical objects: opaque, translucent, and transparent. Figure 2.9 also shows the experimental apparatus set-up for this case. Measure the light intensity on the umbra and penumbra areas for each optical object. Fill out Table 2.2 for each optical object. Illustrate the location of the light source, object, and screen in a diagram.

2.6.5 DATA COLLECTION

Case (a): Formation of Shadows

Measure the distance D, size of the object, distance x, and the size of the umbra and penumbra areas when the object is located in front of the light source at five distances. Fill out Table 2.1.

TABLE 2.1
Shadows Formed by an Object

Distance D from Light Source to Screen (unit)	Size of Object (unit)	Distance x from Light Source to Object (unit)	Size of Umbra (unit)	Size of Penumbra (unit)
		$x_1 =$ ( )	( )	( )
		$x_2 =$ ( )	( )	( )
		$x_3 =$ ( )	( )	( )
		$x_4 =$ ( )	( )	( )
( )	( )	$x_5 =$ ( )	( )	( )

Case (b): Measuring Light Intensity of Shadows Formed by Optical Materials

Measure the distance D, the size of the object, the distance x, and the size of the umbra and penumbra areas when the three optical objects are located in front of the light source at five distances. Measure the light intensity at the umbra and penumbra areas for each optical object. Read the principles and procedure of light intensity, which are presented in this book. Fill out and use Table 2.2 for each optical object.

TABLE 2.2
Light Intensity Measurements

Distance D from Light Source to Screen (unit)	Size of Object (unit)	Distance x from Light Source to Object (unit)	Size of Umbra (unit)	Light Intesity on Umbra Area (unit)	Size of Penumbra (unit)	Light Intesity on Penumbra Area (unit)
		$x_1 =$ ( )	( )	( )	( )	( )
		$x_2 =$ ( )	( )	( )	( )	( )
		$x_3 =$ ( )	( )	( )	( )	( )
		$x_4 =$ ( )	( )	( )	( )	( )
( )	( )	$x_5 =$ ( )	( )	( )	( )	( )

2.6.6 CALCULATIONS AND ANALYSIS

Calculate the areas of the umbra and penumbra for Cases (a) and (b).

2.6.7 RESULTS AND DISCUSSIONS

Case (a): Formation of Shadows

1. Illustrate the location of the light source, object, and screen in a diagram.
2. Present the measurements of the umbra and penumbra areas at five different distances.
3. Create a relation of the umbra and penumbra sizes as a function of the distances.
4. Discuss your results.

Case (b): Measuring Light Intensity of Shadows Formed by Optical Materials

1. Illustrate the location of the light source, object, and screen in a diagram.
2. Present the measurements of the umbra and penumbra areas at five different distances.
3. Present the measurements of light intensity on the umbra and penumbra areas at different distances from the light source.
4. Create a relation of the umbra and penumbra sizes as a function of the distances.
5. Create a relation of light intensity on the umbra and penumbra areas as a function of the distances.
6. Compare among light shadows and light intensities.
7. Discuss your results.

2.6.8 CONCLUSION

Summarize the important observations and findings obtained in this lab experiment.

2.6.9 SUGGESTIONS FOR FUTURE LAB WORK

List any suggestions for improvements using different experimental equipment, procedures, and techniques for any future lab work. These suggestions should be theoretically justified and technically feasible.

2.7 LIST OF REFERENCES

List any references that were used in the report. Use one format in writing the references. Never mix reference formats in a report.

2.8 APPENDICES

List all of the materials and information that are too detailed to be included in the body of the report.

FURTHER READING

Beiser, A., *Physics*, 5th ed., Addison-Wesley Publishing Company, Reading, MA, 1991.

Cutnell, J. D. and Johnson, K. W., *Physics*, 5th ed., Wiley, New York, 2001.

Ewen, D., *Physics for Career Education*, 4th ed., Prentice Hall, Englewood Cliffs, NJ, 1996.

Giancoli, D. C., *Physics*, 5th ed., Prentice Hall, Englewood Cliffs, NJ, 1998.

Halliday, R. W., *Fundamental of Physics*, 6th ed., Wiley, New York, 1997.

Jones, E. and Childers, R., *Contemporary College Physics*, McGraw-Hill Higher Education, New York, 2001.

McDermott, L. C., *Introduction to Physics*, Preliminary Edition, Prentice Hall, Inc., Englewood Cliffs, NJ, 1988.

Nolan, P. J., *Fundamentals of College Physics*, Wm. C. Brown Publishers, Dubuque, IA, 1993.

Robinson, P., *Laboratory Manual to Accompany Conceptual Physics*, 8th ed., Addison-Wesley, Inc., Reading, MA, 1998.

Salah, B. E. A. and Teich, M. C., *Fundamentals of Photonics*, Wiley, New York, 1991.

Sears, F. W., *University Physics—Part II*, 6th ed., Addison-Wesley Publishing Company, Reading, MA, 1998.

Tippens, P. E., *Physics*, 6th ed., Glencoe McGraw-Hill, Westerville, OH, U.S.A., 2001.

Urone, P. P., *College Physics*, Brooks/Cole Publishing Company, Pacific Grove, CA, 1998.

Walker, J. S., *Physics*, Prentice Hall, Englewood Cliffs, NJ, 2002.

White, H. E., *Modern College Physics*, 6th ed., Van Nostrand Reinhold Company, New York, 1972.

Wilson, J. D., *Physics: A Practical and Conceptual Approach*, Saunders College Publishing, Philadelphia, 1989.

Wilson, J. D. and Buffa, A. J., *College Physics*, 5th ed., Prentice Hall, Inc., Englewood Cliffs, NJ, 2000.

Young, H. D. and Freedman, R. A., *University Physics*, 9th ed., Addison-Wesley Publishing Company, Inc., Reading, MA, 1996.

3 Thermal Radiation

3.1 INTRODUCTION

In the absence of heat, life cannot be sustained in the universe. Thermal radiation serves as an essential ingredient for the formation (and existence) of life in the cosmos Thermal radiation represents a finite region of the electromagnetic spectrum.

Thermal imaging cameras permit the ability to observe the flow of thermal energy between two bodies. Thermal images from space have illustrated the flow of energy within the universe.

Thermal radiation is the transfer of energy by electromagnetic waves, which have a range of wavelengths. One example of thermal radiation heat transfer is that of energy transfer between the Sun and the Earth. The emitted radiant heat is the net flow of heat from a higher temperature material to a lower temperature material. The high temperature material emits energy and the low temperature material absorbs this energy. Many developing countries have exploited this principle as a means of renewable solar energy. This transfer (or flow) is dependent on a material's ability to transmit, reflect or absorb thermal energy. In this chapter, the nature of thermal radiation, its characteristics, and properties are presented.

The experiment included in this chapter requires an understanding of the concepts of solar radiation intensity levels and measurements when using a solar radiation meter. The solar radiation variation during the day will be measured and studied. Also the sun positions during the day hours will be studied.

3.2 THERMAL RADIATION

Thermal radiation is the electromagnetic radiation emitted by a body as a result of its temperature. For a net thermal radiation interchange of energy between two bodies to occur, there must be a temperature difference. The rate at which the thermal radiation interchange occurs is proportional to the difference of the fourth powers of their absolute temperatures. For a better understanding of thermal radiation, it will be necessary to study several different disciplines:

1. Electromagnetic Theory: Thermal radiation can be treated mathematically as if it was a traveling wave that contains energy and exerts a pressure. This wave-like nature is described by using electromagnetic wave theory.
2. Statistical Mechanics: Thermal radiation energy is transported over a range of wavelengths. Statistical mechanics help describe how this energy is distributed over the spectrum of wavelengths. For example, most energy coming from the Sun is radiated over many wavelengths.
3. Quantum Mechanics: All substances radiate and receive thermal radiation energy continuously. Associated with these processes are atomic and molecular activities that are modeled with the principles of quantum mechanics. Quantum mechanics gives a description of molecular processes associated with radiation.
4. Thermodynamics: Thermodynamics gives a description of how bulk properties of a material behave in the presence of thermal radiation.

There is only one type of thermal radiation; it propagates at the speed of light $c = 3.00 \times 10^8$ m/s. This speed is equal to the product of the frequency f and wavelength λ of the radiation as given in Equation (1.1). A portion of the electromagnetic spectrum is shown in Figure 3.1. This figure shows that the thermal radiation lies in the range from about 100 to 2000 nm and the very narrow portion of the spectrum is the visible light. Visible light lies in the range from about 450 to 780 nm. The propagation of thermal radiation takes place in the form of discrete quanta; each quantum having the energy of a photon can be defined by Equation (1.8).

By considering the thermal radiation of a gas, the principles of quantum-statistical thermodynamics can be applied to derive an expression for the energy density of radiation per unit volume and per unit wavelength as given in Equation (3.1):

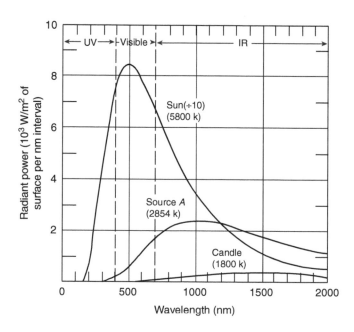

FIGURE 3.1 Blackbody radiation as a function of wavelength and temperature.

$$u_\lambda = \frac{8\pi hc\lambda^{-5}}{e^{hc/\lambda kT} - 1} \tag{3.1}$$

where

> h is Planck's constant $= 6.626\times10^{-34}$ (J s),
> k is Boltzmann's constant $= 1.38066\times10^{-23}$ J/molecule K, and
> T is the temperature in degrees Kelvin.

When the energy density is integrated over all wavelengths, the total energy emitted is proportional to the absolute temperature to the fourth power, which is known as the Stefan–Boltzmann law as given in Equation (3.2):

$$E_b = \sigma T^4 \tag{3.2}$$

where

> E_b is the energy radiated in W/m^2,
> T is the temperature in degree K, and
> σ is the Stefan–Boltzmann constant, $\sigma = 5.669\times10^{-8}$ W/m^2 K^4.

The total energy radiated by the surface increases rapidly with temperature. Figure 3.1 compares the spectrum of blackbody radiation at different temperatures for several light sources. Almost all of the energy at the temperatures shown lies in the infrared regions of the spectrum, where wavelengths are longer than visible light. The wavelength of peak (greatest) emission λ_{peak} (nm) decreases as temperature (T) increases, and is inversely proportional to the temperature in Kelvin. This is known as Wien's displacement law as given in Equation (3.3):

$$\lambda_{peak}T = 2.898 \times 10^{-3} \text{ m K} \tag{3.3}$$

If an object is heated to 2000 K, part of the blackbody radiation spectrum lies in the visible region and is seen as a dull glow, although the wavelength of maximum intensity remains in the infrared. If the temperature increases to 4000 K the glow is a bright orange. At 6000 K (comparable to the temperature of the Sun's surface atmosphere) the maximum peak is in the visible spectrum and the object glows white hot.

In thermodynamics and heat transfer analysis, the energy density is defined as the energy radiated from a surface per unit time and per unit area. Thus the heated interior surface of an enclosure produces a certain energy density of thermal radiation in the enclosure. The thermal radiation energy exchange from a surface is related to its temperature. The subscript b in Equation (3.2) denotes that this is the radiation from a blackbody. This is called blackbody radiation because materials, which obey this law, appear black to the eye. They appear black because they do not reflect any radiation. Thus a blackbody also is considered as a body that absorbs all radiation incident upon it. E_b is called the emissive power of a blackbody.

When radiant energy is incident on a material's surface, part of the radiation is reflected, part is absorbed, and part is transmitted, as shown in Figure 3.2. One defines the reflectivity ρ as the fraction reflected, the absorptivity α as the fraction absorbed, and the transmissivity τ as the fraction transmitted. The total radiant energy distribution can be written in the form:

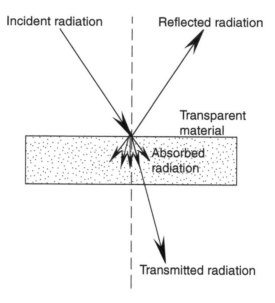

FIGURE 3.2 Incident thermal radiation on a transparent material.

$$\rho + \alpha + \tau = 1 \qquad (3.4)$$

Most solid bodies do not transmit thermal radiation; therefore, the transmissivity may be taken as zero. In this case, Equation (3.4) becomes:

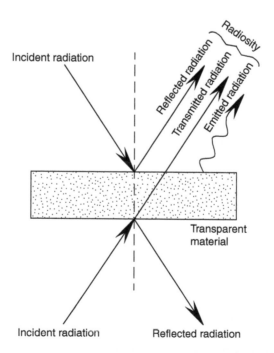

FIGURE 3.3 The definition of radiosity.

$$\rho + \alpha = 1 \tag{3.5}$$

Two types of reflection phenomena may be observed when radiation strikes a surface. If the angle of incidence is equal to the angle of reflection, the reflection is referred to as specular. On the other hand, when the reflected radiation is distributed uniformly in all directions, the reflection is referred to as diffuse. The reflection phenomena are presented in details later in this book.

Figure 3.3 shows the definition of another radiation, referred to as the radiosity. Consider a transparent material receiving radiation on both sides. All the radiation leaving one of its surfaces, which includes emitted, reflected, and transmitted components, is defined as radiosity. Although Figure 3.3 shows the components of radiosity as single rays, it should be remembered that radiosity leaves the surface in all directions and over all wavelengths. Again, if the material is opaque, then the transmitted component is zero and Equation (3.5) applies.

3.3 LIGHT AND ENERGY

The idea that the energy in a light beam consists of numerous amounts of photons simply highlights the fact that light is a form of energy. Energy has the ability to cause transformations from one state to other. Thus ice can be changed to water by heat or a car can be set in motion by the power of fuel energy. Light is just one form of energy. There are many other kinds of energy as well, such as thermal, chemical, electrical, nuclear, and mechanical energy. Whenever a transformation takes place in nature, energy is being changed from one form into another. One particularly important type of transformation involves the conversion of light energy into chemical energy on the retina of the eye or in the film of a camera. In both cases, the energy contained in individual photons is used to produce chemical reactions that are necessary for visual or photographic processes. Both the eye and the camera use visible light photons to initiate the appropriate chemical reactions. Photons from infrared, microwave, and radio wave parts of the spectrum do not contain sufficient energy to be effective. On the other hand, ultraviolet, x-ray, and γ-ray photons carry too much energy, and would be damaging and hazardous, especially to the retina of the eye. The higher energy corresponds to the shorter wavelength of light.

3.4 SOLAR RADIATION ENERGY

The Sun supplies an inexhaustible amount of solar thermal energy to the universe. Solar radiation energy is one of the renewable engines available for many applications. Solar radiation energy reaches the Earth's surface in two forms: (i) direct radiation (the solar parallel rays in a clear sky); and (ii) diffuse radiation (the non-parallel rays of sky radiation, scattered in the atmosphere by a cloudy sky, gases, and atmospheric dust). The average intensity of solar radiation normal to the Sun's rays at the outer edge of the Earth's atmosphere is 1353 W/m^2. This energy is received at the surface of the Earth from the Sun and is subject to variations due to:

1. Variations in tilt of the Earth relative to the Sun. The Earth rotates around its own axis, one complete rotation in 24 h. The axis of this rotation is tilted at an angle of 23.5° to the plane of the Earth's orbit and the direction of this axis is constant, as shown in Figure 3.4. Maximum intensity of solar radiation is received on a plane normal to the direction of radiation. The equatorial regions of the Earth closest to the direction of solar radiation would always receive maximum radiation, if the axis of the Earth were perpendicular to the plane of the orbit. However, due to the tilt of the Earth's axis, the area receiving the maximum solar radiation moves north and south, between the Tropic of Cancer

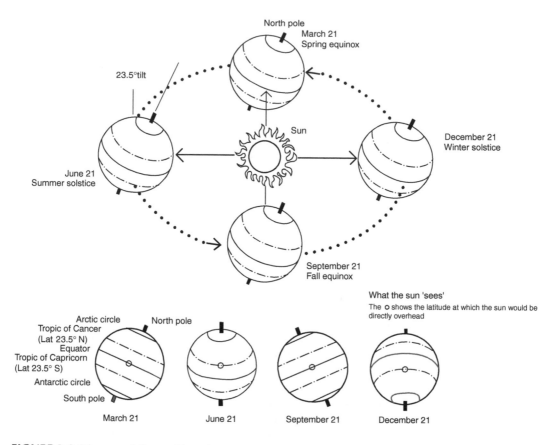

FIGURE 3.4 Diagram of the earth's path.

(latitude 23.5°N) and the Tropic of Capricorn (latitude 23.5°S). This is the primary cause of seasonal change.

2. Variations in atmospheric scattering by air molecules of water vapor, gases, and dust particles.
3. Variations in atmospheric absorption by O_2, O_3, H_2O, and CO_2. Solar radiation incident normal to the Earth's atmosphere has a special distribution. The x-ray and other very short-wave radiation of the solar spectrum are absorbed high in the ionosphere by nitrogen, oxygen, and other atmospheric components; most of the ultra-violet radiation is absorbed by ozone.

At wavelengths longer than 2.5 μm, a combination of low extra-terrestrial radiation and strong absorption by CO_2 and H_2O means that very little energy reaches the ground. For the application of solar energy, only radiation of wavelengths between 0.29 and 2.5 μm need to be considered, as shown in Figure 3.5. This figure shows the intensity of the light from the Sun as a function of wavelength.

Solar radiation is transmitted through the atmosphere, undergoing variations due to scattering and absorption. The other components of the atmosphere scatter a portion of the solar radiation reaching the ground. Thus there is always some diffuse radiation, even in periods of very clear skies. Particles of water and solids in clouds scatter radiation, and in periods of heavy clouds, all of the radiation reaching the ground will be diffused. According to the Rayleigh

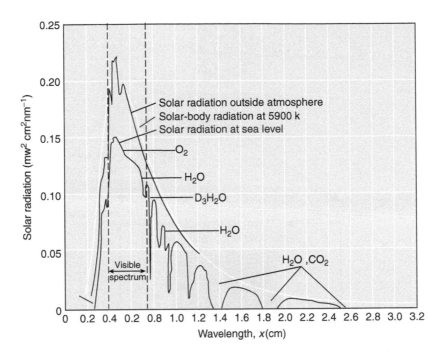

FIGURE 3.5 Spectral radiation distribution curves as a function of wavelength.

theory of scattering, the shorter wavelengths are scattered most and hence, diffuse radiation will tend to be at shorter wavelengths.

Solar radiation measurements often include total (beam and diffuse) radiation, in energy per unit time per unit area, on a horizontal surface. The measurements are made by a pyrheliometer using a collimated detector for measuring solar radiation for a small portion of the sky including the Sun (i.e., beam radiation) at normal incidence. Also the radiation measurements made by a pyranometer measure the total hemispherical solar (beam + diffuse) radiation, usually on a horizontal surface, as shown in Figure 3.6(a). If shaded from the beam radiation by a shade ring, it measures only the diffuse radiation, as shown in Figure 3.6(b). The primary meteorological measurements, done in virtually every weather network stations, are those of global solar radiation and of sunshine

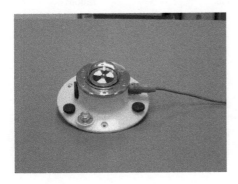

(a)   The black and white pyranometer   (b)   A shadow band stand to shade the
                                                                        pyranometer

FIGURE 3.6 Solar radiation measurements. (a) The black and white pyranometer. (b) A shadow band stand to shade the pyranometer.

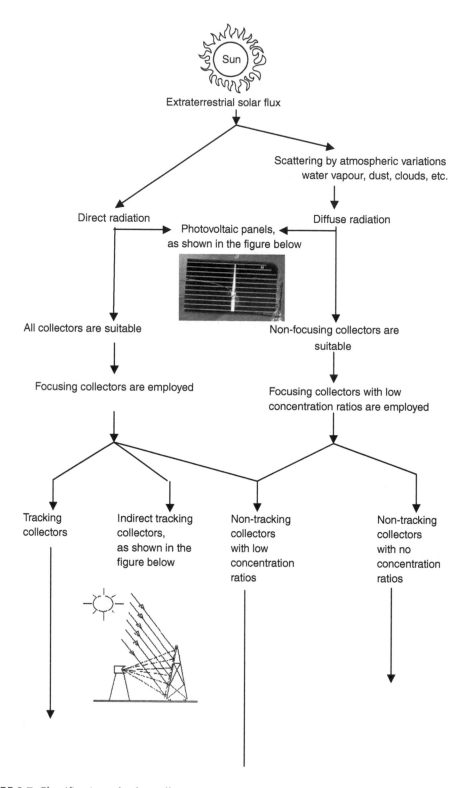

FIGURE 3.7. Classification of solar collectors.

Fresnel lens concentrators
STRA collector
Paraboloidal concentrators
Fresnel reflectors
Heliostat collectors
Parabolic cylindrical collectors,
as shown in the figure below

Honeycomb collectors
Evacuated collectors
Fined flat-plate collectors
Flat-plate collectors,
as shown in the figure
below

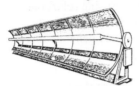
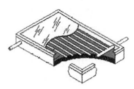

Trapezoidal concentrating collector
Spiral or sea shell collectors
Tabor circular cylinder collector
Axicon (conical) mirror concentrator
AKS composite axicon-lens collector
Trombe-meinel collector
Compound parabolic concentrator,
as shown in the figure below

FIGURE 3.7   *(continued)*

duration, using pyranometers. Pyranometers are radiometers designed for measuring the irradiance on a plane surface, normally from solar radiation and lamps, for measuring photosynthetically active radiation (PAR) in natural daylight or for measurement of illuminance, for the measurement of sunshine duration. There are many types of pyrgeometers designed to measure meteorological radiation parameters. Pyrgeometers are designed for IR (infrared) radiation measurement. Sunshine duration is defined by the World Metrological Organization (WMO) as when the solar radiation exceeds the level of 120 W/m^2.

3.5 CLASSIFICATION OF SOLAR COLLECTORS

Solar collectors may convert solar radiation directly into a useable energy form. A solar collector differs in several respects from more conventional heat exchangers. In solar collectors, energy transfer is from a distant source of radiant energy into thermal energy in a fluid or into electrical energy by photovoltaic panels. Thus, the analysis of solar collectors presents unique problems of low and variable energy fluxes. The type of device used to collect solar radiation energy depends primarily on the application. There are different types of collectors suitable for hot water, hot air, refrigeration, air-conditioning, and electricity generation.

Solar energy collectors are divided into two classes:

1. Liquid-heating solar collectors
2. Air-heating solar collectors

Collectors of solar energy are classified into two main categories. The non-concentrating solar collector category, such as flat plate collectors, is shown in Figure 3.7. In this category the received flux density is low, which makes this type of collector suitable for low temperature applications. On the other hand, in the concentrating solar collectors, such as cylindrical collectors shown in Figure 3.7, the incident flux on a large area is concentrated into a narrow zone with a high intensity of solar flux which generates a high temperature. This correspondingly reduces thermal losses (radiation and convection losses). Thus the concentrating collectors are suitable for high temperature applications. Some applications may require a higher temperature than obtainable by a flat plate collector. However, still a very high temperature is achievable by using solar concentrators with very high concentration ratios, such as cylindrical and heliostat solar concentrators.

Solar electrical collectors are called photovoltaic panels. They generate direct electrical power to customers, as shown in Figure 3.7. All types of solar collectors are a costly purchase, but the cost can be recovered in the long run. Solar collectors also can be used to concentrate solar radiation onto fibre optic cables, which are located in the collector's focusing line. In this arrangement, the fibre optic cables carry solar light into the interior spaces of a building for specular lighting via fibre cables. Fibre optic lighting technology is discussed in detail in this book. Figure 3.7 shows the types of solar collectors, which have been recently built.

3.6 FLAT-PLATE COLLECTORS

The flat-plate solar collector is the basic device used in solar heating and cooling systems. The operation of a flat-plate solar collector is simple. Most of the solar energy incident on the collector is absorbed by the surface, which is black to solar radiation. The essential parts of the collector are:

1. the absorber plate, generally made of metal with a non-reflective black finish to maximize the absorption of solar radiation,

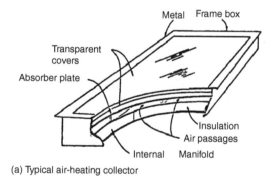

(a) Typical air-heating collector

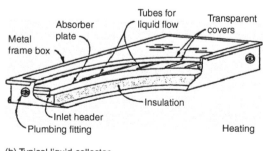

(b) Typical liquid-collector

FIGURE 3.8 Flat-plate collectors. (a) Typical air-heating collector. (b) Typical liquid-heating collector.

2. pipes or ducts to circulate either liquid or air in thermal contact with the absorber plate; thermal insulation for the back side and edges of the plate, and

3. one or more air spaces separated by transparent covers to provide insulation for the top of the plate.

The important parts of a typical flat-plate solar collector are shown in Figure 3.8. The transparent covers usually are made of glass. Glass has excellent weather durability and good mechanical properties. Plastics can be used; however, they generally do not have as high a resistance to weathering as glass; the plastic surface can be scratched and many plastics degrade, yellow with age, and reduce transitivity. The advantage of glass over plastics is that glass absorbs or reflects the entire long-wave radiation incident on it from the solar heated absorber plate. This reduces radiation losses from the absorber plate more effectively than plastics, since plastic transmits part of the long-wave radiation.

Flat-plate collectors absorb both beam and diffuse radiation. Flat-plate collectors can be designed for applications requiring energy delivery at moderate temperatures for such purposes as air heating, water heating for domestic uses, refrigerant systems and heating swimming pools.

Flat-plate collectors are usually mounted in a stationary position with an orientation dependent upon the location and the time of year in which the solar energy system is intended to operate. The cost of energy delivered from a solar collector will depend on the system's thermal performance, installation costs, and maintenance costs.

3.7 SOLAR HEATING SYSTEMS

A solar heating and/or cooling system can be defined as any system, which utilizes solar energy to heat and/or cool a building, although a distinction is often made between active and passive systems. A passive system can be defined as having no moving parts, but it may involve natural circulation of fluid to the heated space. A south-facing window or a skylight, which transmits sunlight can be considered a passive system if it admits more energy than it loses as heat. Another type of passive system may involve movable insulation material, which reduces heat loss from solar absorbing surfaces when there is no sunshine. In contrast to a passive system, an active system involves hardware that collects solar energy, stores heat, and distributes the heat to the rooms in the building. A solar heating system can be classified into a space heating system or a water heating system depending on the fluid employed. In the latter, a liquid is heated in the collector and transferred to a storage unit or directly to the heating loads. The two most common heat transfer working fluids are air and water.

3.7.1 SOLAR AIR HEATING SYSTEMS

An example of a basic solar air heating system is shown in Figure 3.9. This system is called an active solar air heating system. The components of this typical system include:

1. a solar air heating collector, as shown in Figure 3.8(a),
2. a pebble-bed heat storage unit to and from which heat is transferred by circulating air through the bed,
3. a control unit, which includes the sensors and control logic necessary to automatically maintain comfort conditions at all, times,
4. an air heating module comprising of automatic dampers, filters and blowers, and
5. an auxiliary heating unit (usually a warm-air furnace) to provide 100% back-up space heating when storage temperatures are insufficient to meet demands or when the solar system is not operating.

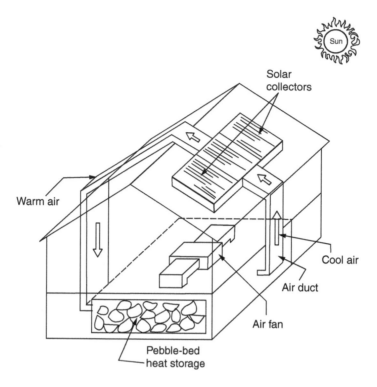

FIGURE 3.9 Basic active air-heating system.

Many solar air-heating designs (e.g., the Denver Solar House, MIT house IV), have been experimentally and commercially employed for space heating of residential buildings in different regions of the world.

There are many designs of passive air solar heating systems used in buildings, some of which are used to create natural ventilation to heat/cool buildings. Figure 3.10 shows two types of the passive air solar systems; direct solar energy collection through the glass roof of a room is one

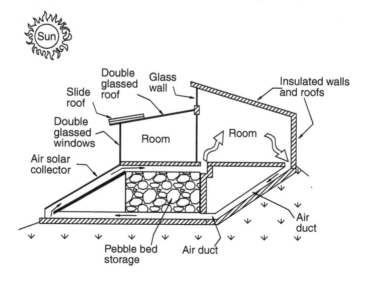

FIGURE 3.10 A passive air heating solar system.

system. The other system is called passive solar air heating through an air solar collector, which is connected to an air heating system.

3.7.2 SOLAR WATER HEATING SYSTEMS

Most of the solar water systems that have been experimentally and commercially used for domestic hot water can be placed in two main groups:

1. Circulating types involving the supply of solar heat to a fluid circulating through a collector, as shown in Figure 3.8(b), and storage of hot water in a separate tank.
2. Non-circulating types involving the use of water containers that serve both as solar collector and storage.

The circulating group may be further divided into the following types and sub-types:

1. Direct heating, single-fluid types in which the water is heated directly in the collector, by:
   (a) thermosiphon circulation between collector and storage and
   (b) pumped circulation between collector, load or storage.
2. Indirect heating, dual-fluid types in which a non-freezing medium is circulated through the collector for subsequent heat exchange with water when:
   (a) heat transfer medium is a non-freezing liquid and
   (b) heat transfer medium is air.

The most common type of direct heating, thermosiphon circulating solar water systems shown in Figure 3.11 utilizes the natural upward movement of heated fluids to circulate water from the collectors to the storage tank without the use of pumps and other components of "active" systems. Location of the storage tank higher than the top of the collector permits circulation of water from the bottom of the storage tank through the collector and back to the top of the storage tank.

Figure 3.12 shows the direct heating, pump circulation type of solar water heater. This arrangement is usually more practical than the thermosiphon type, since the collector would often be located on the roof with a storage tank in the basement. When solar energy is available, a temperature sensor activates a pump, which circulates water through the collector-storage loop. If the solar water heat is used in a cold climate, it may be protected from freeze damage by draining the

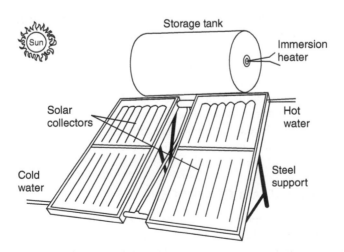

FIGURE 3.11 Passive circulation type of a solar water heater direct heating by thermal themosiphon.

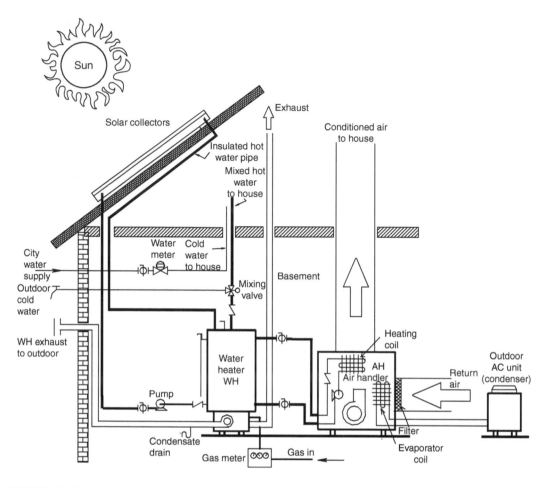

FIGURE 3.12 Active circulation type of a solar water heater direct heating system.

collector when sub-freezing temperatures are encountered. Drainage of the collector can be accomplished by automatic valves, which provide water outflow to a drain and the inflow of air to the collector.

Figure 3.13 illustrates a method for solar water heating using a liquid heat transfer medium in the solar collector used in freezing climates, such as in Canada. The most commonly used liquid is a solution of antifreeze (ethylene glycol) in water. A pump circulates this un-pressurized solution, as in the direct water heating system, and delivers the liquid to and through a liquid-to-liquid heat exchanger. Simultaneously, another pump circulates domestic water from the storage tank through the exchanger, then back to storage. The control system is essentially the same as any design employing water in a direct solar collector system. In a manner similar to that described above, solar energy can be employed in an air heating collector with subsequent transfer to domestic water in an air-to-water heat exchanger, as shown in Figure 3.14.

3.8 HOT WATER AND STEAM GENERATION SYSTEMS

Alternative versions of solar heating systems can be utilized for hot water and steam generation for industrial processes, as shown in Figure 3.15. Water is heated to 60°C in flat-plate solar collectors and then to 88°C in concentrating solar collectors. The hot water flows from the solar collector field into insulated water storage tanks to feed the system by a pump.

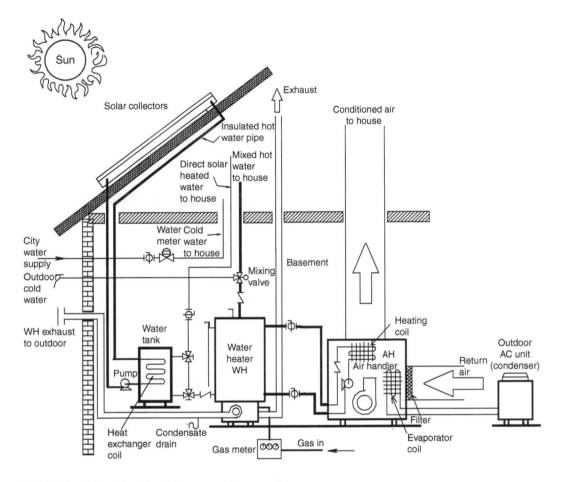

FIGURE 3.13 Dual liquid solar hot water heater.

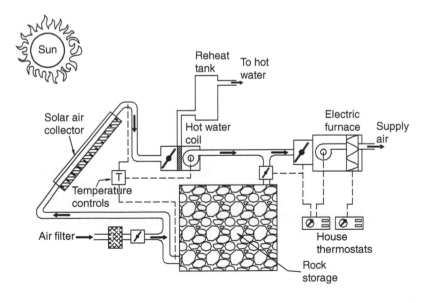

FIGURE 3.14 Hot water heater with solar air collector.

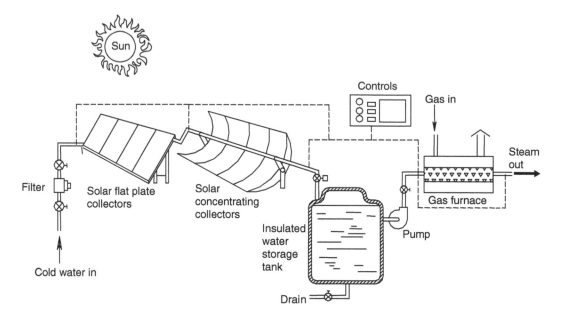

FIGURE 3.15 Hot water and steam generation for industrial processes.

3.9 VAPOUR ABSORPTION REFRIGERATION/AIR CONDITIONING SYSTEMS

The use of solar energy to drive cooling cycles has been considered for two different major applications; the first is in the cooling of buildings and the second is in refrigeration for food preservation. The working fluid is a solution of refrigerant and absorbent. When solar heating

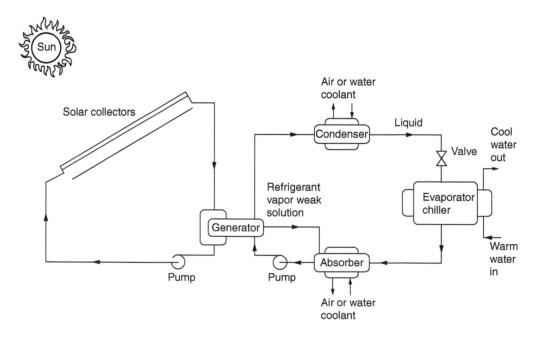

FIGURE 3.16 A solar air conditioner.

is supplied to the generator, some refrigerant is vaporized, and a weak mixture is left behind. The vapor then is condensed, and the liquid is expanded through a throttle valve to the lower pressure evaporator. The spent refrigerant leaving the evaporator next is returned to be absorbed by the weak solution returned from the generator. The resulting strong solution is transferred to the generator. A schematic diagram of one arrangement of such a system is shown in Figure 3.16. The most common refrigerant mixtures used in solar energy absorption refrigeration systems are: ammonia-water and lithiumbromide-water, the former for high temperatures required in the generator while the latter can be operated within the temperature range attainable with simple flat-plate collectors.

3.10 PHOTOVOLTAIC SYSTEMS

The word *photovoltaic* (PV) is a combination of the Greek word for light and the name of the physicist Allesandro Volta. Photovoltaic devices sometimes are called solar cells. They directly convert the incident solar radiation into electrical power. The conversion process is based on the photoelectric effect discovered by Alexander E. Becquerel in 1839. The photoelectric effect describes the release of positive and negative charge carriers in a solid state when light strikes its surface. The solar energy knocks electrons loose from their atoms, allowing the electrons to flow through the semiconductors to produce electricity. This process of converting light (photons) to electricity (voltage) is called the photovoltaic (PV) effect. Solar cells often are used to power calculators, watches, outdoor lighting, remote telecommunication systems, etc. Solar cells are combined typically into modules that hold about 40 cells; about 10 of these modules are mounted in solar cell arrays that can measure up to several metres per side. These flat-plate solar cell arrays can be mounted at a fixed angle facing south, or they can be mounted on a tracking device that follows the sun allowing them to capture the most sunlight over the

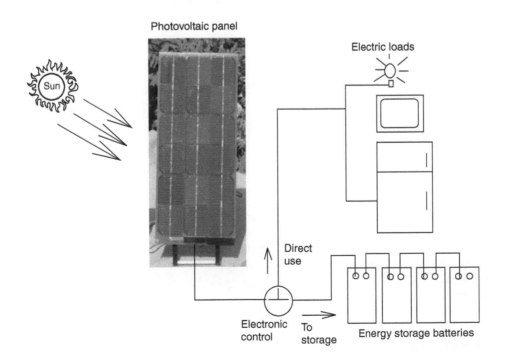

FIGURE 3.17 Major photovoltaic system components.

course of a day. The solar cell arrays are tilted at an angle equal to the geographical location of the site. About 10–20 solar cell arrays can provide enough power for a household. For large electric utility or industrial applications, hundreds of arrays can be interconnected to form a single, large solar cell system. A simplified schematic diagram of a typical photovoltaic system producing electricity is shown in Figure 3.17. More detailed materials and experiments are presented in the semiconductor chapter.

3.11 EXPERIMENTAL WORK

3.11.1 Solar Radiation Measurements

The purpose of this experiment is to measure the solar radiation intensity using a solar radiation meter (black and white pyranometer) as explained above. Students will measure daily solar radiation using a pyranometer connected to a data recorder and a monitor, which displays the data. The students will study the Sun positions during the day hours. Figure 3.18 illustrates outdoor solar radiation measurements arrangement.

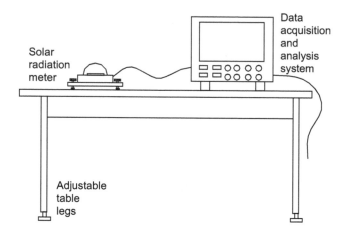

FIGURE 3.18 Solar radiation measurements arrangement.

3.11.2 Technique and Apparatus

Appendix A presents the details of the devices, components, tools, and parts.

1. Solar radiation meter (black and white pyranometer), as shown in Figure 3.6(a).
2. Data acquisition and computer system with display.
3. Clock.
4. Table.

3.11.3 Procedure

Follow the laboratory procedures and instructions given by the professor and/or instructor.

3.11.4 SAFETY PROCEDURE

Follow all safety procedures and regulations regarding the use of pyranometer and working under the sun.

3.11.5 APPARATUS SET-UP

Solar Radiation Measurements

1. Figure 3.19 shows the experimental apparatus set-up.
2. Place the table in a clear area free from shadows generated by trees and buildings during the day.
3. Adjust the legs of the table to level the top surface of the table.
4. Mount a solar radiation meter (black and white pyranometer) on the table. Normally, the solar radiation meter is positioned on a horizontal surface.
5. Place a data acquisition and computer system on the table.
6. Connect the solar radiation meter to the data acquisition system.
7. Measure the solar radiation intensity, time duration, and date using the data acquisition system. Fill out Table 3.1.
8. Keep track of the periods of sunshine during the day of the experiment. Fill out Table 3.2.
9. Repeat the steps 7 and 8 for a few days.

FIGURE 3.19 Solar radiation measurements set-up.

3.11.6 DATA COLLECTION

1. Measure solar radiation intensity, time duration, and date using the data acquisition system. Fill out Table 3.1.
2. Keep track of the periods of sunshine during the day of the experiment. Fill out Table 3.2.

TABLE 3.1
Solar Radiation Measurements

Day:	Month:	Year:
Hour	Solar radiation intensity (unit)	
8:00		
9:00		
10:00		
11:00		
12:00		
13:00		
14:00		
15:00		
16:00		
17:00		

TABLE 3.2
Sunshine Duration

Day:	Month:	Year:
Sunshine duration		
Start time	End time	Duration

3.11.7 CALCULATIONS AND ANALYSIS

No calculations and analysis are required for this case.

3.11.8 RESULTS AND DISCUSSIONS

1. Present solar radiation intensity for each day in a graph.
2. Present sunshine duration for each day in a graph.
3. Discuss solar radiation data collection related to the hours during the day and weather conditions.

3.11.9 CONCLUSION

Summarize the important observations and findings obtained in this lab experiment.

3.11.10 SUGGESTIONS FOR FUTURE LAB WORK

List any suggestions for improvements using different experimental equipment, procedures, and techniques for any future lab work. These suggestions should be theoretically justified and technically feasible.

3.12 LIST OF REFERENCES

List any references that were used in the report. Use one format in writing the references. Never mix reference formats in a report.

3.13 APPENDICES

List all of the materials and information that are too detailed to be included in the body of the report.

3.14 WEATHER STATION

This is a proposal for experimental work using a weather station to measure and study some weather elements, such as solar radiation (direct and diffuse components), temperatures, precipitation, UV index, atmospheric pressure, and wind speed and direction.

FURTHER READING

Adams, R. A., Courage, M. L., and Mercer, M. E., Systematic measurements of human neonatal color vision, *Vision Research*, 34, 1691–1701, 1994.

Air-Conditioning and Refrigeration Institute, *Refrigeration and Air-Conditioning*, Prentice-Hall, Inc., Englewood Cliffs, NJ, 1979.

Al-Azzawi, A. R., The use of solar energy to provide air conditioning and hot water in buildings. M.Sc., Thesis, University of Strathclyde, Glasgow, Scotland, U.K., 1980.

Duffie, J. A. and Beckman, W. A., *Solar Energy Thermal Processes*, Wiley-Interscience Publication, New York, 1974.

EPLAB, *EPLAB Catalogue 2002*, Eppley Laboratory, Inc., Newport, RI, 2002.

Halliday, D., Resnick, R., and Walker, J., *Fundamentals of Physics*, 6th ed., Wiley, New York, 1997.

Holman, J. P., *Thermodynamics*, 4th ed., McGraw Hill, New York, 1988.

Holman, J. P., *Heat Transfer*, 9th ed., McGraw Hill Higher Education, New York, 2002.

Janna, W. S., *Engineering Heat Transfer*, PWS Publishers, Boston, MA, 1986.

Kreith, F. and Kreider, J. F., *Principles of Solar Energy*, McGraw-Hill Book Co., New York, 1980.

Lerner, R. G. and Trigg, L. G., *Encyclopedia of Physics*, 2nd ed., VCH Publishers, Inc., New York, 1991.

McDaniels, D. K., *The Sun: Our Future Energy Source*, 2nd ed., Wiley, New York, 1984.

McDermott, L. C. and Shaffer, P. S., *Tutorials in Introductory Physics, Preliminary Edition*, Prentice Hall Series in Educational Innovation, Upper Saddle River, New Jersey, U.S.A., 1988.

Melles Griot. The practical application of light. *Melles Griot Catalog, Rochester*, NY, U.S.A., 2001.

National Material Advisory Board, National Research Council, *Harnessing Light Optical Science and Engineering for the 21ˢᵗ Century*, National Academy Press, Washington, DC, 1998.

National Resources Canada NRC, *Advanced Houses Testing New Ideas for Energy-Efficient Environmentally Responsible Homes*, CANMET's Building Group, NRC, Ottawa, On, Canada, 1993.

National Resources Canada NRC, Canada advances on the home front. Canada's advance houses program: Leading the way in energy efficiency, indoor air quality, and environmental responsibility. *Home Energy Magazine, NRC, Ottawa, On, Canada*, 1996.

Ontario Ministry of Environment and Energy, *Home Heating and Cooling: A Consumer's Guide, Featuring the Energy Calculator*, Ontario Ministry of Environment and Energy, Toronto, On., 1994.

SEPA, *Solar Power*, Solar Electric Power Association, Washington, DC, 2002.

Smith, W. J., *Modern Optical Engineering*, McGraw-Hill Book Co., New York, 1966.

Tao, W. K. Y. and Janis, R. R., *Mechanical and Electrical Systems In Buildings*, Prentice Hall, Englewood Cliffs, NJ, 2001.

Wilson, J. D., *Physics-A Practical and Conceptual Approach*, Saunders College Publishing, London, 1989.

Young, H. D. and Freedman, R. A., *University Physics*, 9th ed., Addison-Wesley Publishing Co., Inc., Reading, MA, 1996.

4 Light Production

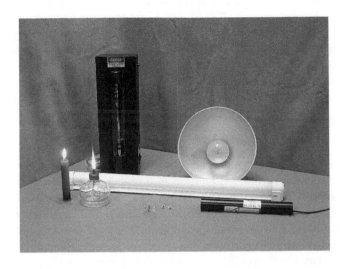

4.1 INTRODUCTION

For many thousands of years, humans relied on the sun as their only source of light. The ability to create fire, and its by-product, light, led to a profound change in the way humans managed their time. Today, there are many options for creating light.

Light can be produced through the rapid change of state of an electron from a state of relatively high energy to a ground state of lower energy. The energy of the electron has to leave the atom somehow, and it is often emitted in the form of a photon of light. This light emission takes place in some optical materials. A series of controlled rapid transitions will produce a stream of photons sufficient to provide illumination. This principle is utilized in the design of flashlights, light emitting diodes, and lasers.

Production of artificial light can be achieved by two types of emissions: either by spontaneous emission or stimulated emission. The spontaneous emission of light serves as the basis for most lighting systems. By passing an electric current through a metal wire (filament), the filament will begin to glow. The chemical composition and temperature of the wire will determine the wavelength of the light being generated. As such, light that is generated by this method is composed of many different wavelengths. The stimulated emission of light occurs in special materials when the energy transfer is stimulated by monochromatic wave light. As a result, the monochromatic light is amplified to produce the lasing light. Stimulated emission can be achieved using either light, electricity, or both as the catalyst. Lasers are the most common example of stimulated light emission. Gas and semiconductor lasers offer the coherent and monochromatic light required to re-shape a cornea, remove a pigment from skin, or transmit signals efficiently through optical fibre. Light production is explained in detail in this chapter.

This chapter also presents the types of energy sources used as a catalyst to produce light. The energy transfers involved vary in conversion efficiency. Experiments at the end of this chapter involve testing different light sources and calculating the conversion efficiency of the light sources.

4.2 SPONTANEOUS LIGHT EMISSION

Spontaneous light emission is the most common process of light production. When an electron is elevated to a high-energy state. This state is usually unstable. The electron will return spontaneously to the ground state, which is a more stable state, immediately emitting a photon. When light is emitted spontaneously, the light propagates in all directions. Light emission from an incandescent light bulb is an example of spontaneous emission. The glow of the thin wire filament of a light bulb is caused by the electrical current passing through it. The electrical current is transferred to thermal energy by the collisions of the excited electrons in the atoms of the wire. This causes the wire's temperature to elevate to a high level so that it emits light. There are many different types of materials with several energy states that can be used to produce spontaneous light at various wavelengths.

4.3 STIMULATED LIGHT EMISSION

Stimulated light emission is another process of light production. This process happens in laser light production. Once the lasing medium is pumped to a higher energy level by an external power source, it contains atoms with many electrons sitting in excited energy levels. The excited electrons occupy higher energy levels than the more relaxed electrons. The higher energy levels are meta-stable states. Just as the electron absorbed some amount of energy to reach this excited level, it can also release this energy. The electron can simply relax, emitting some energy in the form of a photon. This photon has a very specific wavelength that depends on the change of the electron's energy when the photon is released. While the excited electron is in the metastable state, an incoming photon can stimulate it to emit an identical photon with the same wavelength and phase. Thus, laser light is produced. This process is called stimulated light emission.

4.4 LIGHT PRODUCTION BY DIFFERENT ENERGY SOURCES

As explained in light emission processes, energy from an external source must be supplied to a material to boost the electron from its low energy state to a higher energy state before emitting a photon. The following are the types of energy sources used to produce light.

4.4.1 HEAT ENERGY

The most common way of providing an external energy to boost an electron into a higher energy state is to apply heat energy. The electron cannot stay in this excited state for very long. It immediately returns to a lower energy state and gives off its energy in the form of a photon. One example is light emission from a hot metal.

4.4.2 ELECTRICAL DISCHARGE

When an electrical current passes through a gas, such as neon, energy from the current ionizes the gas particles. This ionizing process injects energy into the electrons in the gas. When this ejected electron is reclaimed into a molecule, energy is given off in the form of a photon. One example is the emission of light from an ordinary fluorescent tube.

4.4.3 ELECTRICAL CURRENT

Another way to produce light is by applying electrical power to semiconductors, such as lasers and light emitting diodes. An electrical current is applied to a semiconductor's p–n junction to produce pairs of electrons and holes, as shown in Figure 4.1. When the positive pole is connected to the p-type material and the negative pole connected to the n-type; then the junction conducts. On the n-type side, free electrons are repelled from the electrical contact and pushed towards the junction.

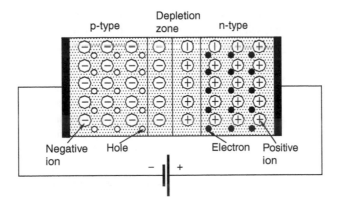

FIGURE 4.1 A P–N junction in a semiconductor.

FIGURE 4.2 Light emitting and laser diodes.

On the p-type side, holes are repelled from the positively charged contact towards the junction. At the junction, electrons will cross from the n-type side to the p-type side, and holes will cross from the p-type side to the n-type side. As soon as they cross the junction, most holes and electrons will recombine and eliminate each other. During this recombination process, each free electron loses a discrete packet of energy called a quantum of energy. This quantum of energy is radiated in the form of photons as electromagnetic waves.

Light emitting diodes and laser diodes are used to produce light in various wavelengths. Light emitting diodes have a similar but simpler structure than laser diodes but they have a lot in common. Almost all light sources used in communication systems are made from semiconductors. Figure 4.2 shows different types of light emitting diodes, and a pigtail fibre connected to a side emitting laser diode commonly used in telecommunication systems.

4.4.4 ABSORPTION OF LIGHT

Some materials, such as phosphorus materials absorb light at a particular wavelength. During light absorption, the electrons move to an excited energy state. They then give off light at a different wavelength. Phosphorus material is used to coat the inside surface of the fluorescent tube. The phosphor absorbs the UV light, which is produced by electrical discharge within the tube. The UV light is then re-emitted as visible light.

4.4.5 CHEMICAL REACTION

There are many chemical materials that emit light when they are mixed together during chemical reactions. This process occurs without adding heat energy; it can even occur at room temperature. During a chemical reaction, atoms and molecules are restructured. Often as a result of the restructuring process, electrons are left in high-energy unstable states. The electrons will emit some of the energy while returning to the lower energy state. This excess energy is either emitted in the form of light or heat. Figure 4.3 shows an example of a chemical reaction taking place between two components in plastic tubes.

4.4.6 BIOLOGICAL REACTIONS

Light production by biological reactions is simply chemical reactions taking place within living organisms. Many biological organisms use bioluminescence for attracting prey and mates or to scare predators. Within these organisms, light is generated by initiating chemical reactions that leave electrons in a high energy state. These excited electrons subsequently decay to a more stable energy state by emitting their energy as light. This occurs in luminous insects, such as the glowworm and the North American firefly. Many fish and sea animals are also capable of this light emission.

4.4.7 NUCLEAR RADIATION

Light production by nuclear radiation can result when there is a nuclear reaction. A nuclear reaction occurs within the unstable nucleus of an atom of a nuclear-reactive (radioactive) material. Similar to chemical reactions, nuclear reactions are accompanied by a release of energy that was stored in the nucleus. This released energy is partially emitted as stream of photons of light in the form of electromagnetic radiation. The nuclear reaction produces a major fragmentation of the particles in the nucleus and releases high-energy nuclear radiations called alpha, beta, and gamma radiations.

FIGURE 4.3 Light production by chemical reaction.

An alpha (α) radiation particle (the nucleus of the He atom) is heavy and moves at very high speeds (high energy) after a nuclear reaction. A beta (β) radiation particle is simply an electron moving at high speeds after a nuclear reaction. Gamma (γ) radiation rays are electromagnetic radiation of very short wavelength (very high frequency) light; therefore, they have very high energy.

Light emission can be a by-product of a chain reaction of nuclear radiation. The main nuclear decay creates nuclear radiation moving at excessive speed. Nuclear radiation will eject electrons from their orbit in the atoms in almost any material it encounters. This process is called ionization. The encounter transfers energy to the ejected electron, exciting the electron to a higher energy state. When the electron in the ionized atom then relaxes to the lower energy state, it gives off the excess energy, which can be emitted as a photon of light. Gamma rays are capable of ionizing more atoms in their path than alpha or beta particles.

Depending on the material involved in the nuclear reaction, the emitted light can be visible. For example, cobalt (a radioactive material) placed in a storage pool of heavy water emits intense cobalt blue light.

4.4.8 Electrical Current

Electrical current can be applied through metal or gas to produce light. The following describes the methods of light production using different lamps. These are reasonably efficient light sources

4.4.8.1 Incandescent Light Lamps

When a solid material is heated, it emits radiant energy in the form of light. The light power at each wavelength depends on the temperature of the material. Figure 4.4 shows the various continuous spectra produced by a black body radiator at increasing temperatures. The amount of energy that falls within the visible portion of the spectrum is a function of temperature. The wavelengths of the visible light range from approximately 400–700 nm. The ideal light source would need to operate around 6000 K.

An incandescent light bulb has a filament of tungsten (a high melting point metal) that is heated to a high temperature by the passage of electric current and emits visible light of all wavelengths.

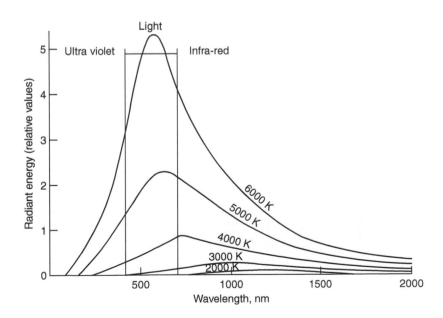

FIGURE 4.4 The radiant energy from a black body at various temperatures.

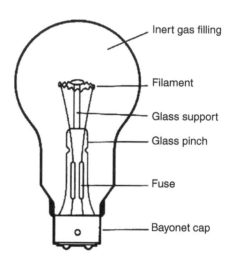

FIGURE 4.5 Incandescent lighting bulbs.

Figure 4.5 shows incandescent bulbs for general lighting service. The filaments are constructed of drawn tungsten wire, which is then coiled. The glass envelopes are spherical, elliptical or mushroom in shape and can be clear, pearl, or white. Clear and pearl bulbs have the same efficiency. The glass envelopes are filled with gases that are chemically inactive with hot tungsten filaments. At the bottom of the glass envelope, a lamp cap holds the electric terminal contacts.

Most of the power supplied to an incandescent light bulb is converted to light and heat. This bulb increases in efficiency with input electrical power, since the higher the filament temperature, the greater the proportion of visible light in the total radiation. The light emitted by an incandescent bulb is called white light, which is a mixture of all visible light wavelengths. Light with a single wavelength is called monochromatic or one colour. Figure 4.6 shows the radiant energy spectrum for an incandescent lamp. Incandescent bulbs are very inefficient light sources. Appendix C.1. presents shape codes and lamp designations for most common incandescent bulbs that are used in different lighting systems.

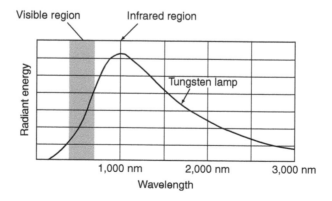

FIGURE 4.6 The radiant energy spectrum for an incandescent lamp.

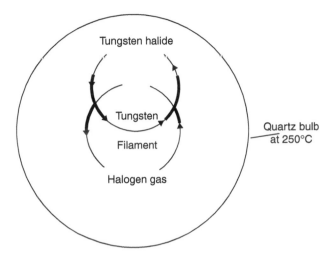

FIGURE 4.7 The halogen regenerative cycle.

4.4.8.2 Tungsten Halogen Lamps

The temperature of a filament is limited by the rate of gas evaporation. The introduction of a halogen (iodine, chlorine, bromine, or fluorine) gas sets up a regenerative cycle as illustrated in Figure 4.7. The evaporated tungsten associates with the halogen at the bulb wall temperature so that the tungsten does not get deposited on the bulb wall blackening it. Tungsten halide vapour forms and when it diffuses back to the hot filament, the vapour dissociates, and the tungsten is redeposited on the filament. This cycle enables the filament to run at temperatures above 3000 K. Appendix C.2. presents shape codes for most common tungsten lamps that are used in different lighting systems.

4.4.8.3 Fluorescent Light Lamps

The history of the electric gas-discharge lamp dates back to before the electric filament lamp, but it was only in 1939 that the fluorescent tube became generally available as an alternative to the filament

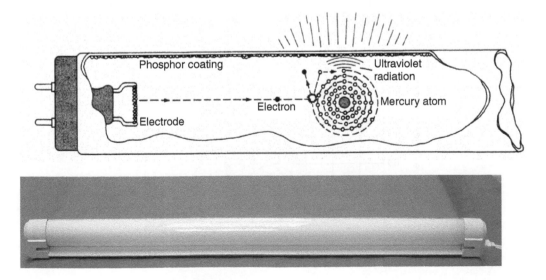

FIGURE 4.8 The tubular fluorescent lamp.

lamp. The fluorescent tube development has been so rapid since then that if a fitting installed in 1940 was relamped today, it could give three times the illumination for the same filament lamp wattage. In the same period, there has been very little equivalent development in the filament lamp output.

A fluorescent light lamp consists of a glass tube filled with a mixture of mercury vapour and an inert gas, such as argon, as shown in Figure 4.8. When a current passes through the mixture, ultraviolet radiation is produced. The inside of the tube is coated with a phosphorus material. This material absorbs the ultraviolet radiation and emits visible light.

Figure 4.9 shows that most of the emitted light is in the visible region of the light spectrum. Fluorescent light lamps produce little heat, so they are very efficient, more so than incandescent lamps. Appendix C.3. presents shape codes and lamp designations for most common fluorescent light lamps that are used in different lighting systems.

4.4.8.4 Black Lights

The fluorescent excitation by ultraviolet (UV) radiation has another common use. Black light used at nightclubs emits radiation in the violet and near-ultraviolet regions. Ultraviolet black lights cause fluorescent paints and dyes on signs, posters, and performers' clothes to fluoresce with brilliant colours. Many products, such as laundry detergents, are packaged in boxes that have bright colours containing phosphors in the ink. These fluoresce somewhat under the store's fluorescent lamps and appear brighter to get your attention. In the dark, when illuminated with black light, the boxes glow brilliantly. Some natural minerals are also fluorescent. Illumination with black lights is a method of identifying these materials.

4.4.8.5 Phosphorescent Materials

Phosphorescent materials, such as zinc sulfide products have numerous applications in safety signs, luminous watch dials, and military and novelty applications. They can be produced as paints, tapes, ropes, vinyl, plastisol, inks, pigments, and varnishes to fit many applications.

Phosphors absorb UV light and re-emit this energy at longer UV wavelengths or as visible light. When an atom is raised to an excited state, it drops back down to a relaxed state within a fraction of a second. However, in phosphorescent substances, atoms excited by photon absorption to high-energy states, called metastable states, can remain there for a few seconds. In a collection of such atoms, many of the atoms will descend to the lower energy state fairly soon, however many will remain in the excited state for over an hour. Hence, light will be emitted even after long periods of time.

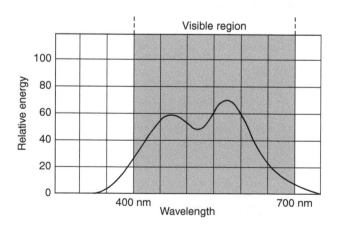

FIGURE 4.9 The radiant energy spectrum for a tubular fluorescent lamp.

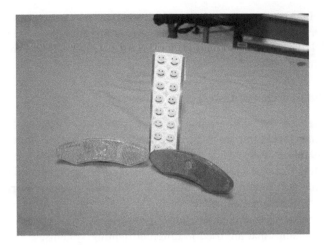

FIGURE 4.10 Phosphorescent materials emitting light.

Figure 4.10 shows phosphorescent materials emitting light. There are also super phosphorescent products based on strontium, an earth alkaline. This self-emitting light technology allows after glow time to exceed 12 h. Safety and novelty applications use this luminous pigment.

4.4.8.6 High-Pressure Mercury Discharge Lamps

A high-pressure discharge lamp contains a gas or vapour that causes light to radiate due to the passage of an electric current. The colour of the light depends on the nature of the gas or vapour. In a high-pressure mercury discharge lamp, the mercury gas discharge is increased, the resonant light radiation at 253.7 nm is absorbed and other radiations occur, many within the visible part of the spectrum. The colour of light from a high-pressure lamp is much whiter than that from a low-pressure one and the ultraviolet content is reduced. Figure 4.11 shows a high-pressure mercury

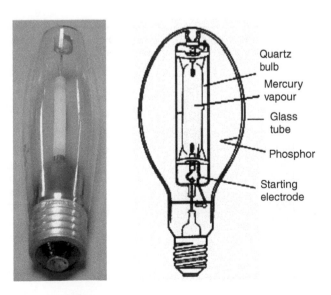

FIGURE 4.11 A high-pressure mercury discharge lamp.

discharge lamp. The lamp operates on a choke circuit and incorporates its own starting electrode; no circuit transformer or switch is necessary. High-pressure lamps are more efficient than incandescent lamps. Appendix C.4. presents shape codes for most common high-pressure mercury lamps that are used in different lighting systems.

4.4.8.7 Metal Halide Discharge Lamps

The high-pressure mercury discharge lamp emits a rather cold white light, as the colour distribution is strong in blues and greens but weak in reds. Better colour distribution can be achieved by introducing other metals in their halide form into the discharge tube. This produces additional radiation. The wavelength and colour depend on the metal used. Two examples of these metals are the iodides of thallium and sodium. Metal halide lamps require a higher starting voltage than the high-pressure mercury discharge lamps, but operate at higher efficiencies, and their colour appearance is suitable for industrial and commercial uses. Appendix C.5. presents shape code for most common metal halide lamps that are used in different lighting systems.

4.4.8.8 Sodium Lamps

Sodium lamps are a very efficient light source. However, the radiation is concentrated into two wavelengths around 589 nm resulting in a strong yellow light, monochromatic in nature. This is inappropriate where any reasonable colour is required. The lamp produces light by discharging an electrical current through sodium vapour. A control circuit limits the current and provides a high starting voltage. There are a number of different types of low-pressure sodium lamps which are illustrated in Figure 4.12. High-pressure sodium lamps are used in lighting systems. Appendix C.6. presents shape codes for most common high-pressure sodium lamps that are used in different lighting systems.

4.4.8.9 Energy Efficient Light Bulbs

An energy-efficient lighting system uses less natural resources and reduces the impact on the environment. Energy efficiency in lighting systems requires careful attention to the type of lighting and fixture purchased and to their location in the building. Energy efficient bulbs will provide energy savings and improve the quality of lighting.

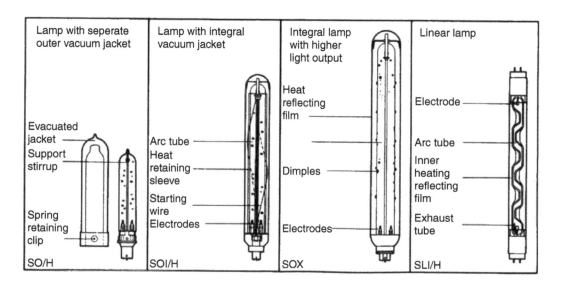

FIGURE 4.12 Low-pressure sodium lamps.

A bewildering array of light fixtures and lamps are available many of which have changed dramatically in recent years. The choices made when choosing new lighting products can lower the energy spending. Every lighting product can be thought of as having two costs: the initial cost is the purchase price, and the secondary cost is the ongoing cost of the energy used by the light bulb. If an energy efficient light bulb is chosen, the initial cost will be high, but the ongoing cost will be far less than any ordinary light bulb with the same power.

The first consideration is the amount of artificial light that is required. There is a natural tendency to provide too much light. The lighting selection also depends on the application. General lighting provides moderate light throughout a room. Several light sources can create uniform lighting and minimize glare and contrasts. Task lighting facilitates activities, such as reading and cooking; it requires more focused light in addition to general lighting. Protective and safety lighting helps prevent accidents, especially in stairwells, and discourages prowlers. Decorative lighting highlights room features, such as drapes, a fireplace, or a piece of art.

The second consideration is the type of fixture to be used. The three main types of household lighting available are standard incandescent, fluorescent, and tungsten-halogen fixtures. With incandescent bulbs, a very small percentage of the electricity used actually becomes light. Fluorescent lights and tungsten-halogen bulbs, on the other hand, are much more energy-efficient.

Whatever type of lighting is selected, the choice of fixtures and the range of efficiencies that they offer is wide. To select the most efficient fixture for the application, the following should be considered:

- Fixtures are specifically designed for the purpose of the application. For task lighting, such as for reading, the lamp should provide highly directional lighting. For general room lighting, choose a fixture that provides light over a broad area.
- Some fixtures have features that limit light output. Heavy shades and bowls, for example, can reduce light levels significantly. Features that will enhance useful light output, such as reflectors, will direct the light outwards.
- A fixture with a single bulb gives more useful light than one with several bulbs having the same total wattage. For example, four 25 W bulbs give little more than half the light of one 100 W bulb.

Energy Efficient Incandescent Bulbs

Energy efficient incandescent bulbs are regular incandescent light bulbs that have been improved to use less energy, but with slightly less light output. They do not offer energy savings as large as the compact fluorescents; however they are compatible with dimmers, work well outdoors, fit any light fixtures that take a regular incandescent bulb, and cost less. Long-life or extended-life incandescent bulbs last longer than regular bulbs, but output 30 per cent less light while using the same amount of energy. Bulbs with a higher than normal voltage rating are also available (typically, 130 V instead of the standard 120 V). These are intended for use where the voltage supply fluctuates (as in some rural areas). These bulbs are less efficient than standard incandescent bulbs.

Energy Efficient Fluorescent Tubes

Traditionally, fluorescent lighting has had limited use in most homes; the older fixtures were big, and the light quality was poor for most home needs. Today, however, new types of fluorescent tubes and bulbs produce light comparable to incandescent lighting, and a new generation of compact fluorescent tubes and fixtures are available. Fluorescent lighting now can be used in almost any area of a building and is best suited for utility areas. Though the initial cost may be higher, the greatly reduced energy costs and the very long tube life can make fluorescent fixtures an economical choice. Fluorescent tubes use 60–80% less energy and last 10–20 times longer than incandescent bulbs. They also work with conventional switches.

Energy Efficient Compact Fluorescent Tubes

Fluorescent lighting now can fit conveniently into most standard light fixtures by using compact fluorescent tubes. Compact fluorescents use about 25% of the energy that incandescent bulbs use and last up to 10 times longer. This makes them ideal for lights that are hard to reach or difficult to change, such as stairwells and recessed luminaires. To maximize the efficiencies of compact fluorescent tubes, they are best suited for lights that are used for long periods of time (three or more hours per day).

Compact fluorescents work most efficiently when the lamp is oriented downwards, with the base up. This is because the efficiency of the tube depends on the temperature of the coldest part of the lamp, which is the end most remote from the ballast. As heat rises, a base-up lamp will be coolest at the bottom, and therefore produce the greatest amount of light. Compact fluorescents emit the same kind of warm, natural light as regular incandescent bulbs. Two types can be installed together in the same room to produce a balanced, even illumination. Most fluorescent tubes are not compatible with dimmer switches and are not recommended for cold outdoor applications.

Energy Efficient Tungsten-Halogen Bulbs

Tungsten-halogen parabolic aluminum reflectors (PAR) type bulbs are low wattage floodlights (halogen-type incandescent bulbs). A standard 150 W incandescent spotlight can be replaced by a 90 W PAR bulb; this will cut down on electricity consumption by up to 40%. Tungsten-halogen bulbs can be used outdoors as well as indoors and are suitable for small parking lots, gardens, and marking pathways. No warm-up or waiting is required after a momentary power interruption.

Energy Efficient High Intensity Discharge (HID) Bulbs

Efficient HID light sources include high-pressure sodium (HPS) and metal halide types. HPS bulbs are an efficient source for exterior lighting. They provide bright light, which is ideal to ensure safety around buildings. HPS bulbs use 70% less energy than standard floodlights and last up to eight

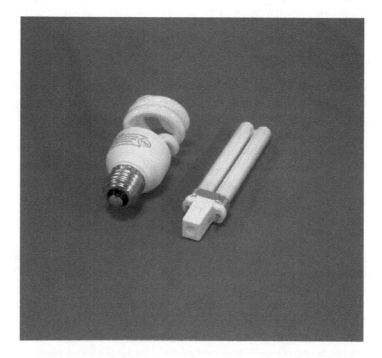

FIGURE 4.13 Energy efficient light lamps.

times longer. This type of fixture is estimated to have a lifetime of approximately 10 years. Metal halides provide a blue–white light source and can be used to highlight plants in gardens.

Figure 4.13 shows light output that is measured in lumens. A standard 110 W bulb produces about 1,680 lumens; whereas a 26 W compact fluorescent tube produces about 1,800 lumens. This figure can be used when replacing regular bulbs with energy efficient compact fluorescents.

Solar Energy Powered Lighting

Solar energy is an alternative energy that is useful for lighting to reduce the energy cost. Solar lighting systems are suitable for small garden and sign displays. Figure 4.14 shows a solar powered system. It's fixture has a re-chargeable battery to store energy generated by solar photovoltaic cells. The solar photovoltaic cells generate energy from solar radiation during day hours for use at night. The fixture uses a fluorescent lamp powered by the stored energy.

4.4.8.10 Lasers

One of the most fascinating inventions in the second half of the twentieth century is the laser. Laser is an acronym for Light Amplification by Stimulated Emission of Radiation. Laser devices produce a very narrow intense beam of monochromatic and coherent light. Laser beams are used in many applications such as for aligning subway tunnels, precisely adjusting integrated circuit chips, boring holes in steel, sending signals into fibre optic cables in communication systems, performing surgery on the retina of the eye, activating remote control devices, playing CDs, printing materials in laser printers, reading bar codes. Lasers send thousands of signals through telephone lines and many television signals over fibre optic cables. There are other applications listed in details in the laser chapter.

The beam emitted from the laser is coherent and directional, while an ordinary light source emits incoherent light in all directions. The action of laser production is based on electrons existing at specific energy levels or states characteristic of that particular atom. Electrons can be pumped up to higher energy levels by the external energy. These electrons may jump spontaneously to the lower state, resulting in the emission of energy in the form of a photon. However, if a photon with

FIGURE 4.14 Solar powered outdoor lighting.

FIGURE 4.15 Helium–Neon (He–Ne) laser device.

this same energy strikes the excited atom, it can stimulate the atom to make the transition to the lower state. In this phenomenon, called stimulated emission, the new photon has the same frequency as the original one. Also, these two photons are in phase and moving in the same direction. This is how coherent light is produced in a laser. Depending on the particular lasing material being used, specific wavelengths of light are emitted.

There are many types of lasers, such as the gas laser, Ruby laser, Helium–Neon (He–Ne) laser, Fabry-Perot laser, CO_2 gas laser, and laser diode. Figure 4.15 shows the Helium-Neon (He–Ne) laser device with internal glass tube that has two parallel mirrors, one at each end. More details on laser principles, operations, and devices are presented laser chapter.

4.5 EXPERIMENTAL WORK

This experiment studies light emission from different types of light sources. The student will study power output from different light sources, such as a fluorescent tube, incandescent bulb, halogen lamp, mercury lamp, He–Ne laser, LED, and laser diode. The student will measure the power

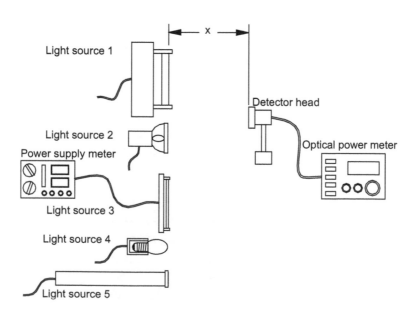

FIGURE 4.16 Light emission from different light sources.

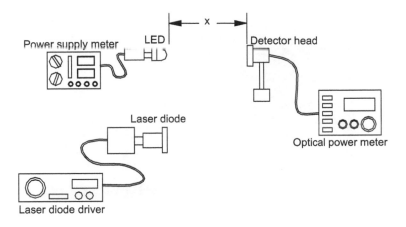

FIGURE 4.17 Light emission from laser diode and light emitting diode.

consumption and the energy output from these light sources. Then the efficiency of a light source will be calculated. An efficiency comparison among each light source will be carried out.

This experiment is arranged into two cases:

a. Light emission from five different light sources shown in Figure 4.16.
b. Light emission from an LED and laser diode shown in Figure 4.17.

4.5.1 TECHNIQUE AND APPARATUS

Appendix A presents the details of the devices, components, tools, and parts.

1. He–Ne laser source, as shown in Figure 4.16 and Figure 4.18. Always follow laser operation procedure and safety instructions when using a laser light source.
2. Laser power supply.
3. Fluorescent tube, incandescent lamp, halogen lamp, mercury lamp, as shown in Figure 4.18.
4. LED and laser diode/laser diode drive, as shown in Figure 4.2.
5. Hardware assembly (clamps, posts, screw kits, screwdriver kits, sundry positioners, etc.).

FIGURE 4.18 Five different light sources.

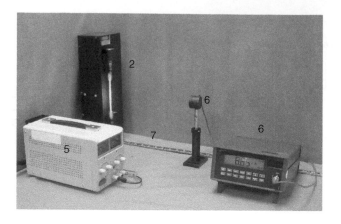

FIGURE 4.19 Light emission from five different light sources.

6. Electrical power supply meter, as shown in Figure 4.19.
7. Optical detector head and optical power meter, as shown in Figure 4.19.
8. Ruler.

4.5.2 PROCEDURE

Follow the laboratory procedures and instructions given by the professor and/or instructor.

4.5.3 SAFETY PROCEDURE

Follow all safety procedures and regulations regarding the use of light source devices, measurement instruments, and power supplies.

4.5.4 APPARATUS SET-UP

4.5.4.1 Light Emission from Five Different Light Sources

1. Figure 4.19 shows the experimental apparatus set-up.
2. Prepare five types of light sources on the table.
3. Connect light source 1 to the power supply meter.
4. Mount an optical detector head connected to an optical power meter at a distance of x cm facing the light source.
5. Align the optical detector and light source 1 to face each other.
6. Measure the distance x between light source 1 and the optical detector head. Fill out Table 4.1.
7. Turn on the power supply to provide electrical power to light source 1. Be sure not to exceed the maximum voltage and current for each light source.
8. Turn off the lights of the lab.
9. Measure the voltage (V, in Volts) and the current (I, in Amperes) to calculate the power (W, in Watts $= V \times I$) consumed by light source 1. Fill out Table 4.1.
10. Measure light power emitted by light source 1 using the optical power meter. Fill out Table 4.1.
11. Repeat steps 3–10 for the four remaining light sources.
12. Illustrate the location of the light sources and the optical detector head in a diagram.

FIGURE 4.20 Light emission from a light emitting diode and laser diode.

4.5.4.2 Light Emission from an LED and Laser Diode

1. Figure 4.20 shows the experimental apparatus set-up.
2. Repeat the procedure explained in Case (a).
3. Use an LED and laser diode/laser diode drive as light sources. Be sure not to exceed the maximum voltage and current for the LED and laser.
4. Measure the voltage V and the current I for the LED and laser diode, and the distance x. Fill out Table 4.2.
5. Measure the light power emitted by the LED and laser diode. Fill out Table 4.2.
6. Illustrate the location of the LED and laser diode and the optical detector head in a diagram.

4.5.5 DATA COLLECTION

4.5.5.1 Light Emission from Five Different Light Sources

1. Measure the voltage (V), the current (I), and the distance (x). Fill out Table 4.1.
2. Measure the light power emitted by the light sources. Fill out Table 4.1.

TABLE 4.1
Light Emission from Five Different Light Sources

Light	Light Source	Volts	Amperes	Input Power = ($V \times I$)	Optical Power (Output Power)	Distance	Efficiency = (Output Power/Input Power)×100
Source	Type	(V)	(I)	(W)	(W)	(x)	(%)
No. 1							
No. 2							
No. 3							
No. 4							
No. 5							

4.5.5.2 Light Emission from an LED and Laser Diode

1. Measure the voltage (V), the current (I), and the distance (x). Fill out Table 4.2.
2. Measure the light power emitted by the LED and laser diode. Fill out Table 4.2.

TABLE 4.2
Light Emission from a Laser Diode and LED

Light Source	Light Source Type	Volts (V)	Amperes (I)	Input Power = (V × I) (W)	Optical Power (Output Power) (W)	Distance (x)	Efficiency = (Output Power/Input Power)×100 (%)
No. 1							
No. 2							

4.5.6 CALCULATIONS AND ANALYSIS

4.5.6.1 Light Emission from Five Different Light Sources

1. Calculate the power (W, in Watts $= V \times I$) consumed by the light sources. Fill out Table 4.1.
2. Calculate the efficiency of each light source, by dividing power output by power input. Fill out Table 4.1.
3. Illustrate the location of the light sources and the optical detector head in a diagram.

4.5.6.2 Light Emission from an LED and Laser Diode

1. Calculate the power (W, in Watts $= V \times I$) consumed by the LED and laser diode. Fill out Table 4.2.
2. Calculate the efficiency of the LED and laser diode, by dividing the power output by power input. Fill out Table 4.2.
3. Illustrate the location of the LED and laser diode and the optical detector head in a diagram.

4.5.7 RESULTS AND DISCUSSIONS

4.5.7.1 Light Emission from Five Different Light Sources

1. Discuss the measurements of the voltage and current, the calculation of the power consumed by the light sources, and the optical power measurements.
2. Compare the efficiency of the light sources.
3. Verify the calculated efficiency with the efficiency provided by the manufacturer.

4.5.7.2 Light Emission from an LED and Laser Diode

1. Discuss the measurements of the voltage and current, the calculation of the power consumed by the LED and laser diode, and the optical power measurements.
2. Compare the efficiency of the LED and laser diode.
3. Verify the calculated efficiency with the efficiency provided by the manufacturer.

4.5.8 CONCLUSION

Summarize the important observations and findings obtained in this lab experiment.

4.5.9 Suggestions for Future Lab Work

List any suggestions for improvements using different experimental equipment, procedures, and techniques for any future lab work. These suggestions should be theoretically justified and technically feasible.

4.6 LIST OF REFERENCES

List any references that were used in the report. Use one format in writing the references. Never mix reference formats in a report.

4.7 APPENDICES

List all of the materials and information that are too detailed to be included in the body of the report.

FURTHER READING

Cornsweet, T. N., *Visual Perception*, Academic Press, New York, 1970.

Deveau, R. L., *Fiber Optic Lighting: A Guide for Specifiers*, 2nd ed., Prentice Hall PTR, Upper Saddle River, NJ, 2000.

Falk, D. et al., *Seeing the Light Optics in Nature, Photography, Color, Vision, and Holography*, Wiley, New York, 1986.

Grove, A. S., *Physics and Technology of Semiconductor Devices*, Wiley, New York, 1967.

Hood, D. C. and Finkelstein, M. A., Sensitivity to light handbook of perception and human performance, In *Sensory Processes and Perception*, Boff, K. R., Kaufman, L., and Thomas, J. P., Eds., Vol. 1, John Wiley & Sons, Toronto, 1986.

Illuminating Engineering Society of North America, *IESNA lighting education—Intermediate level*, IESNA ED-150.5A, 1993.

Kuhn, K., *Laser Engineering*, Prentice Hall, Inc, Upper Saddle River, NJ, 1998.

Lengyel, B., *Lasers*, Wiley, New York, 1971.

Lerner, R. G. and Trigg, G. L., *Encyclopedia of Physics*, 2nd ed., VCH Publishers, Inc, New York, 1991.

McComb, G., *The laser cookbook—88 practical projects Tab Book*, Division of McGraw-Hill, Inc, Blue Ridge Summit, PA, 1988.

Melles Griot, The practical application of light, *Melles Griot Catalog*, Irvine, California, U.S.A., 2001.

Overheim, R. D. and Wagner, D. L., *Light and Color*, John Wiley and Sons, Inc, New York, 1982.

Pritchard, D. C., *Environmental Physics: Lighting*, Longmans, Green & Co, London, 1969.

Product Knowledge, Lighting reference guide, *Product Development*. 2nd ed., Ontario Hydro, Toronto, ON, Canada, 1988.

Salah, B. E. A. and Teich, M. C., *Fundamentals of Photonics*, Wiley, New York, 1991.

Schildegen, T. E., *Pocket Guide to Color With Digital Applications,* Delmar Publishers, Albany, New York, U.S.A., 1998.

SCIENCETECH, *Designers and Manufacturers of Scientific Instruments Catalog*, SCIENCETECH, London, Ontario, Canada, 2003.

Serway, R. A., *Physics for Scientists and Engineers*, 3rd ed., Saunders Golden Sunburst Series, Philadelphia, 1990.

Weisskopf, V. F., How light interacts with matter, *Scientific American*, 219, (3), pp. 60–71, U.S.A., September, 1968.

Williamson, S. J. and Cummins, H. Z., *Light and Color in Nature and Art*, Wiley, New York, 1983.

Yariv, A., *Optical Electronics*, Wiley, New York, 1997.

Yeh, C., *Applied Photonics*, Academic Press, San Diego, CA, 1994.

5 Light Intensity

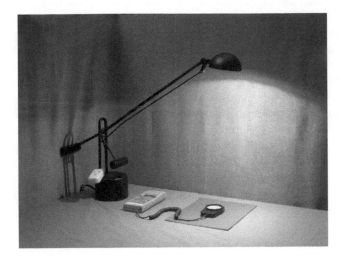

5.1 INTRODUCTION

The light from a fire or flame will propagate spherically outwards from the source. In general, the apparent intensity of the light will vary inversely with the distance from the observer to the source. The earliest unit for measuring light, the foot-candle, was based on the amount of light being emitted from a single candle at a distance of one foot. The more candles lit, the brighter the radiant light.

Since light waves propagate spherically from a candle, the number of photons per unit area will vary based on the distance to the candle. The measurement of light intensity is called luminous flux.

Experimental work presented in this chapter involves measuring light intensity of five different light sources at five distances. It also involves measuring the distribution of light intensity on a floor, to see if the light is evenly distributed throughout the floor and if it complies with the recommended indoor lighting levels.

5.2 LIGHT INTENSITY

Electromagnetic wave motion involves the propagation of energy from a source to a detector. The rate of energy transfer is expressed as intensity. The intensity is defined as the energy transported per time per unit area as given in the following equation:

$$\text{Intensity} = \frac{\frac{\text{Energy}}{\text{Time}}}{\text{Area}} = \frac{\text{Power}}{\text{Area}} \tag{5.1}$$

The standard units of intensity are watts per square metre (W/m^2).

5.3 LUMINOUS FLUX

Light sources (except coherent and directional light sources such as lasers) emit electromagnetic waves in all directions. A lamp emits radiation when electric energy is supplied to the lamp. The radiation energy emitted per unit of time by the lamp is called the radiant power or sometimes referred to as the radiant flux. Radiant energy covers a wide range of the light spectrum. The visible light is a small range of the light spectrum. The range of the visible light is approximately between wavelengths of 400 and 700 nm. The visual sensing of the eye perceives the visible light and luminous energy radiation. Therefore, the luminous flux is defined as that part of the total radiant power emitted from a light source that is capable of affecting the sense of sight.

The human eye is not equally sensitive to all colours. In other words, equal radiant power of different wavelengths does not produce equal perceived brightness. Each colour has perceived brightness different from other colours. For example, a 40-W green light bulb appears brighter than a 40-W blue light bulb. A relative sensitivity graph of the visible light spectrum is shown in Figure 5.1. The sensitivity curve is a bell-shape with the centre of the visible spectrum at wavelength of 555 nm. Under normal conditions, the human eye is most sensitive to yellow–green light of wavelength 555 nm. This curve also shows the sensitivity falling off rapidly for longer and shorter wavelengths.

5.4 LUMINOUS INTENSITY

The brightness of a light source is referred to as luminous intensity I, for which the candela (cd) is the SI unit (metric system). The candela is defined in terms of the light emitted by a small pool of platinum at its melting point. A candle has a luminous intensity of about 1 cd, and in fact the former standard of this quantity was an actual candle of specified composition and dimensions.

Consider the definition of a solid angle θ in terms of the radius and the circular arc length in two-dimensional analysis. Figure 5.2 shows the angle θ, measured in radians, is defined as:

$$\theta = \frac{S}{R} \tag{5.2}$$

where S is the arc length and R is the radius of the circle.

Similarly, the solid angle Ω in space analysis is explained in Figure 5.3. Point C represents the centre of the sphere. A steradian sr is defined as conical in shape. The solid angle (conical) Ω, representing one steradian, such that the area A of the subtended portion of the sphere is equal to R^2,

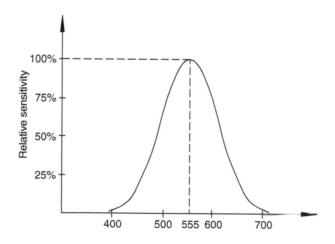

FIGURE 5.1 A relative sensitivity curve for visible light perceived by the eye.

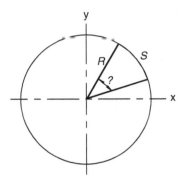

FIGURE 5.2 An angle θ expressed in radians in a plane.

where R is the radius of the sphere. The total solid angle of a sphere is 4π steradians. The number of steradians in a given solid angle can be determined by dividing the area on the surface of a sphere (lying within the intersection of that solid angle with the surface of the sphere) by the square of the radius R of the sphere.

Therefore, the solid angle in steradians sr is defined as:

$$\Omega = \frac{A}{R^2} \, (\text{sr}) \tag{5.3}$$

Notice that, the steradian is a unitless quantity in Equation (5.3). The solid angle is independent of the radius R of a sphere. There are 4π sr in a sphere, regardless of the length of its radius as given in the following equation:

$$\Omega = \frac{A}{R^2} = \frac{4\pi R^2}{R^2} = 4\pi (\text{sr}) \tag{5.4}$$

The luminous flux F emitted by a light source describes the total amount of visible light that it emits. A small bright source may give off less light than a large diameter source, just as a small hot object may give off less heat than a large warm one. The unit of luminous flux is the lumen lm. Consider a 1-candela light source at the centre of the sphere, as shown in Figure 5.3. Such a light source is referred to as isotropic because it radiates equally in all directions. One lm equals the luminous flux falling on each square metre of a 1 m radius sphere. A lumen of flux is equivalent to about 0.0015 W of yellow–green light of wavelength 555 nm. Since the area of a sphere of radius R

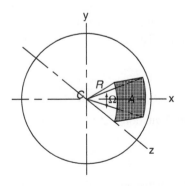

FIGURE 5.3 A solid angle Ω expressed in steradians in space.

is $4\pi R^2$, the above sphere has a total surface area of 4π m^2, and the total luminous flux radiated by a 1-cd source is thus 4π lm.

Light travels radially outward in straight lines from a point source that is small compared with its surroundings. For a point light source, the luminous flux included in a solid angle Ω remains the same in all directions from the source. Therefore, the luminous flux per unit solid angle is simply expressed as the total flux. The physical quantity that expresses this relationship is called the luminous intensity I. The luminous intensity I of a source of light is the luminous flux F emitted per unit solid angle Ω, as given in the following equations:

$$I = \frac{F}{\Omega}. \tag{5.5}$$

or,

$$F = I\Omega = \tag{5.6}$$

The unit for luminous intensity I is the lumen per steradian (lm/sr), which is called a candela (cd). The candela, or candle as it was sometimes called, originated when the international standard was defined in terms of the quantity of light emitted by the flame of a certain make of candle. This standard was found unsatisfactory and eventually was replaced by the platinum standard. The total solid angle Ω for an isotropic source is 4π sr. Thus, the luminous flux is given by:

$$F = 4\pi I \tag{5.7}$$

5.5 ILLUMINATION

If the intensity of a source is increased, the luminous flux F transmitted to each unit of surface area in the vicinity of the source is also increased. The surface appears brighter. The illumination E of a surface A is defined as the luminous flux F per unit area A, as given in the following equation:

$$E = \frac{F}{A} \tag{5.8}$$

When the flux F is measured in lumens and the area A in square metres, the illumination E has units of lumens per square metre or lux (lx). When the area A is expressed in square feet, the E is expressed in lumens per square foot. The lumen per square foot is sometimes referred to as the footcandle, where 1 footcandle $= 0.093$ lux.

Figure 5.3 shows a point light source emitting the light energy in all directions. To see the relations between intensity and illumination, consider a surface A at a distance R from a point source at C of intensity I, as shown in Figure 5.3. The solid angle subtended by the surface at the source is given by Equation (5.3) as where the area A is perpendicular to the emitted light. If the luminous flux makes an angle θ with the normal to the surface, then the projected area, $A \cos\theta$, is the perpendicular area to the normal. Therefore, the solid angle in Equation (5.3) can be written in general form as follows:

$$\Omega = \frac{A \cos\theta}{R^2} \tag{5.9}$$

Substitute Equation (5.9) into Equation (5.6) to obtain:

$$F = I\Omega = \frac{IA \cos\theta}{R^2} \tag{5.10}$$

In this equation, θ is the angle between the direction of the light and a normal to the surface where the light strikes. When $\theta = 0°$, then $\cos\theta = 1$. To express the illumination as a function of

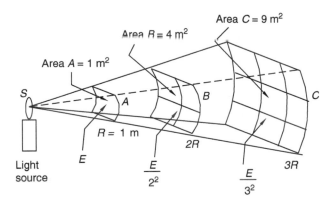

FIGURE 5.4 Inverse-square law of light.

intensity, substitute Equation (5.10) into Equation (5.8) to give the following equation:

$$E = \frac{F}{A} = \frac{IA \cos \theta}{AR^2} = \frac{I \cos \theta}{R^2} \qquad (5.11)$$

For the special case in which the surface is normal to the flux, $\theta = 0°$, Equation (5.11) is simplified to:

$$E = \frac{I}{R^2} \qquad (5.12)$$

Equation (5.12) involving illumination and luminous intensity is a mathematical formulation of the inverse-square law of light waves. This law can be stated as: The illumination of a surface is perpendicular to the luminous intensity of a point light source and is inversely proportional to the square of the distance. This describes the fundamental law of light intensity variation with distance from a point light source.

Figure 5.4 illustrates this inverse-square law of light. The figure shows light emitted from a point source S becomes distributed over 1 m² at the first distance R. If the distance is doubled to $2R$, the light is distributed over 4 m², and the illumination per square metre is one-quarter of what it was at the distance R. If the distance is tripled to $3R$, the illumination per square metre will be one-ninth of what is was at distance R. Therefore, the intensity of light emitted from the point light source varies inversely as the square of the distance from the light source. This is called the inverse square law for electromagnetic light waves.

5.6 EXPERIMENTAL WORK

This experiment studies the light intensity from different types of light sources. These different light sources could be a fluorescent tube, incandescent bulb, halogen lamp, mercury lamp, LED, and laser diode. The student will measure the light intensity for at least five different distances from these light sources. Then intensity of a light source will be drawn as a function of distance. Light intensity comparison among each light source will be carried out. The student will also measure light intensity on a flat area and compare it with the recommended lighting levels related to the area under consideration according to indoor lighting system codes.

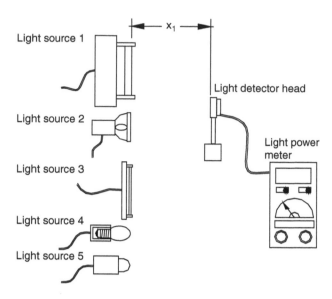

FIGURE 5.5 Light intensity measurements from different light sources.

This experiment covers two cases:

a. Light intensity measurements of five different light sources are shown in Figure 5.5. Each light source is placed in front of a light meter at five distances between them. Light intensity is measured using a light meter.

b. Distribution of light intensity on a flat surface is shown in Figure 5.6. A light meter is used to measure the distribution of the intensity of the light from the ceiling lighting system projected onto the floor of a room. The measurements will show if the light is evenly distributed throughout the floor and also if the measurements comply with the recommended indoor lighting levels.

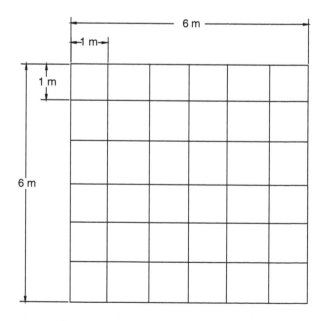

FIGURE 5.6 Distribution of light intensity on a flat surface.

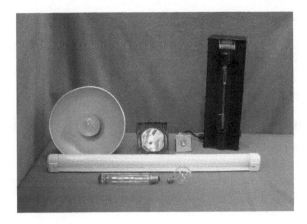

FIGURE 5.7 Five different light sources.

5.6.1 TECHNIQUE AND APPARATUS

Appendix A presents the details of the devices, components, tools, and parts.

1. Fluorescent tube, high power incandescent lamp, low power incandescent lamp, halogen lamp, and mercury lamp, as shown in Figure 5.7.
2. Hardware assembly (clamps, posts, positioners, etc.).
3. Light meter, as shown in Figure 5.6.2.
4. Ruler.

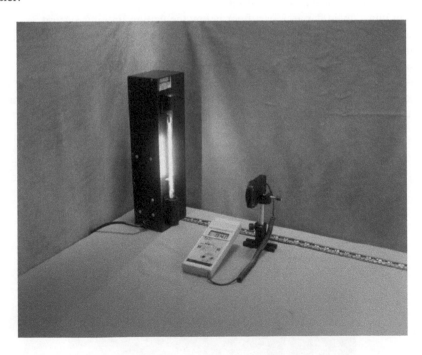

FIGURE 5.8 Light intensity from five different light sources set-up.

5.6.2 PROCEDURE

Follow the laboratory procedures and instructions given by the professor and/or instructor.

5.6.3 Safety Procedure

Follow all safety procedures and regulations regarding the use of optical instruments and measurements, and light source devices.

5.6.4 Apparatus Set-up

5.6.4.1 Light Intensity From Five Different Light Sources

1. Figure 5.8 shows the experimental apparatus set-up.
2. Prepare five types of light sources on the table.
3. Connect light source 1 to the power outlet.
4. Mount a light meter at a distance x_1 from light source 1.
5. Align the light meter and light source 1 to face each other.
6. Measure the distance x_1 between light source 1 and the light meter. Fill out Table 5.1.
7. Switch on the power supply of light source 1.
8. Turn off the lights of the lab.
9. Measure the intensity of light source 1 using the light meter. Fill out Table 5.1.
10. Switch off the power supply of light source 1.
11. Repeat steps 3 to 10 for four different light sources.
12. Turn on the lights of the lab.
13. Illustrate the location of the light sources and the light meter in a diagram.

5.6.4.2 Distribution of Light Intensity on a Flat Surface

1. Figure 5.9 shows the experimental set-up.
2. Find an unoccupied flat area of a floor below the ceiling lighting system.
3. Create a grid pattern on the flat area in 1 square metre sections. Use masking tape to map out the flat floor.
4. Draw the flat area layout on a paper using a suitable scale.
5. Measure the light intensity at the centre of each square in the grid. Be careful as not to shadow the light meter with your body during the measurements. Fill out Table 5.2.
6. Illustrate the distribution of the light intensity on the flat floor in a diagram.

FIGURE 5.9 Distribution of light intensity on a flat floor.

5.6.5 Data Collection

5.6.5.1 Light Intensity from Five Different Light Sources

Measure the intensity of each of the five light sources, at five distances. Fill out Table 5.1.

TABLE 5.1
Light Intensity from Five Different Light Sources

Distance between Light Source and Light Meter (unit:)	Intensity of the (unit:)				
	Light Source Type 1	Light Source Type 2	Light Source Type 3	Light Source Type 4	Light Source Type 5
$x_1 =$					
$x_2 =$					
$x_3 =$					
$x_4 =$					
$x_5 =$					

5.6.5.2 Distribution of Light Intensity on a Flat Surface

Measure the intensity of light at the centre of each grid on the floor. Fill out Table 5.2.

TABLE 5.2
Distribution of Light Intensity on a Flat Surface

	1	2	3	4	5	6 (meter)
1	× (unit:)	×	×	×	×	×
2	×	×	×	×	×	×
3	×	×	×	×	×	×
4	×	×	×	×	×	×
5	×	×	×	×	×	×
6 (meter)	×	×	×	×	×	×

5.6.6 Calculations and Analysis

5.6.6.1 Light Intensity from Five Different Light Sources

No calculations are required for this case.

5.6.6.2 Distribution of Light Intensity on a Flat Surface

No calculations are required for this case.

5.6.7 Results and Discussions

5.6.7.1 Light Intensity from Five Different Light Sources

1. Discuss the measurements of the intensity of light sources at different distances.
2. Compare the intensity of the light sources.
3. See if the inverse square law of light is applicable with your measurements.
4. Verify the intensity of the light sources with the intensity that is provided by the manufacturer.

5.6.7.2 Distribution of Light Intensity on a Flat Surface

1. Discuss the measurements of the intensity of light at the centre of each grid of the flat floor under consideration.
2. Present the light intensity distribution of the flat floor under consideration.
3. Compare your measurements of the intensity of light with indoor lighting system codes.

5.6.8 Conclusion

Summarize the important observations and findings obtained in this lab experiment.

5.6.9 Suggestions for Future Lab Work

List any suggestions for improvements using different experimental equipment, procedures, and techniques for any future lab work. These suggestions should be theoretically justified and technically feasible.

5.7 LIST OF REFERENCES

List any references that were used in the report. Use one format in writing the references. Never mix reference formats in a report.

5.8 APPENDICES

List all of the materials and information that are too detailed to be included in the body of the report.

FURTHER READING

Adams, R. A., et al., Systematic measurements of human neonatal color vision, *Vision Research*, 34, 1691–1701, 1994.

Alda, J., Laser and Gaussian beam propagation and transformation, in *Encyclopedia of Optical Engineering*, Barry Johnson, R., Ed., Marcel Dekker, Inc., New York, 2002.

Beiser, A., *Physics*, 5th ed., Addison-Wesley Publishing Company, Reading, MA, 1991.

Chen, K. P., In-Fiber light powers active fiber optical components, *Phontonics Spectra*, 78–90, 2005.

Cutnell, J. D. and Johnson, K. W., *Physics*, 5th ed., Wiley, New York, 2001.

Digonnet, M. J. F., *Rare-Earth Fiber Lasers and Amplifiers*, Marcel Dekker, NY, U.S.A., 2001.

Falk, D., et al., *Seeing the Light Optics in Nature, Photography, Color, Vision, and Holography*, Wiley, New York, 1986.

Hood, D. C. and Finkelstein, M. A., Sensitivity to light handbook of perception and human performance, in *Sensory Processes and Perception*, Boff, K. R., Kaufman, L., and Thomas, J. P., Eds., Vol. 1, Wiley, Toronto, 1986.

Jameson, D. and Hurvich, L. M., Theory of brightness and color contrast in human vision, *Vision Research*, 4, 135–154, 1964.

Jenkins, F. W. and White, H. E., *Fundamentals of Optics*, McGraw Hill, New York, 1957.

Jones, E. R. and Childers, R., *Contemporary college physics*, McGraw-Hill Higher Education, Maidenhead, U.K., 2001.

Kuhn, K., *Laser engineering Englewood, Chffs*, Prentice Hall, Inc, NJ, 1998.

Lerner, R. G. and Trigg, G. L., *Encyclopedia of Physics*, 2nd ed., VCH Publishers, Inc, New York, 1991.

Melles Griot, The practical application of light, *Melles Griot Catalog*, Irvine, California, U.S.A., 2001.

Pritchard, D. C., *Environmental Physics: Lighting*, Longmans, Green & Co, London, 1969.

Product Knowledge, Lighting reference guide, *Product development*, 2nd ed., Feb. 1988.

Robinson, P., *Laboratory Manual to Accompany Conceptual Physics*, 8th ed., Addison-Wesley, Inc., Reading, MA, 1998.

Romine, G. S., *Applied physics concepts into practice Englewood, Chffs*, Prentice Hall, Inc, NJ, 2001.

Shen, L. P., Huang, W. P., Chen, G. X., et al., Design and optimization of photonic crystal fibers for broad-band dispersion compensation, *IEEE Photonics Technology Letter*, 15, 540–542, 2003.

Urone, P. P., *College Physics Belmont*, Brooks/Cole Publishing Company, CA, U.S.A, 1998.

Warren, M. L., *Introduction to Physics*, W.H. Freeman and Company, San Francisco, CA, U.S.A., 2001.

Weisskopf, V. F., How light interacts with matter, *Scientific American*, 60–71, 219(3), U.S.A., Sept. 1968.

6 Light and Colour

6.1 INTRODUCTION

Visible light radiating from the sun consists of a broad spectrum of wavelengths ranging from violet to red. When light strikes a material, some of the wavelengths are reflected and others are absorbed. The characteristics of a material and the incident light striking them determine the apparent colour of an object. Light can be used to measure an object's temperature; for example, blue stars are hotter than red stars.

Display devices mix red, green, and blue (RGB) light to form different coloured pixels in an image. Likewise, digital cameras separate and measure the RGB light from a scene to form digital images.

Experimental work presented in this chapter will demonstrate mixing coloured light to produce different colours. Newton's colour wheel and black and white colour wheels are used in the experiment.

6.2 COLOURS

Colour is a phenomenon of light, and a visual perception using the physical sense of human sight. As with most human senses, vision is susceptible to sensory adaptation. It has been known at least since the early days of the ancient Egyptians that fragments of clear, colourless glass and precious stones emit the colours of the rainbow when placed in the path of a beam of white light. It was not until 1666, however, that this phenomenon called desperation was systematically investigated by scientists. The refracting telescope was invented by a Dutch eyeglass maker named Hans Lippershey (1570–1619). Later, the English physicist and mathematician, Isaac Newton (1642–1727), was starting to search for a technique for removing colouration from the images seen through reflecting telescopes. In 1672, Newton described his experiments to the Royal Society in London. His theory, that white light was made up of many colours, was revolutionary and it was greeted with skepticism. Indeed Newton and another English physicist, Robert Hooke (1635–1703), became involved in a bitter debate, and Newton refused to publish his conclusions until after Hooke's death, 32 years later.

As with all human sense organs, the eye is sensitive to physical energy. The human eye is sensitive to a very small portion of the electromagnetic light spectrum. This portion of the spectrum is referred to as light energy covering the visible range. The human eye only sees colours starting from blue light waves that measure about 400 nm and ending with the longer red light waves of about 700 nm, as shown in Figure 6.1.

Light is one of the four essential elements in the visual experience of colour. For the human sensation of colour vision, the following four elements are required:

1. A coloured object
2. A light source to illuminate the coloured object
3. The human eye as a receptor of the light energy reflected from the object
4. The human brain to interpret the electrochemical neural impulses sent from the eye

There are many different pigments or colourants, and these can be organic or inorganic in nature. The coloured object either absorbs or reflects the incident light that illuminates it. Some objects, such as coloured glass, can allow light to be transmitted through them, filtering all colours other than their own.

The eye is the physical receptor of light energy. The major parts of the human eye are shown in Figure 6.2. As the eye focuses on a coloured object, light passes through the cornea and is regulated by the pupil opening. The lens of the eye focuses the light on the retina in the back of the eye. The retina is the light-sensitive surface located around the back of the eye. The retina is made up of rods and cones, which are the photosensitive cells that convert the light energy into different nerve impulses. The retina also has ganglion cells and bipolar cells. Vision is a function of light energy reaching the rods and cones of the eye. The bulbous cones can see chromatic vision or colour, such as red, green, and blue. The cones can also see achromatic colour, such as white, grey, and black. There are more than six million bulbous cones distributed in the retina of the eye. The cones need higher levels of illumination to produce colour vision. There are around 100 million rods in the retina, and they function in dim light conditions and produce monochromatic vision. The rods produce vision in times of low illumination. The rods see black and white or grey tonal values. If you look at a landscape during the daylight, the grass and shrubs are green and the sky is blue in

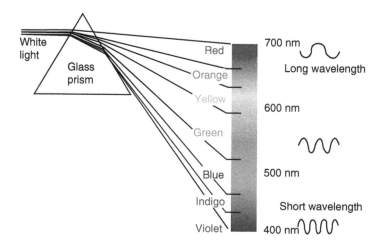

FIGURE 6.1 (See colour insert following page 200.) Dispersion of white light into the visible spectrum by a prism.

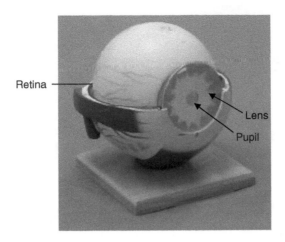

Retina — Lens Pupil

FIGURE 6.2 (See colour insert following page 200.) The major parts of the human eye.

colour as a result of the different wavelengths of energy that stimulate the bulbous cones. Looking at the landscape at dusk requires the use of the rods in the eye and the green grass or blue sky appears to be a dark grey due to limited illumination. The fovea is the most sensitive part of the retina in daylight conditions. The fovea is made up of only bulbous cones and is centrally located in the retina. The light reflected from a coloured object is seen best when it strikes the fovea, because these cones are connected directly to the optic nerve which transmits the data to the brain. Other cones and rods in the retina require a relay of nerves to the brain. There are several theories regarding colour vision. The most common theory regarding colour vision was presented by the Young-Helmholtz theory, composed by Thomas Young (1773–1829), a British physician and Egyptologist and Hermann Helmholtz (1821–1894), a German physicist and physiologist. Their theory of colour vision proposed that the eye has three different colour receptors, each sensitive to different wavelengths of the spectrum. Because the eye receives light energy it was proposed that the bulbous cones are either sensitive to red, green, or blue light energy.

Another more modern colour theory includes elements from previous models of colour vision. In 1955, this theory was called either the zone or the opponent-process, and was proposed by Leo Hurvich and Dorothea Jameson. In the opponent-process theory, the eye has three colour receptors (cones with broad bands of sensitivity) that peak at three different portions of the visible spectrum. These different wavelength receptors in the eye are connected to the ganglion cells, some of which are always active regardless of light energy. The ganglion cells that receive light energy pulsate faster, and if the cell is inhibited in some way the pulsation diminishes. In this theory the ganglion cells can provide two opposing signals to the brain, which better explains differences in colour vision and negative afterimages that we experience.

Normal colour vision is often called trichromat, which refers to a person having the ability to sense or discriminate light from dark, red from green, and yellow from blue. Colour blindness is when an individual is limited in his/her ability to discriminate colour, such as red from green and yellow from blue. If a person has the ability to see contrast from light to dark and cannot discriminate one colour pattern from another, that individual has dichromat vision. If the individual has only the ability to see contrast of light to dark, the person has the vision of a monochromat.

Each colour of light covers a specific range of wavelengths. The spectral power distribution of the colour is measured by a spectroscope, which is an instrument that works on the principle of white light dispersion in a prism to indicate the power of the light's components at various wavelengths. Light entering the spectroscope is dispersed through a prism, such as in Newton's

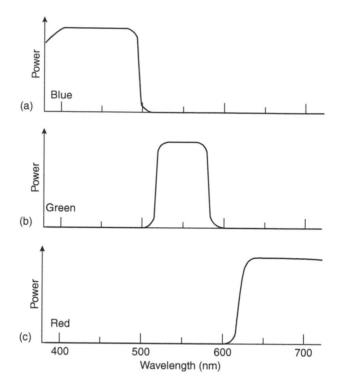

FIGURE 6.3 Spectral power distributions for the (a) blue, (b) green, and (c) red colours of the white light dispersion in a prism.

experiment. As a detector moves across the screen, for each wavelength, it measures the power of light received from the light source. The resulting curve giving the power to each wavelength is called the spectral power distribution. Figure 6.3 shows the spectral power distributions (SPD) for the blue, green, and red portions of the visible light spectrum. For blue, the power is found concentrated at short wavelengths; for green, at intermediate wavelengths; and for red, at long wavelengths.

6.3 MIXING LIGHT COLOURS

All colour reproduction is accomplished through either additive or subtractive colour theory. Additive theory is synonymous with light energy and the direct experience of light. Subtractive theory describes the experience of light reflected from a coloured object, such as paper coated with printing inks.

6.3.1 ADDITIVE METHOD OF COLOUR MIXING

Newton's edition of *Optics* in 1704 described a number of experiments in which he arranged mirrors to direct two different monochromatic components of the spectrum to the same location on a white screen. When he observed the colour of the mixed beams, Newton noticed that he could not always produce white light by mixing two monochromatic colours together. His systematic studies revealed the general rules of colour addition. As described above, each colour has a spectral power distribution. The spectral power distribution of one colour is added to the spectral power distribution of the other colour to produce the net spectral power distribution of the new colour. The rules of additive colour mixing are valid whether or not the light that produces a given colour is

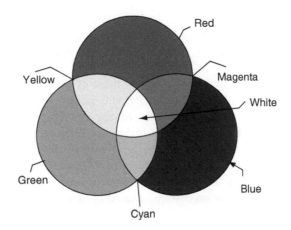

FIGURE 6.4 (See colour insert following page 200.) Additive colour mixing.

monochromatic. For simplicity, it is assumed that the brightness of each original is the same. Mixing green light with red light produces a yellow light. A second example of addition is when blue light and red light are mixed to produce magenta light (a purple). For a third example, blue light and green light are added together to produce cyan light (a turquoise). When the three colours blue, green, and red are mixed additively in various proportions a wider range of colours can be produced. If these three colours have the same proportion and brightness when added, they produce white colour. Therefore, the three colours the red, green, and blue (R, G, and B, respectively) are called primary colours for additive colour mixing. Adding the three primary colours of red, green, and blue lights will create white light, as shown in Figure 6.4.

Any two colours that produce white colour when added together are called complementary. The complement of a primary colour is called a secondary colour. Therefore, cyan is a secondary colour and is the complement of red colour. The complement of green is blue plus red colours, or magenta, so magenta is also a secondary colour. Also, the complementary of blue is green plus red colours, or yellow. Thus, yellow is a secondary colour and is the complementary of blue. As shown in Figure 6.4, the outside colours are the primary colours and the interior colours are the secondary colours.

6.3.2 Subtractive Method of Colour Mixing

The idea of primary colours can be applied to a different set of situations where the spectral power distribution is altered as light passes through a medium, which absorbs differing amounts of power across the spectrum. This occurs when a beam of light passes through one or more coloured filters (such as a piece of coloured glass or a plastic containing a dye). The more coloured filters we insert into the path of the beam, the more light is absorbed and the dimmer is the emerging light. A similar effect is produced when paints of various colours are mixed together. These are examples of subtractive colour mixing. The rules of subtractive mixing are more complicated than those of additive mixing. If several filters are placed in a beam of white light, only the first filter receives all of the white light. The others receive variously coloured light depending on the filters that precede them.

An understanding of the theory of filters is a good introduction to subtractive mixing of colours. By convention, a filter is named by the colour of the light that emerges from it when it is placed in a white beam. A red filter allows long wavelengths to pass and absorbs the remaining wavelengths. A common yellow filter allows medium through long wavelengths to pass and absorbs short wavelengths. The process where a material absorbs light in only certain regions of the spectrum is called selective absorption. It is the primary cause of most colours that we see. In terms of our additive primaries, a red filter allows only the red component of the incident light to pass. A common yellow filter transmits green and red. Similarly, a magenta filter transmits blue and red, and a cyan filter transmits blue and green.

Consider the combination of two filters, such as those of magenta and yellow. The magenta filter allows blue and red to pass. The yellow filter would transmit any incident green and red, but since green is absorbed by the magenta filter, only red emerges. The combination of the two filters therefore has the same effect as a red filter. It makes no difference which of the two filters is inserted into the beam first. To figure out what happens we must follow the path of light filtering, as shown in Figure 6.5. This figure shows the effect of (a) Magenta, (b) Yellow, (c) Cyan filters on white light. Also, (d–f) illustrates how the three additive primary colours can be produced from white light by using two filters in succession. The SPD of the emerging light are shown for each case.

The three subtractive primary colours are yellow, magenta, and cyan, as shown in Figure 6.6. The primary subtractive colours receive their name from the fact that each colour absorbs one-third of the white light spectrum. These subtractive primaries will make suitable ink colours

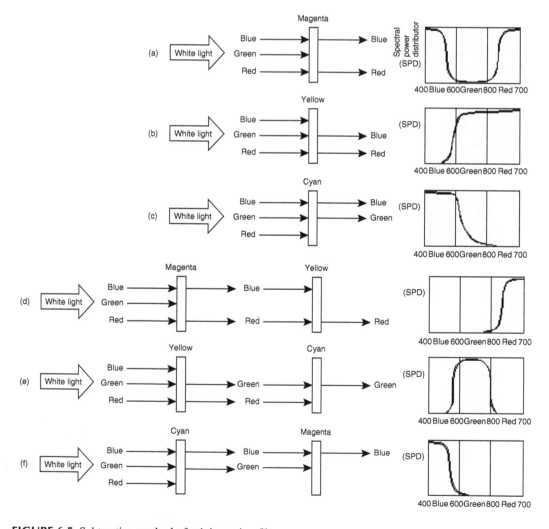

FIGURE 6.5 Subtractive method of mixing using filters.

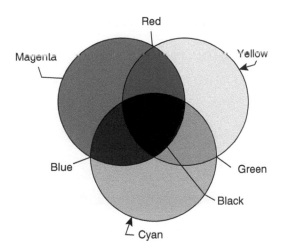

FIGURE 6.6 (See colour insert following page 200.) Subtractive colour mixing.

because when all are mixed they create black, or the total absence of colour. Printers subtract colour from the white sheet of paper. Basically, all the colours we see in a printed separation are contained in the red, green, and blue light energy reflected from a white sheet of paper. To produce the image, different degrees of red, green, and blue light are subtracted from the white light spectrum or paper.

The relationship between the additive primary colours and the subtractive primaries is considered complementary, as shown in Figure 6.7. The term complementary comes from the paired relationship where each subtractive colour absorbs its complementary one-third of the light spectrum. Yellow ink on white paper absorbs its complementary colour of blue light. Magenta ink on white paper will absorb one-third of the light spectrum, or its complementary colour of green. Finally, the cyan ink will absorb its complement of red light.

The colour wheel is used as a graphic aid in understanding basic colour theory. Though simple in design, the colour wheel represents a significant amount of information. The printing of colour reproductions involves creating colour separations of the primary colours contained in the original image, and reproducing them. When you mixed red and green tempera paints they created brown paint. The colour wheel used in graphic arts is built on the physical sciences; when red and green light are mixed we see the yellow. Using the colour wheel shown in Figure 6.8, note that the three additive primary colours are separated by the subtractive primary colours. Also, the

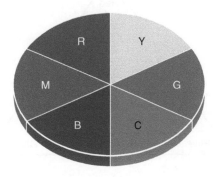

FIGURE 6.7 (See colour insert following page 200.) The Relationship between the additive and subtractive primary's complement.

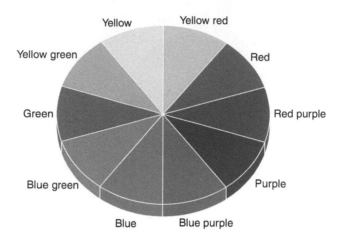

FIGURE 6.8 (**See colour insert following page 200**.) The colour wheel (the three primary additive colours sandwiched between variations of coloured light).

complementary colour pairs are opposite each other in the colour wheel. Finally, note that the two additive colours of light that make up a particular subtractive primary are located on each side of that ink colour.

6.4 THE COLOUR TRIANGLE

The colour triangle, as shown in Figure 6.9, is a triangular arrangement of the additive and subtractive primaries with white at the centre. Red, green, and blue are located at the corners, while magenta, yellow, and cyan are located at the sides. The order of the colours is such that the sum of any two additive primaries at the corners gives the subtractive primary between them on the sides, and the sum of all three primaries gives white at the centre.

Colours opposite each other on the colour triangle are complementary. Two colours are said to be complementary if, when added together, they produce white. Magenta and green are complementary, for when added together, they contain all of the spectrum colours of white light. Similarly red and cyan, as well as yellow and blue, are complementary.

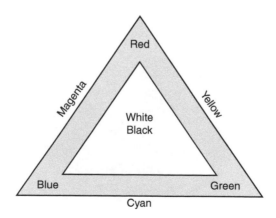

FIGURE 6.9 Diagram of the colour triangle, with the additive primaries at the corners and the subtractive primaries at the sides.

6.5 THE C.I.E. CHROMATICITY DIAGRAM

Well-planned steps toward a quantitative measurement of colour were taken by the Commission Internationale de I' Eclairage C.I.E. (International Commission on Illumination I.C.I.) in 1931. The three primaries—red, green, and blue—were adopted, in which the visible spectrum was divided into three overlapping spectral curves. It should be mentioned here that any given colour sample can be measured with a spectroscope in terms of the three adopted primaries, and the results of the measurements can be expressed by two numbers. These two numbers can then be plotted on a graph. When the pure spectrum colours (red R, orange O, yellow Y, green G, blue B, and violet V) are matched against a mixture of the standard primaries, a smooth curve is obtained, as shown in Figure 6.10. The figure shows white colour at the centre, the complete array of all possible colour mixtures lies within the enclosed area RGBVW, with the purples P and magentas M confined to the region RWV between the two ends of the spectrum.

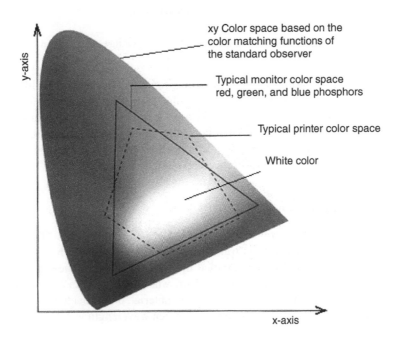

FIGURE 6.10 (**See colour insert following page 200**.) The C. I. E. Chromaticity diagram.

6.6 COLOUR TELEVISIONS

The use of artificial light sources to reproduce colour, such as a colour television, must also use the blue, green, and red portions of the spectrum to produce a white image. The television set has three cathode ray guns that project different degrees of red, green, and blue onto the phosphor coating on the inside of the picture tube to create different colours in the spectrum. The mixing of light energy to create a colour image is additive colour theory. In a dark room, the overlapping of red, green, and blue light beams will create additional colours, such as magenta, cyan, and yellow. Additive colour can also be seen in a television studio where coloured lights are used to create different effects on a studio backdrop. The various light adjustments on a control panel allow for different amounts of red, green, and blue filtered lights to create any colour desired, as shown in Figure 6.11.

The colour monitor on a desktop computer or colour workstation also uses red, green, and blue dots of phosphor on a black background to produce images. When electronically activated, the phosphor dots emit light energy. Unfortunately, the range of light energy emitted from various

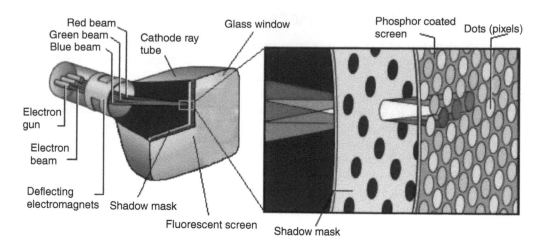

FIGURE 6.11 (**See colour insert following page 200.**) The additive theory is used in colour television.

monitors differ just like in various colour televisions for the home market. With this in mind, it becomes critical to understand that one colour monitor on a desktop workstation will vary from another. Furthermore, as a colour monitor and phosphor coating age the colour output will shift. There are some more expensive colour monitors that have been colour-calibrated and can be adjusted to a production standard every day. The need to calibrate colour devices is becoming more critical as the hardware and software become more sophisticated.

6.7 SPECTRAL TRANSMITTANCE CURVES

Figure 6.12 shows the spectral transmitted curve for different colours. The curve specifies the percentage of incident light that is transmitted for each wavelength across the spectrum. This percentage at a given wavelength is called the spectral transmittance for that wavelength. The spectral transmittance can range from 0 to 1 or, when expressed as a percentage, from 0 to 100%. The rest of the colour is absorbed by the filter or material, or is reflected back from one or both surfaces. The value of the spectral transmittance at a given wavelength is defined in Equation (6.1):

$$\text{Spectral Transmittance} = \frac{\text{Transmitted Power}}{\text{Incident Power}} \tag{6.1}$$

6.8 COLOUR TEMPERATURE

Colour temperature is a method of specifying the colour of a light source, but it should only be used when the source emits a continuous spectrum if it is to indicate the colour-rendering properties of the source. The colour temperature is measured in absolute temperature in Kelvin (K). Figure 6.13 shows some examples of colour temperature.

Although white light contains all the colours of the spectrum, they need not be present with the same intensity. For example, sunlight, often used as a reference standard for white light, does not have the same intensity at all wavelengths. The intensity of sunlight reaching the earth's surface peaks near a wavelength of 474 nm and falls off sharply in the ultraviolet. A comparison of the solar data with the Planck theory of blackbody radiation can be used to determine the sun's temperature, giving a result of about 6000 K. Incandescent lamps are good approximations of blackbody

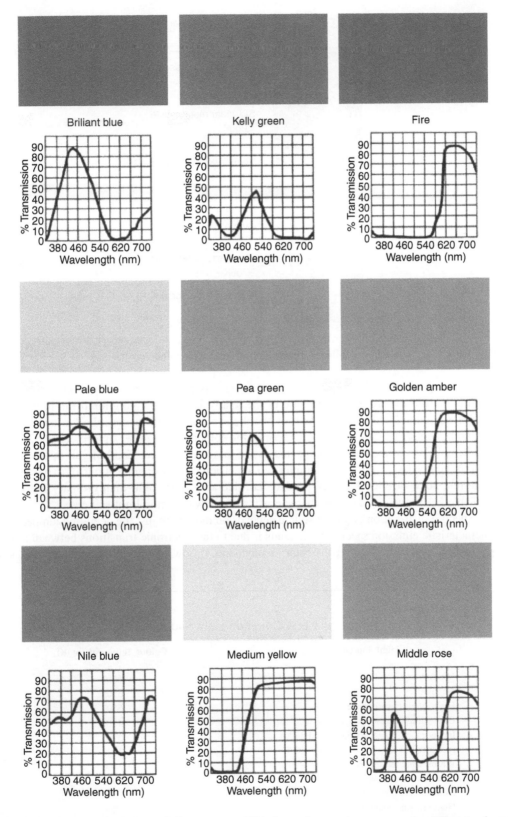

FIGURE 6.12 (See colour insert following page 200.) Spectral transmittance curve for different colours.

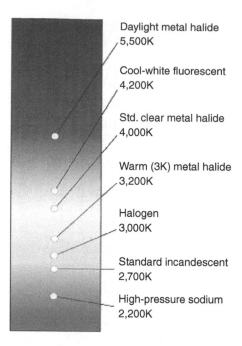

FIGURE 6.13 (See colour insert following page 200.) Colour temperature chart.

radiations and give off white light; however, this light is deficient in the blue and violet. The spectrum is skewed toward the red because of the relatively low operating temperature of incandescent lamps, about 2900 K. The spectrum distribution of the light is described in terms of the corresponding temperature of a blackbody, termed the colour temperature. At a colour temperature of 2900 K, only 3% of the energy dissipated in the lamp emerges as visible light. Special high-temperature lamps designed for photography and television usually operate at one of two reference temperatures, 3000 or 3400 K. They produce considerably more blue than ordinary household lamps. Table 6.1 shows some of the characteristics of a few common light sources.

The spectral distribution of a light source depends on more than just the operating temperature. The characteristic emission spectra of elements is due to the electronic transitions between atomic energy levels. Thus, the light from each element has its own characteristic set of emission

TABLE 6.1
Colour Temperature of a Few Common Light Sources

Light Source	Colour Temperature (K)
Mercury arc	6000
Daylight	5500
High-intensity carbon arc	5500
Cool white fluorescent	4200
Incandescent tungsten (Photoflood 3400)	3400
Incandescent tungsten (Photoflood 3200)	3200
Warm white fluorescent	3000
Incandescent tungsten (100 W)	2900
Incandescent tungsten (40 W)	2650
High pressure sodium arc	2200

wavelengths and intensities, which give it a characteristic colour. An electric discharge in hydrogen produces light of the Balmer series. We perceive this mixture of wavelengths as pink. Sodium vapour lamps produce a characteristic yellow light. Similarly, neon lamps produce a red orange light. The fluorescent lamp is a gas-discharge tube containing mercury vapour and a coating of fluorescent powder on the inside of the tube. Fluorescent materials absorb light at some wavelengths and then radiate light at longer wavelengths. Generally, both the absorption and emission states are broad bands rather than sharply-defined energy levels of the type observed in free atoms. When the mercury vapour in a fluorescent lamp is excited by an electric discharge, it emits its characteristic spectral radiation. Part of this radiation extends beyond the visible range into the ultraviolet. The fluorescent material absorbs the ultraviolet radiation and then radiates light in the visible spectrum. Fluorescent lamps of this type are considerably more efficient than incandescent lamps, converting about 20% of the electrical energy into visible light.

6.9 NEWTON'S COLOUR WHEEL

In 1672, Newton described his experiment to the Royal Society in London. His theory that white light was made up of many colours was revolutionary, and it was greeted with skepticism. The demonstration of recomposition with Newton's colour wheel is only possible because of the persistence of vision. The image of a colour produced on the retina of the eye is retained for a fraction of a second. If the wheel is rotated fast enough, the image of one colour is still present on the retina when the image of the next colour is formed. The brain sums up and blends together the rapidly changed coloured images on the retina, producing the effect of a white image.

The seven colours in Newton's optical spectrum (red, orange, yellow, green, blue, indigo, and violet) may be recombined, by rotating Newton's colour wheel, as shown Figure 6.14. When the wheel is rotating with enough speed, the colours blur together; the eye, unable to respond rapidly enough, sees the colours mixed together to form white colour.

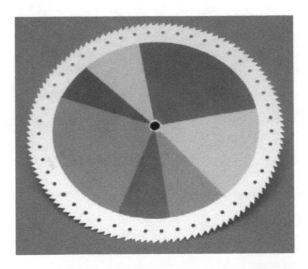

FIGURE 6.14 (See colour insert following page 200.) Newton's colour wheel.

6.10 BLACK AND WHITE COLOUR STRIP INTERSECTION WHEEL

The black and white colour strip intersection wheel works on the same principle as Newton's colour wheel. Figure 6.15 shows the picture of this wheel. If the wheel is rotated fast enough, the image produces

FIGURE 6.15 Black and white colour strip intersection wheel.

a combination of white and black bands of grey colours. The brain sums up and blends together the rapidly changing coloured images on the retina, producing the effect of a white, black, or grey image.

6.11 BLACK AND WHITE COLOUR STRIP WHEEL

The black and white colour strip wheel works as explained in Case (d). Figure 6.16 shows the picture of this wheel. If the wheel is rotated fast enough, the image produces a combination of white and black bands of grey colours.

FIGURE 6.16 Black and white colour strip wheel.

6.12 EXPERIMENTAL WORK

This experiment is a study of the colour mixing principles. The students will experience colour mixing for the following cases:

    a. Additive method of colour mixing
    b. Subtractive method of colour mixing
    c. Newton's colour wheel
    d. Black and white colour strip intersection wheel
    e. Black and white colour strip wheel

6.12.1 TECHNIQUE AND APPARATUS

Appendix A presents the details of the devices, components, tools, and parts comprised in the experiment:

1. Newton's colour wheel, as shown in Figure 6.14
2. Black and white colour strip intersection wheel, as shown in Figure 6.15
3. Black and white colour strip wheel, as shown in Figure 6.16
4. Blue, green, and red slides, as shown in Figure 6.17
5. Three slide projectors, as shown in Figure 6.18
6. White screen and stand
7. Subtractive colour filter set, as shown in Figure 6.19
8. Motor-driven rotator, as shown in Figure 6.20
9. Stroboscope with a filter in 35 mm slide format window, and power supply, as shown in Figure 6.21
10. White cardboard

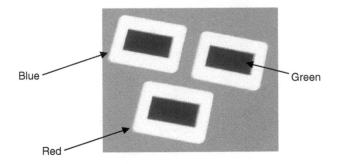

FIGURE 6.17 Blue, green, and red slides.

FIGURE 6.18 Slide projector.

FIGURE 6.19 (**See colour insert following page 200**.) Subtractive colour filter set.

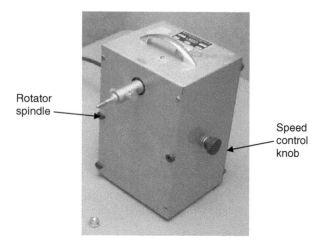

Rotator
spindle

Speed
control
knob

FIGURE 6.20 Motor-driven rotator.

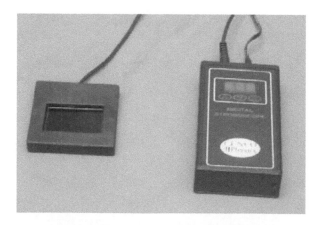

FIGURE 6.21 Stroboscope with a filter in 35 mm slide format window.

6.12.2 Procedure

Follow the laboratory procedures and instructions given by the professor and/or instructor.

6.12.3 Safety Procedure

Follow all safety procedures and regulations regarding the use of electric and optical devices and test measurement instruments.

6.12.4 Apparatus Set-up

6.12.4.1 Additive Method of Colour Mixing

1. Figure 6.22 shows the experimental apparatus set-up.
2. Place the white cardboard against the wall.
3. Mount three slide projectors on a working bench. Locate the three slide projectors facing the white cardboard.
4. Turn on the slide projectors individually and focus the projected images on the white cardboard.

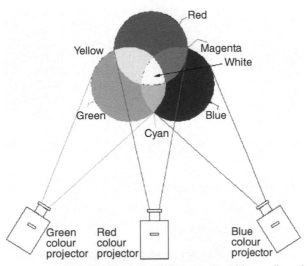

(a) Three slide projectors project lights onto the white cardboard.

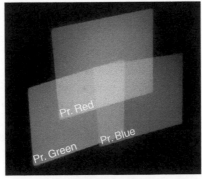

(b) Mixing three colors

FIGURE 6.22 (See colour insert following page 200.) Additive colour mixing. (a) Three slide projectors project lights onto the white cardboard. (b) Mixing three colours.

5. Turn off the laboratory light to be able to see the colours.
6. Insert the green, red, and blue slide colours into the slide projectors, with one-slide colour in each slide projector.
7. Align the images on the white cardboard. Try to mix three colours at a time, as shown in Figure 6.22.
8. Report colour observations from additive method of colour mixing when two or three colours are mixed at a time.
9. For more experience in mixing colours, try to insert two slide colours in one projector at the same time. Try to achieve the colour mixing shown in Figure 6.23.
10. Turn on the laboratory light.

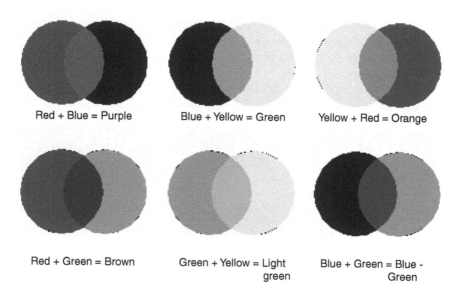

FIGURE 6.23 (**See colour insert following page 200.**) Two colour additive mixing.

6.12.4.2 Subtractive Method of Colour Mixing

1. Hold the subtractive colour filter set between your hands, as shown in Figure 6.24.
2. Try to mix two or three colours at a time.
3. Report colour observations from subtractive method of colour mixing when two or three colours are mixed at a time.

6.12.4.3 Newton's Colour Wheel

1. Figure 6.25 shows the experimental apparatus set-up.
2. Take the colour wheel and mount it on the motor-driven rotator and plug it into a power source.
3. Turn on the colour wheel using the speed control knob and set it to its lowest rotational speed: "Speed 1."
4. Measure the rotational speed of the colour wheel (Speed 1) using the stroboscope with a filter in 35 mm slide format window, as shown in Figure 6.21.
5. Record what colour is observed on the wheel in Table 6.2.
6. Increase the rotational speed to setting "Speed 2" using speed control knob.
7. Repeat steps 4 and 5 for four different speeds and fill out Table 6.2.

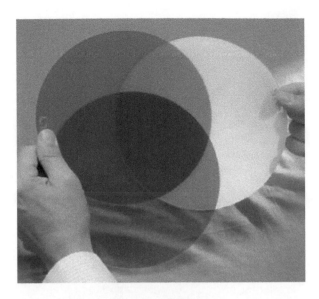

FIGURE 6.24 (**See colour insert following page 200.**) Subtractive colour mixing.

6.12.4.4 Black and White Colour Strip Intersection Wheel

1. Figure 6.26 shows the experimental apparatus set-up.
2. Take the black and white colour strip intersection wheel and mount it on the motor-driven rotator and plug it into a power source.

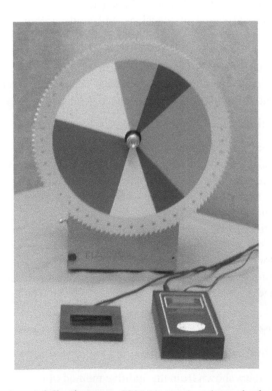

FIGURE 6.25 (**See colour insert following page 200.**) Newton's colour wheel apparatus set-up.

FIGURE 6.26 Black and white colour strip intersection wheel apparatus set-up.

3. Turn on the black and white colour strip intersection wheel using the speed control knob and set it to its lowest rotational speed: "Speed 1."
4. Measure the rotational speed of the black and white colour strip intersection wheel (Speed 1) using the stroboscope with a filter in 35 mm slide format window, as shown in Figure 6.21.
5. Record what colour is observed on the wheel in Table 6.3.
6. Increase the rotational speed to setting "Speed 2" using the speed control knob.
7. Repeat steps 4 and 5 for four different speeds and fill out Table 6.3.

6.12.4.5 Black and White Colour Strip Wheel

1. Figure 6.27 shows the experimental apparatus set-up.
2. Take the black and white colour strip wheel and mount it on the motor-driven rotator and plug it into a power source.
3. Turn on the black and white colour strip wheel using the speed control knob and set it to its lowest rotational speed: "Speed 1."
4. Measure the rotational speed of the black and white colour strip wheel (Speed 1) using the stroboscope with a filter in 35 mm slide format window, as shown in Figure 6.21.
5. Record what colour is observed on the wheel in Table 6.4.
6. Increase the rotational speed to setting "Speed 2" using the speed control knob.
7. Repeat steps 4 and 5 for four different speeds and fill out Table 6.4.

6.12.5 DATA COLLECTION

6.12.5.1 Additive Method of Colour Mixing

Observe and report the colours shown from the additive method of colour mixing when two or three colours are mixed at a time.

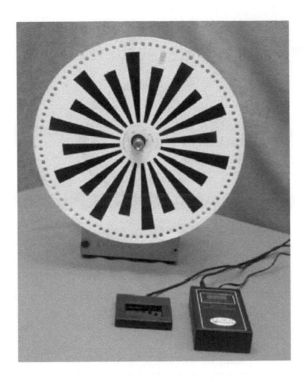

FIGURE 6.27 Black and white colour strip wheel apparatus set-up.

6.12.5.2 Subtractive Method of Colour Mixing

Observe and report the colours shown from the subtractive method of colour mixing when two or three colours are mixed at a time.

6.12.5.3 Newton's Colour Wheel

Fill out Table 6.2 with the observed colour and wheel rotational speed.

TABLE 6.2
Newton's Colour Wheel

Wheel Speed	Measured Wheel Speed (RPM)	Observed Colour
Speed 1		
Speed 2		
Speed 3		
Speed 4		
Speed 5		

6.12.5.4 Black and White Colour Strip Intersection Wheel

Fill out Table 6.3 with the observed colour and wheel rotational speed.

TABLE 6.3
Black and White Colour Strip
Intersection Wheel

Wheel Speed	Measured Wheel Speed (RPM)	Observed Colour
Speed 1		
Speed 2		
Speed 3		
Speed 4		
Speed 5		

6.12.5.5 Black and White Colour Strip Wheel

Fill out Table 6.4 with the observed colour and the wheel rotational speed.

TABLE 6.4
Black and White Colour Strip
Wheel

Wheel Speed	Measured Wheel Speed (RPM)	Observed Colour
Speed 1		
Speed 2		
Speed 3		
Speed 4		
Speed 5		

6.12.6 CALCULATIONS AND ANALYSIS

Not applicable in this lab.

6.12.7 RESULTS AND DISCUSSIONS

6.12.7.1 Additive Method of Colour Mixing

1. Report and discuss the additive method of colour mixing, when two or three colours are mixed at a time.

6.12.7.2 Subtractive Method of Colour Mixing

1. Report and discuss the subtractive method of colour mixing, when two or three colours are mixed at a time.

6.12.7.3 Newton's Colour Wheel

1. Report and discuss the observed colour when the Newton colour wheel rotates at different rotational speeds.

6.12.7.4 Black and White Colour Strip Intersection Wheel

1. Report and discuss the observed colour when the black and white colour strip intersection wheel rotates at different rotational speeds.

6.12.7.5 Black and White Colour Strip Wheel

1. Report and discuss the observed colour when the black and white colour strip wheel rotates at different rotational speeds.

6.12.8 CONCLUSION

Summarize the important observations and findings obtained in this lab experiment.

6.12.9 SUGGESTIONS FOR FUTURE LAB WORK

List any suggestions for improvements using different experimental equipment, procedures, and techniques for any future lab work. These suggestions should be theoretically justified and technically feasible.

6.13 LIST OF REFERENCES

List any references that were used in the report. Use one format in writing the references. Never mix reference formats in a report.

6.14 APPENDICES

List all of the materials and information that are too detailed to be included in the body of the report.

FURTHER READING

Adams, R. A., Courage, M. L., and Mercer, M. E., Systematic measurements of human neonatal color vision, *Vision Res.*, 34, 1691–1701, 1994.

Beiser, A., *Physics*, 5th ed., Addison-Wesley, New York, 1991.

Bise, R. T., Windeler, R. S., Kranz, K. S., Kerbage, C., Eggleton, B. J., and Trevor, D. J., Tunable Photonic Bandgap Fiber, Paper presented at Proc. Optical Fiber Communication Conference and Exhibit, Murray Hill, NJ, 466–468, March 17, 2002.

Bowmaker, J. K. and Dartnall, H. J. A., Visual pigments of rods and cones in human retina, *J. Physiol.*, 298, 501–511, 1980.

Cornsweet, T. N., *Visual Perception*, Academic Press, New York, 1970.

Cox, A., *Photographic Optics*, 15th ed., Focal Press, London, 1974.

Ewen, D., Nelson, R. J., and Schurter, D., *Physics for Career Education*, 4th ed., Prentice Hall, Upper Saddle River, NJ, 1996.

Falk, D., Brill, D., and Stork, D., *Seeing the Light Optics in Nature, Photography, Color, Vision, and Holography*, Wiley, New York, 1986.

Fehrman, K. R. and Fehrman, C., *Color the Secret Influence*, Prentice Hall, Inc., Upper Saddle River, 2000.

Ghatak, A. K., *An Introduction to Modern Optics*, McGraw-Hill, New York, 1972.

Giancoli, D. C., *Physics*, 5th ed., Prentice Hall, Upper Saddle River, NJ, 1998.

Griot, M., The practical application of light, *Melles Griot Catalog*, Wiley, New York, 2001.

Halliday, D., Resnick, R., and Walker, J., *Fundamental of Physics*, 6th ed., Wiley, New York, 1997.

Hood, D. C. and Finkelstein, M. A., Sensitivity to light handbook of perception and human performance, In *Sensory Processes and Perception*, Boff, K. R., Kaufman, L., and Thomas, J. P., Eds., Vol. 1, Wiley, Toronto, 1986.

Hurvich, L. M. and Jameson, D., A psychophysical study of white I natural adoption, *J. Opt. Soc. Am.*, 41, 521–527, 1951a.

Hurvich, L. M. and Jameson, D., A psychophysical study of white III adoption as variant, *J. Opt. Soc. Am.*, 41, 701–709, 1951b.

Hurvich, L. M. and Jameson, D., Some quantitative aspects of an opponent colors theory, II. Brightness, saturation, and hue in normal and dichromatic vision, *J. Opt. Soc. Am.*, 45, 602–616, 1955.

Jameson, D. and Hurvich, L. M., Theory of brightness and color contrast in human vision, *Vision Res.*, 4, 135–154, 1964.

Jenkins, F. W. and White, H. E., *Fundamentals of Optics*, McGraw Hill, New York, 1957.

Jones, E. and Childers, R., *Contemporary College Physics*, McGraw-Hill Higher Education, New York, 2001.

Kaiser, P. K. and Boynton, R. M., *Human Color Vision*, 2nd ed., Optical Society of America, Washington, DC, 1996.

Keuffel & Esser Co., *Physics*, Keuffel & Esser Audiovisual Educator—Approved Diazo Transparency Masters, Audiovisual Division, Keuffel & Esser Co., U.S.A., 1989.

Lerner, R. G. and Trigg, G. L., *Encyclopedia of Physics*, 2nd ed., VCH Publishers, Inc., New York, 1991.

Manngeim, L. A., *Photography Theory and Practice*, Focal Press Ltd., Boston, 1970.

Nassau, K., *Experimenting with Color*, Franklin Watts A Division of Grolier Publishing, New York, 1997.

Newport Corporation, *Photonics Section, Newport Resource 2004 Catalog*, Newport Corporation, Irvine, CA, 2004.

Nolan, P. J., *Fundamentals of College Physics*, Wm. C. Brown Publishers, Dubuque, Iowa, 1993.

Overheim, R. D. and Wagner, D. L., *Light and Color*, Wiley, New York, 1982.

Pritchard, D. C., *Environmental Physics: Lighting*, Longmans, Green & Co. Ltd., London and Harlow, Great Britain, 1969.

Product Knowledge, Lighting reference guide, *Product Development*, 2nd ed., Ontario Hydro, Toronto, ON, Canada, 1988.

Schildegen, T. E., *Pocket Guide to Color with Digital Applications*, Delmar Publishers, Albany, NY, 1998.

Stroebel, L., Compton, J., Current, I., and Zakia, R., *Photographic Materials and Processes*, Focal Press Ltd., Boston, 1986.

Tippens, P. E., *Physics*, 6th ed., Glencoe McGraw-Hill, Westerville, OH, 2001.

Williamson, S. J. and Cummins, H. Z., *Light and Color in Nature and Art*, Wiley, New York, 1983.

Wilson, J. D., *Physics—A Practical and Conceptual Approach Saunders Golden Sunburst Series*, Saunders College Publishing, Philadelphia, 1989.

Wilson, J. D. and Buffa, A. J., *College Physics*, 5th ed., Prentice Hall, Inc., Upper Saddle River, 2000.

7 Laws of Light

7.1 INTRODUCTION

Basic understanding of the reflection and refraction properties of light paved the way for many of the optical measuring devices in use today. When light strikes a surface, a portion of the incident ray is reflected from the surface. Depending on the surface characteristics, either specular reflection or diffuse reflection is observed. When light passes through an optically dense media, such as glass or water, the incident light ray bends as it enters the media. The characteristics of the material (air, glass, water, etc.) and the orientation of the ray determine the path of a reflected or refracted ray.

The laws of reflection and refraction govern the behaviour of light incident on a flat surface, separating two optical media. Light propagation through optical components is explained by these laws, as well as by critical and total internal reflection.

Four experimental lab cases presented in this chapter cover light passing through a water layer as an optical component. The experimental cases include observing laser light passing through a water layer, measuring the angle of incidence and angle of refraction, measuring and calculating the critical angle at the water–air interface, and measuring the total internal reflection occurring at the water–air interface.

7.2 LAW OF REFLECTION

Light travels in a straight line at a constant speed in a uniform optical medium. When a light ray traveling in a medium meets a boundary leading into a second medium, part of the incident ray reflects back to the first medium. Consider a light ray traveling in air incident at an angle (θ_i) on a flat mirror surface, as shown in Figure 7.1. A light ray incident on a mirror surface dividing two uniform media is described by an angle of incidence (θ_i). This angle is measured relative to a line perpendicular to the reflecting surface, as shown in Figure 7.1. Similarly, the reflected ray is

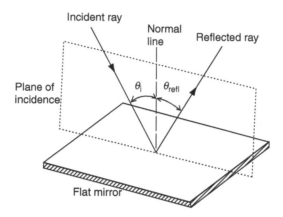

FIGURE 7.1 Light ray reflection.

described by an angle of reflection (θ_{refl}). The incident ray, the normal line, and the reflected ray lie in the plane of incidence.

Figure 7.2 shows three parallel light rays incident on and reflected off a reflecting surface. This figure also illustrates how to derive the law of reflection. The line AB lies along an ingoing wavefront, while CD lies on an outgoing wavefront. In effect, AB transforms upon reflection into CD. The wavelet emitted from A will arrive at C in-phase with the wavelet just being emitted from D, as long as the distances AC and BD are equal. In other words, if all the wavelets emitted from the surface overlap in-phase and form a single reflected plane wave, it must be that $AC = BD$. The two right triangles ABD and ACD are congruent, and they have the same hypotenuse AD. From Figure 7.2, the hypotenuse AD can be obtained:

$$\frac{\sin \theta_i}{BD} = \frac{\sin \theta_{refl}}{AC} \tag{7.1}$$

All the waves travel in the incident medium with the same speed (v_i). In the same time (Δt) that it takes the wavelet from point B on the wavefront to reach point D on the surface, the wavelet emitted from A reaches point C. Therefore, $BD = v_i \Delta t = AC$, and Equation (7.1) can be

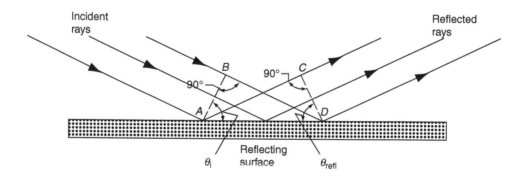

FIGURE 7.2 Light rays incident on and reflected off a reflecting surface.

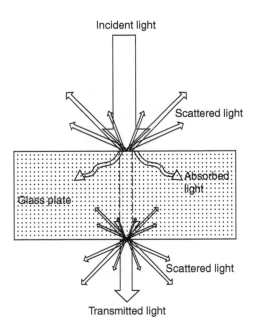

FIGURE 7.3 Transmission of light through a glass plate.

re-written as:

$$\sin \theta_i = \sin \theta_{refl} \tag{7.2}$$

which means that

$$\theta_i = \theta_{refl} \tag{7.3}$$

Therefore, the angle of incidence equals the angle of reflection. The relationship between these angles is given by the law of reflection.

Figure 7.3 shows light passing through an optical medium, for example, a glass plate. When light rays are incident perpendicular to an optical medium surface, part of the light is scattered at the surface of the medium. One part of the light rays is absorbed by the optical medium, whereas the other part is transmitted through the optical medium. Light reflection, scattering, and absorption are important factors in the calculation of light loss in an optical medium.

7.2.1 FRESNEL REFLECTION

There are two common types of reflection that occur in optical components. Fresnel reflection occurs when a light ray passes through materials, which have different refractive indices, as shown in Figure 7.4. A portion of the incident light ray is subjected to multiple reflections between two parallel optical surfaces. The multiple reflections are called Fresnel reflection, named after the French physicist, Augustin Fresnel (1788–1827). The portion of the incident light ray reflected is dependent on the angle of incidence (measured from the normal line) and the polarization of the

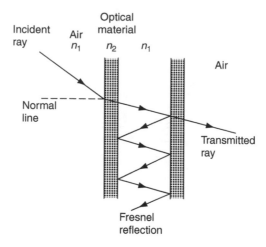

FIGURE 7.4 Fresnel reflection between parallel optical surface.

electric field relative to the plane of incidence. This kind of reflection normally takes place at the ends of mated fibre optic connectors.

7.2.2 Back Reflection

Back reflection, which occurs at the end of the fibre optic cables, is shown in Figure 7.5. Additional losses can occur if the fibre optic cable ends are cut and polished at an angle that does not match the connector angle. Strong back reflection losses can provide feedback, which, in some cases, introduces a spurious signal and increased noise levels.

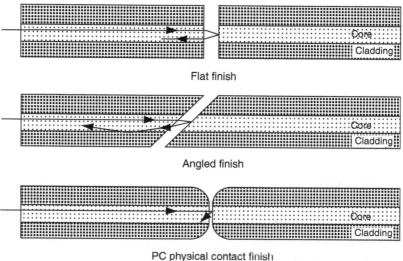

FIGURE 7.5 Back reflection in fibre optic cable ends.

7.3 LAW OF REFRACTION

Light travels in a straight line at a constant speed in a uniform optical medium. If the properties of the optical medium change, then the speed of light within the medium will also change. The velocity of light in an optical medium is less than the velocity of light in a vacuum of 3×10^8 m/s. The ratio of the velocity of light in a vacuum (c) to the velocity of light in a particular medium (v) is called the index of refraction (n) for that medium. The index of refraction of a medium is defined by Equation (7.4).

$$n = \frac{c}{v} \tag{7.4}$$

The index of refraction is a unitless quantity that is always greater than unity. Table 7.1 lists the index of refraction for several common substances.

Consider a ray of light traveling from one medium to a second medium, as shown in Figure 7.6. The ray that enters the second medium is bent at the interface between the two mediums; this phenomenon is called refraction of light. The incident ray, the normal line, and the refracted ray lie on the plane of incidence. The sine of the angle of refraction (θ_{refr}) is directly proportional to the sine of the angle of incidence (θ_i), as shown in Figure 7.6; this relation is called Snell's law of refraction.

Figure 7.7 shows how to derive Snell's law of refraction. Two parallel light rays in a medium with index of refraction n_1 meet at the interface of a medium with index of refraction n_2. It is assumed that the second medium has a greater index of refraction than the first ($n_2 > n_1$). Both rays bend toward the normal as they pass into the denser medium.

In the time interval Δt, ray light ray 1 propagates from A to B and light ray 2 propagates from C to D. Since the velocity v_2 in the second medium is less than the velocity v_1 in the first medium, the distance AB will be shorter than the distance CD. These lengths are given by:

$$AB = v_2 \Delta t \tag{7.5}$$
$$CD = v_1 \Delta t \tag{7.6}$$

TABLE 7.1
Index of Refraction for Several Common Substances

Substance	Index of Refraction (n)
Air	1.00029
Diamond	2.24
Ethyl alcohol	1.36
Fluorite	1.43
Fused quartz	1.46
Glass (by type)	
Crown	1.52
Flint	1.66
Glycerin	1.47
Ice	1.31
Polystyrene	1.49
Rock salt	1.54
Water	1.33

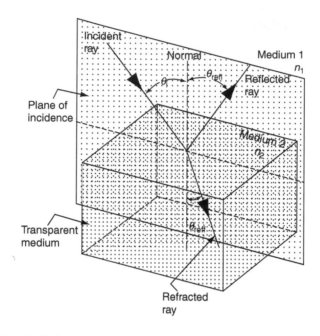

FIGURE 7.6 Refraction of a light ray.

The geometry in Figure 7.7 shows that the angle *CAD* is equal to θ_i, and the angle *ADB* is equal to θ_{refr}. The line *AD* forms a hypotenuse that is common to the two triangles, *ACD* and *ABD*. From these triangles, we find that

$$\sin \theta_i = \frac{v_1 \Delta t}{AD} \tag{7.7}$$

$$\sin \theta_{refr} = \frac{v_2 \Delta t}{AD} \tag{7.8}$$

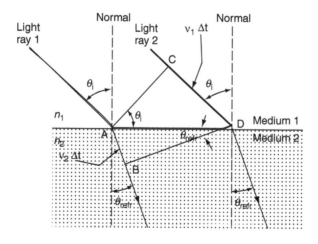

FIGURE 7.7 Deriving Snell's law.

Dividing Equation (7.7) by Equation (7.8) gives:

$$\frac{\sin \theta_i}{\sin \theta_{refr}} = \frac{v_1}{v_2} \qquad (7.9)$$

Equation (7.9) shows that the ratio of the sine of the angle of incidence to the sine of the angle of refraction is equal to the ratio of the velocity of light in the incident medium to the velocity of light in the refracted medium.

Using Equation (7.4) in conjunction with Equation (7.9), gives the following equation:

$$n_1 \sin \theta_i = n_2 \sin \theta_{refr} \qquad (7.10)$$

where n_1 and n_2 are the index of refraction for medium 1 and 2, respectively. The relationship between these angles is known as the law of refraction, or Snell's law.

More generally, the following relationships for light refraction in an optical medium can easily be determined from Snell's law:

1. When a light ray enters a denser medium, its speed decreases, and the refracted ray is bent towards the normal. For example, this occurs when the light ray moves from air into glass, as shown in Figure 7.8.
2. When a light ray enters a less dense medium, its speed increases, and the refracted ray is bent away from the normal. For example, this occurs when the light ray moves from glass into air, as shown in Figure 7.9.
3. There is no change in the direction of propagation, if there is no change in index of refraction. The greater the change is in the index of refraction the greater the change in the propagation direction will be.
4. If a light ray goes from one medium to another along the normal, it is undeflected, regardless of the change in the index of refraction, as shown in Figure 7.10. Although the angle does not change, the speed of light does change as the light travels from one medium to another.

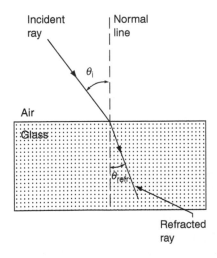

FIGURE 7.8 Light ray propagates from air into glass.

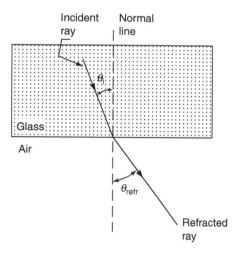

FIGURE 7.9 Light ray propagates from glass in to air.

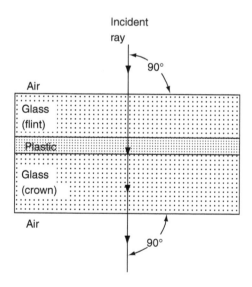

FIGURE 7.10 Normal incident light ray passes through differenct optical materials.

The principles of reflection and refraction are illustrated in Figure 7.6. An incident light ray is partially reflected and partially refracted, and then transmitted at the interface plane, separating two transparent media.

Figure 7.11 shows a light ray passing through two media (air and glass). The change in index of refraction at each interface causes the incident ray to bend twice. This figure also shows that the angle θ_1 equals the angle θ_3, and the ray exiting the glass plate is parallel to the incident ray.

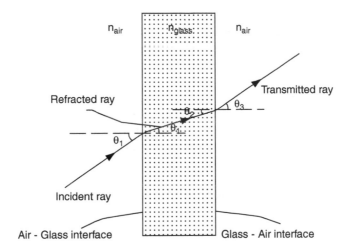

FIGURE 7.11 Light ray refraction through a glass plate.

7.3.1 Critical Angle and Total Internal Reflection

Total internal reflection can occur when light, traveling in a more optically dense medium, is incident on the boundary with a less optically dense medium, as shown in Figure 7.12. The index of refraction (n_1) of medium 1 is greater than index of refraction (n_2) of medium 2. A good example is when light goes from water into air. Consider a light beam traveling in medium 1 and incident on the boundary interface between medium 1 and medium 2. Various possible directions of the light ray are indicated by rays 1 through 5, as illustrated in

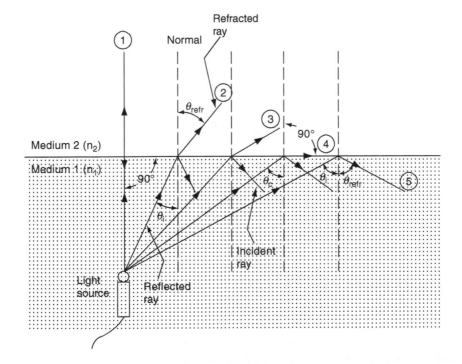

FIGURE 7.12 Critical angle and total internal reflection.

Figure 7.12. The refracted rays are bent away from the normal, according to Snell's law because n_1 is greater than n_2. Furthermore, when the light ray refracts at the interface between the two media, it partially reflects back into medium 1. Figure 7.12 shows that rays 1 through 4 are partially reflected back into medium 1, and ray 5 is totally reflected back into medium 1.

Consider individual light rays in Figure 7.12:

1. Ray 1 is incident perpendicular to the interface between the two media, and passes straight through from medium 1 to medium 2. Part of ray 1 reflects back in the same direction into medium 1; this type of reflection is called a back reflection. Back reflection is one type of loss in optical devices.
2. Ray 2 is an incident ray in medium 1 to medium 2, where $n_1 > n_2$. A portion of ray 2 refracts to medium 2, as given by Snell's law, and a portion of ray 2 reflects back into medium 1 (since the angle of incidence equals the angle of reflection).
3. Similarly, for ray 3, as the angle of incidence increases, the angle of refraction also increases.
4. For ray 4, as the angle of incidence increases, the angle of refraction increases until the angle of refraction is 90°. The angle of incidence is called the critical angle (θ_c), when the refracted angle is 90°. Ray 4 refracts and propagates parallel to the interface between the two media.
5. For angles of incidence greater than the critical angle (θ_c), ray 5 is entirely reflected at the interface. All the intensity of incident ray 5 goes into the reflected ray without any loss; this is called total internal reflection of light.

The critical angle for total internal reflection is found by setting $\theta_{\text{refr}} = 90°$ and applying Snell's law, Equation (7.10):

$$n_1 \sin \theta_c = n_2 \sin 90 = n_2 \times 1$$
$$\because \sin 90 = 1 \tag{7.11}$$

Therefore, the critical angle (θ_c) is given by the following equations:

$$\sin \theta_c = \frac{n_2}{n_1} \tag{7.12}$$

$$\theta_c = \sin^{-1}\left(\frac{n_2}{n_1}\right) \tag{7.13}$$

Since $\sin \theta_c$ is always less than or equal to 1, the index of refraction (n_1) of medium 1 must be larger than the index of refraction (n_2) of medium 2. Thus, total internal reflection can occur only when light in one medium encounters an interface with another medium that has a lower index of refraction (i.e., speed of light is greater in the second medium).

7.4 EXPERIMENTAL WORK

The cases in this experiment use the theoretical principles of light reflection, refraction, and total internal reflection for light passing through optical components. The following cases examine the behaviour of light passing through a water layer in a water tank; these principles are explained by the laws of the reflection and refraction of light. Students will observe the effects of a light ray as it propagates through a tank filled with water for the following cases:

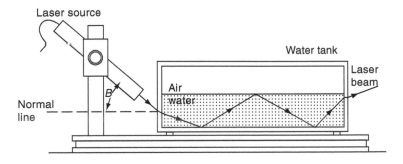

FIGURE 7.13 Incident laser beam of an angle θ into water layer.

7.4.1 Laser Light Passes through a Water Layer

The students will observe a laser beam bouncing through the water layer in a water tank. The students will observe the reflection of a laser beam incident at different angles on the side of the water tank, as illustrated in Figure 7.13.

7.4.1.1 Law of Refraction at Air–Water Interface

The students will calculate the angles of incidence and refraction of a laser beam incident on the top water surface of a water tank. Refraction occurs when light passes from air to water. Air has a lower index of refraction than water. Therefore, the refracted ray bends toward the normal as the ray passes through water. The students will calculate the angle of incidence (θ_i) and the angle of refraction (θ_{refr}) at the air–water interface using the measurement of horizontal and vertical distances, as illustrated in Figure 7.14.

Using the geometrical relations in this figure, the following equations can be written to calculate the angles of incidence (θ_i) and refraction (θ_{refr}).

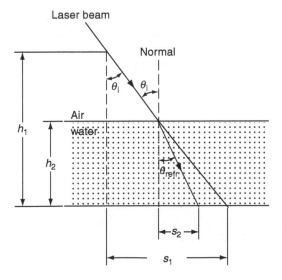

FIGURE 7.14 Refraction angle of light at air–water interface.

$$\tan \theta_i = \frac{s_1}{h_1} \tag{7.14}$$

$$\theta_i = \tan^{-1} \frac{s_1}{h_1} \tag{7.15}$$

$$\tan\theta_{\text{refr}} = \frac{s_2}{h_2} \tag{7.16}$$

$$\theta_{\text{refr}} = \tan^{-1} \frac{s_2}{h_2} \tag{7.17}$$

7.4.1.2 Critical Angle at Water–Air Interface

The students will perform an experiment to calculate the critical incident angle (θ_c) when the refracted angle is 90° with the normal, as illustrated in Figure 7.15. Total internal reflection can occur when light propagates from water to air and the incident angle is greater than the critical angle. In this case, the refracted ray reflects inside the water, and the law of reflection is applicable. When the critical angle occurs, light incident on the underside of a water–air interface produces a laser beam that lies on the water–air interface and shows a bright spot on the target card. The students will find the experimental and theoretical values of the critical angle at the water–air interface.

Using the geometrical relations in this figure, the following equations can be written to calculate the critical angles (θ_c):

$$\tan \theta_c = \frac{X}{Y} \tag{7.18}$$

$$\theta_c = \tan^{-1} \frac{X}{Y} \tag{7.19}$$

7.4.1.3 Total Internal Reflection at Water–Air Interface

The students will observe the phenomena of the total internal reflection of light rays when the incidence angle is greater than the critical angle. The students will observe light rays shining from the bottom of a water tank filled with water, as illustrated in Figure 7.16. Specifically, they will measure the distance from the light source to the bottom of the water tank in relation to the diameter of the light spots on the top and bottom surfaces of the water tank. The students will determine the critical angle experimentally and theoretically at the water–air interface.

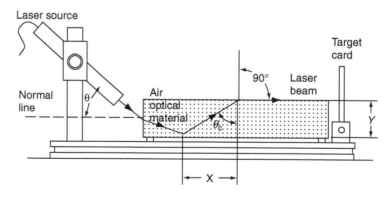

FIGURE 7.15 Critical angle of light at water–air interface.

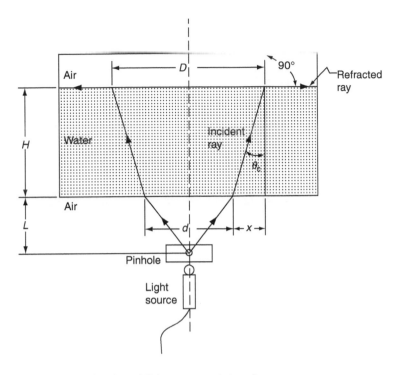

FIGURE 7.16 Total internal reflection of light at water-air interface.

Using the geometrical relations in Figure 7.16, the following equations can be written to calculate the critical angle (θ_c) and x:

$$x = \frac{(D-d)}{2} \tag{7.20}$$

$$\tan \theta_c = \frac{x}{H} \tag{7.21}$$

$$\theta_c = \tan^{-1}\left(\frac{x}{H}\right) \tag{7.22}$$

7.4.2 TECHNIQUE AND APPARATUS

Appendix A presents the details of the devices, components, tools, and parts.

1. 2×2 ft. optical breadboard
2. HeNe laser light source and power supply
3. Laser mount assembly
4. Incandescent or halogen light source
5. Light source positioners
6. A water tank, as shown in Figure 7.17
7. Hardware assembly (clamps, posts, screw kits, screwdriver kits, sundry positioners, etc.)
8. Adjustable pinhole, as shown in Figure 7.18
9. Rotation Stage
10. Black/white card and cardholder
11. Target card

FIGURE 7.17 Water tank.

    12. Vernier
    13. Ruler

7.4.3 PROCEDURE

Follow the laboratory procedures and instructions given by the professor and/or instructor.

7.4.4 SAFETY PROCEDURE

Follow all safety procedures and regulations regarding the use of laser light and light source devices.

FIGURE 7.18 Adjustable pinhole.

7.4.5 Apparatus Set-up

7.4.5.1 Laser Light Passes through a Water Layer

1. Figure 7.19 shows the apparatus set-up.
2. Bolt the laser short rod to the breadboard.
3. Bolt the laser mount to the clamp using bolts from the screw kit.
4. Put the clamp on the short rod.
5. Place the HeNe laser into the laser mount and tighten the screw. Turn on the laser device. Follow the operation and safety procedures of the laser device in use.
6. Mount the water tank and fill it with water to a depth of about 10 cm.
7. Place the water tank near the laser source.
8. Turn off the laboratory light to be able to see the laser beam path.
9. Direct the laser beam in the perpendicular position to the left side of the water tank (set the first angle equal to zero degrees with the horizontal level).
10. Observe the laser beam path. Report your observation in Table 7.2.
11. Align the laser source at an angle so the laser beam enters the left side of the water tank. Observe the laser beam path when the laser beam bounces between the bottom of the water tank and water surface. Report your observation in Table 7.2.
12. Repeat step 11 for another two different angles of the laser beam entering the left side of the water tank. Report your observations in Table 7.2.
13. Turn on the laboratory light.

7.4.5.2 Law of Refraction at Air–Water Interface

1. Figure 7.20 shows the apparatus set-up.
2. Bolt the laser short rod to the breadboard.
3. Bolt the laser mount to the clamp using bolts from the screw kit.
4. Put the clamp on the short rod.
5. Place the HeNe laser into the laser mount and tighten the screw. Turn on the laser device. Follow the operation and safety procedures of the laser in use.

FIGURE 7.19 Laser light passes through a water layer.

6. Place the water tank on the breadboard near the laser source.
7. Place a lined paper under the water tank, for ease of marking the laser beam spots.
8. Turn off the laboratory light to be able to see the laser beam path.
9. Direct the laser beam to be perpendicular to the bottom of the water tank.

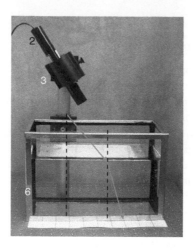

FIGURE 7.20 Refraction angle of light at air–water interface.

10. Mark the laser beam spot on the lined paper, when the laser beam is positioned vertically to the bottom of the water tank.
11. Rotate the laser source to direct the laser beam by an angle less the 90° with the normal. For example, 30°. Refer to Figure 7.14 for distances to be measured.
12. Measure the vertical distance (h_1) from the laser source to the spot where the laser beam is incident on the lined paper. This is the reference spot. Fill out Table 7.3.
13. Measure the distance (s_1) from the reference spot to where the laser beam is incident on the lined paper. Fill out Table 7.3.
14. Fill the water tank with water to about 5 cm below the top. (Note: It is preferable to add sugar or salt to the water to enhance the viewing of the light shining through the water tank. In this case, you will need to find the index of refraction (n) of the solution).
15. Mark the laser beam spot on the lined paper when the laser beam is incident on it.
16. Make a vertical line from where the laser beam intersects the water surface to the lined paper. Mark this spot on the lined paper.
17. Measure the vertical distance (h_2) and the horizontal distance (s_2), as shown in Default 7.14 and 7.20. Fill out Table 7.3.
18. Turn on the laboratory light.

7.4.5.3 Critical Angle at Water–Air Interface

1. Figure 7.21 shows the apparatus set-up.
2. Bolt the laser short rod to the breadboard.
3. Bolt the laser mount to the clamp using bolts from the screw kit.
4. Put the clamp on the short rod.
5. Place the HeNe laser into the laser mount and tighten the screw. Turn on the laser device. Follow the operation and safety procedures of the laser in use.
6. Mount the water tank on the breadboard and fill it with water to a depth of about eight to ten centimetres. (Note: It is preferable to add sugar or salt to the water to enhance the viewing of the light shining through the water tank. In this case, you will need to find the index of refraction (n) of the solution).
7. Place the water tank near the laser source.
8. Place a target card on the side of the water tank.
9. Turn off the laboratory light to be able to see the laser beam path.

FIGURE 7.21 Critical angle at water–air interface.

10. Align the laser source at an angle of incident on the side of the water tank so the laser beam enters the left side of the water tank.
11. Observe the laser beam path when it bounces between the bottom of the water tank and the water surface. Keep aligning the laser beam angle until you see the refracted laser beam lies on the water surface and produces a red spot on the target card (i.e. the aligned laser beam is at the critical angle and the refracted angle is 90°).
12. Measure the vertical and the horizontal distances (Y) and (X), as shown in Figure 7.15. Fill out Table 7.4.
13. Turn on the laboratory light.

7.4.5.4 Total Internal Reflection at Water–Air Interface

1. Figure 7.22 shows the apparatus set-up.
2. Mount the water tank on the stand. Fill it with water to within a few centimetres below the top of the water tank. (Note: It is preferable to add sugar or salt to the water to

FIGURE 7.22 Total internal reflection at water–air interface.

enhance the viewing of the light shining through the water tank. In this case, you will need to find the index of refraction (n) of the solution).

3. Measure water depth (H) in the water tank.
4. Place the light source underneath the water tank so the light shines up towards the bottom of the water tank. Place the light source at a distance (L_1) from the bottom of the water tank.
5. Place the pinhole on the top of the light source.
6. Connect the light source to the power supply.
7. Turn off the laboratory light to be able to see the light shining through the water.
8. Adjust the distance (L_1) until you see a bright circular pattern on the top water surface in the water tank.
9. Measure the distance (L_1), the diameter of the bright circle (D_1) on the top water surface, and the diameter (d_1) of the bright circle on the bottom water surface, as shown in Figure 7.16. Fill out Table 7.5.
10. Repeat steps 8 and 9 for another five distances. Fill out Table 7.5.
11. Turn on the laboratory light.

7.4.6 DATA COLLECTION

7.4.6.1 Laser Light Passes through a Water Layer

Observe the laser beam path for four angles of incidence (θ) on the left side of the water tank. Draw the laser beam path and report your observation in Table 7.2.

TABLE 7.2
Observation of the Laser Beam Path

Laser Beam Angle (Degrees)	Observation of the Laser Beam Path
$\theta_1 = 0°$	
$\theta_2 =$	
$\theta_3 =$	
$\theta_4 =$	

7.4.6.2 Law of Refraction at Air–Water Interface

1. Measure the vertical distances (h_1 and h_2) and horizontal distances (s_1 and s_2).
2. Report your measurements in Table 7.3.

TABLE 7.3
Law of Refraction at Water–Air Interface

Vertical Distances (cm)		Horizontal Distances (cm)		Calculated Incident Angle (θ_i) (degrees)	Calculated Refraction Angle (θ_{refr}) (degrees)
h_1	h_2	s_1	s_2		

7.4.6.3 Critical Angle at Water–Air Interface

1. Measure the vertical and horizontal distances (Y and X).
2. Report your measurements in Table 7.4.

TABLE 7.4
Critical Angle at Water–Air Interface

Vertical Distance Y	Horizontal Distance X	Calculated Critical Angle (θ_c) Using Equation (32.18) (degrees)	Refraction Angle (θ_{refr}) (degrees)	Calculated Critical Angle (θ_c) Using Equation (32.12) (degrees)
			$90°$	

7.4.6.4 Total Internal Reflection at Water–Air Interface

1. Measure the distances (L) between the pinhole and the tank bottom, the light spot diameters (D) at the top water surface, and the light spot diameters (d) at the bottom water surface. Report your measurements in Table 7.5.
2. Measure water depth (H) in the water tank.

TABLE 7.5
Total Internal Reflection at Water–Air Interface

Measured Distance (L) (cm)	Measured Top Diameter (D) (cm)	Measured Bottom Diameter (d) (cm)	x (cm)	Determined Critical Angle (θ_c) (degree)	Calculated Critical Angle by Snell's Law (θ_c) (degree)
$L_1 =$	$D_1 =$	$d_1 =$	$x_1 =$	$\theta_{c1} =$	
$L_2 =$	$D_2 =$	$d_2 =$	$x_2 =$	$\theta_{c2} =$	
$L_3 =$	$D_3 =$	$d_3 =$	$x_3 =$	$\theta_{c3} =$	$\theta_c =$
$L_4 =$	$D_4 =$	$d_4 =$	$x_4 =$	$\theta_{c4} =$	
$L_5 =$	$D_5 =$	$d_5 =$	$x_5 =$	$\theta_{c5} =$	
$L_6 =$	$D_6 =$	$d_6 =$	$x_6 =$	$\theta_{c6} =$	

7.4.7 CALCULATIONS AND ANALYSIS

7.4.7.1 Laser Light Passes through a Water Layer

No calculations and analysis are required in this case.

7.4.7.2 Law of Refraction at Air–Water Interface

1. Calculate the angles of incidence and refraction, using Equation (7.15) and Equation (7.17). Report your results in Table 7.3.
2. Verify Snell's law, given by Equation (7.10), using your calculated angles and the index of refraction for air and water as 1.0003 and 1.33, respectively.

7.4.7.3 Critical Angle at Water–Air Interface

1. Calculate the experimental critical angle using Equation (7.19). Report your results in Table 7.4.
2. Calculate the theoretical critical angle using Equation (7.13) and the index of refraction for air and water as 1.0003 and 1.33, respectively. Report your results in Table 7.4.
3. Verify the experimental and theoretical critical angles.

7.4.7.4 Total Internal Reflection at Water–Air Interface

1. Draw the geometry of the light propagation from the light source through the water for each distance (L). Determine the angles of incidence (θ_{in}) of the light beam path through the water, from the geometry of the light propagation. Report your results in Table 7.5.
2. Plot a graph of distance (L) verses diameters (D) and (d).
3. Calculate the critical angle (θ_c) using Snell's Law in Equation (7.10).
4. Calculate the critical angles (θ_c) obtained from the geometry of the light propagation in the water using Equation (7.20) and Equation (7.22).
5. Compare the critical angle obtained from Snell's law to the critical angles obtained from the geometry of the light propagation in the water.

7.4.8 Results and Discussions

7.4.8.1 Laser Light Passes through a Water Layer

Discuss your observations of the laser beam path at four angles of incidence on the left side of the water tank.

7.4.8.2 Law of Refraction at Air–Water Interface

Discuss Snell's law verification. Try to explain any discrepancies between your measurements and Snell's law.

7.4.8.3 Critical Angle at Water–Air Interface

1. Present a geometric drawing of the laser beam propagation through the water.
2. Discuss the calculated and measured critical angles.
3. Explain the variation between the critical angles obtained by Snell's law and the critical angle obtained from the geometry of the light beam passing through the water layer.

7.4.8.4 Total Internal Reflection at Water–Air Interface

1. Present a ray drawing of the light propagation from the light source through the water for each distance (L).

2. Find a relation between the distance (L) and diameters (D) and (d).
3. Discuss the critical angles for each distance (L).
4. Explain the variation between the critical angle obtained by Snell's law and the critical angles obtained from the geometry of the light propagation in the water.

7.4.9 CONCLUSION

Summarize the important observations and findings obtained in this lab experiment.

7.4.10 SUGGESTIONS FOR FUTURE LAB WORK

List any suggestions for improvements using different experimental equipment, procedures, and techniques for any future lab work. These suggestions should be theoretically justified and technically feasible.

7.4.11 LIST OF REFERENCES

List any references that were used in the report. Use one format in writing the references. Never mix reference formats in a report.

7.4.12 APPENDICES

List all of the materials and information that are too detailed to be included in the body of the report.

FURTHER READING

Beiser, Arthur, *Physics*, 5th ed., Addison-Wesley, Reading, MA, 1991.

Born, M. and, Wolf, E., Elements of the theory of diffraction and rigorous diffraction theory, *Principles of Optics: Electromagnetic Theory of Propagation, Interference, and Diffraction of Light*, Cambridge University Press, Cambridge, England, pp. 370–458 and 556–592, 1999.

Bouwkamp, C. J., Diffraction theory, *Rep. Prog. Phys.*, 17 35, 100, 1949.

Bromwich, T. J. I'A, Diffraction of waves by a wedge, *Proc. London Math. Soc.*, 14, 450–468, 1916.

Cutnell, J. D. and Johnson, K. W., *Physics*, 5th ed., John Wiley & Sons, Inc., New York, New York, U.S.A., 2001a.

Cutnell, J. D. and Johnson, K. W., *Student Study Guide—Physics*, 5th ed., John Wiley & Sons, Inc., New York, U.S.A., 2001b.

Francon, M., *Optical Interferometry*, Academic Press, New York, pp. 97–99, 1966.

Ghatak, K., *An Introduction to Modern Optics*, McGraw-Hill, New York, 1972.

Griot, Melles, *The Practical Application of Light*, Melles Griot Catalog, Rochester, NY, U.S.A., 2001.

Halliday, D., Resnick, R., and Walker, J., *Fundamental of Physics*, 6th ed., John Wiley & Sons, New York, U.S.A., 1997.

Heath, R. W., Macnaughton, R. R., and Martindale, D. G., *Fundamentals of Physics*, D.C. Heath Canada, Ltd., Canada, 1979.

Hecht, Eugene, *Optics*, 4th ed., Addison-Wesley Longman, Inc., Reading, MA, 2002.

Hewitt, P. G., *Conceptual Physics*, 8th ed., Addison-Wesley, Inc., Reading, MA, U.S.A., 1998.

Hoss, R. J., *Fiber Optic Communications—Design Handbook*, Prentice Hall Pub. Co., Englewood, Chffs. NJ, 1990.

Jenkins, F. W. and White, H. E., *Fundamentals of Optics*, McGraw Hill, New York, 1957.

Lambda, Lambda Catalog, Lambda, Research Optics, Inc., Costa Mesa, California, U.S.A., 2004.

Lehrman, R. L., *Physics-The Easy Way*, 3rd ed., Barron's Educational Series, Inc., U.S.A., 1998.

Lerner, R. G. and Trigg, G. L., *Encyclopedia of Physics*, 2nd ed., VCH Publishers, Inc., New York, 1991.

Levine, H. and Schwinger, J., On the theory of diffraction by an aperture in an infinite plane screen II, *Phys. Rev.*, 75, 1423–432, 1949

McDermott, L. C., *Introduction to Physics*, Prentice Hall, Inc., Englewood, Chffs. NJ, U.S.A., 1988.

Naess, R. O., *Optics for Technology Students*, Prentice Hall, Englewood, Chffs. NJ, 2001.

Newport Corporation, *Optics and Mechanics Section*, the Newport Resources 1999/2000 Catalog, Newport Corporation, Irvine, CA, U.S.A., 1999/2000.

Newport Corporation, *Photonics Section*, the Newport Resources 2004 Catalog, Newport Corporation, Irvine CA, U.S.A., 2004.

Nolan, P. J., *Fundamentals of College Physics*, Wm. C. Brown Publishers, Dubuque, IA, U.S.A., 1993.

Otto, F. B. and Wilson, J. D., *Physics—a Practical and Conceptual Approach*, 3rd ed., Saunders Goldern Sunburst Series, Saunders College Publishing, Harcourt Brace College Publishers, Orlando, Florida, U.S.A., 1993.

Pedrotti, F. L. and Pedrotti, L. S., *Introduction to Optics*, 2th ed., Prentice Hall, Inc., Englewood, Chffs. NJ, 1993.

Plamer, Christopher, *Diffraction Grating Handbook*, 5th ed., Thermo RGL, Richardson Grating Laboratory, New York, 2002.

Robinson, P., *Laboratory Manual to Accompany Conceptual Physics*, 8th ed., Addison-Wesley, Inc., Reading, MA, U.S.A., 1998.

Serway, R. A. and Jewett, J. W., *Physics for Scientists and Engineers with Modern Physics*, 6th ed., Vol. 2, Thomson Brooks/Cole, U.S.A., 2004.

Smith, W. J., *Modern Optical Engineering*, McGraw-Hill Book Co., New York City, NY, 1966.

Tippens, P. E., *Physics*, 6th ed., Glencoe McGraw-Hill, Westerville, OH, U.S.A., 2001.

Urone, P. P., *College Physics*, Thomson Brooks/Cole Publishing, U.S.A., 1993.

Walker, J. S., *Physics*, Prentice Hall, Upper Saddle River, New Jersey, U.S.A., 2002.

Warren, M. L., *Introduction to Physics*, W.H. Freeman and Company, San Francisco, CA, U.S.A., 1979.

White, H. E., *Modern College Physics*, 6th ed., Van Nostrand Reinhold Company, New York, 1972.

Williams, J. E., *Teacher's Edition Modern Physics*, Holt, Rinehart and Winston, Inc., New York, U.S.A., 1968.

Woods, N., *Instruction's Manual to Beiser Physics*, 5th ed., Addison-Wesley Publishing, Reading, MA, 1991.

Yeh, Chai, *Handbook of Fiber Optics: Theory and Applications*, Academic Press, San Diego, 1990.

8 Plane Mirrors

8.1 INTRODUCTION

Plane mirrors are made from a piece of glass polished on one side. If one stands before a plane mirror, one sees an upright image of oneself as far "behind" the mirror as one's position in front of it. All plane mirrors form an upright image equal to the height of the object placed in front them. If one raises one's right hand, the mirror image raises its left hand. Lettering read in a plane mirror is reversed right to left as well. This is the reason why writing on emergency vehicles, such as ambulances and police cars, is reversed right to left. Thus, when a driver looks into the mirror of his car, the writing can easily be read, and he can give room to emergency vehicles passing through.

Image formation by plane mirrors follows the law of reflection, as explained in the chapter on the laws of light. The ray-tracing method is used to show the image formation by a plane mirror. The reflection of light from a smooth shiny surface is more clear and focused than the reflection of light from a rough surface.

Plane mirrors are used in simple reflection applications, such as building optic and optical fibre devices.

Seven experimental lab cases presented in this chapter study the formation of images when an object or a laser beam source is placed in front of a plane mirror. The cases are: a candle placed in front of a fixed plane mirror and between two plane mirrors at right angles, tracing a laser beam passing between two plane mirrors at different angles, and tracing a laser beam incident on a rotating plane mirror.

8.2 THE REFLECTION OF LIGHT

When a light ray traveling in a medium comes across a boundary leading into a second medium, such as that between air and glass, part of the incident light ray is reflected back into the first

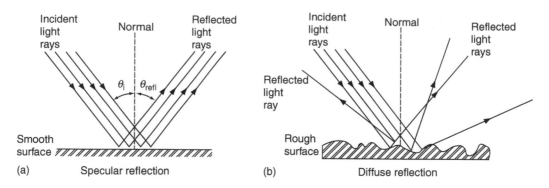

FIGURE 8.1 Light reflection from smooth and rough surfaces.

medium. The light entering the glass is partially absorbed and partially transmitted. The transmitted light is refracted. These phenomena are called the laws of reflection and refraction, and have been presented in detail in the laws of light chapter.

Figure 8.1(a) shows two types of reflection: specular and diffuse reflection. Light reflection from a smooth surface is called regular or specular reflection. The figure also shows that all the reflected light moves in parallel. Reflection of light from mirrors and smooth surfaces is an example of specular reflection. In contrast, if a surface is rough, as shown in Figure 8.1(b), the reflected light is spread out and scattered in all directions. Light reflection from a rough surface is called diffuse reflection. Reflection of light from brick, concrete, and other rough surfaces is an example of diffuse reflection.

Consider a point source of light placed a distance S in front of a plane mirror at a location O, as shown in Figure 8.2. The distance S is called the object distance. Light rays leaving the light source are incident on the mirror and reflect from the mirror. After reflection, the rays diverge, but they appear to the viewer to come from a point I located "behind" the mirror at distance S'. Point I is called the image of the object at O. The distance S' is called the image distance. Images are formed in the same way for all optical components, including mirrors and lenses. Images are formed at the point where rays of light actually intersect at the point from which they appear to originate. Solid lines represent the light coming from an object or a real image. Dashed lines are drawn to represent the light coming from an imaginary image. Figure 8.2 shows that the rays appear to originate at I, which is behind the mirror. This is the location of the image. Images are classified as real or virtual. A real image is one in which light reflects from a mirror or passes through a lens and the image can be captured on a screen. A virtual image is one

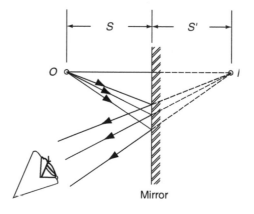

FIGURE 8.2 Image formation by a plane mirror.

in which the light does not really pass through a mirror or lens but appears either behind the mirror or in front of the lens. The virtual image cannot be displayed on a screen.

8.2.1 An Object Placed in Front of a Plane Mirror

Figure 8.3 shows a virtual image formed by a plane mirror. The images seen in plane mirrors are always virtual for real objects. Real images can usually be displayed on a screen, but virtual images cannot. Some of the properties of the images formed by plane mirrors can be examined using simple geometric techniques shown in Figure 8.3. In order to find out where an image is formed, it is always necessary to follow at least two rays of light as they reflect from the mirror. One of those rays starts at O, follows a horizontal path to the mirror, and reflects back on itself, OA. The second ray follows the oblique path OB and reflects, as shown in Figure 8.3. An observer to the left of the mirror, as shown in this figure, would trace the two reflected rays back to the point from which they appear to originate, point I. A continuation of this process for points on the object other than I would result in a virtual image to the right of the mirror, as shown in this figure. Since triangles OAB and IAB are congruent, $OA = IA$. Therefore, the image formed by an object placed in front of a plane mirror is the same distance behind the mirror as the object is in front of the mirror.

The geometry of Figure 8.3 also shows that the object height, h, equals the image height, h'. Let lateral magnification, M, of a plane mirror be defined as given in Equation (8.1):

$$M = \frac{h'}{h} \tag{8.1}$$

Equation (8.1) gives a general definition of the lateral magnification of any type of plane mirror. $M = 1$ for a plane mirror because $h' = h$, as shown in the geometry of Figure 8.3.

The image formed by a plane mirror has one more important property: that of right-left reversal between image and object. This reversal can be seen by standing in front of a mirror and raising one's right hand. The image seen raises its left hand.

In summary, the image formed by a plane mirror has the following properties:

1. The image is as far "behind" the plane mirror as the object location is in front of the mirror.
2. The image is unmagnified, virtual, and erect, as shown in Figure 8.3.
3. The image has right–left reversal.

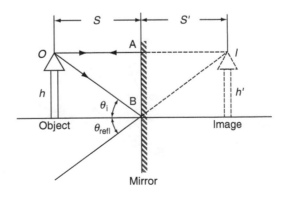

FIGURE 8.3 Geometric construction used to locate the image of an object placed in front of a plane mirror.

8.2.2 MULTIPLE IMAGES FORMED BY AN OBJECT PLACED BETWEEN TWO PLANE MIRRORS AT RIGHT ANGLES

Figure 8.4 shows an object at point O placed between two plane mirrors at right angles to each other. In this situation, multiple images are formed. The positions of the multiple images that formed by the two mirrors can be located using the simple geometric techniques shown in Figure 8.4. The image of the object is at image$_1$, behind mirror 1, and at image$_2$, behind mirror 2. In addition, a third image is formed at image$_3$, which will be the image of image$_1$ in mirror 2 or, equivalently, the image of image$_2$ in mirror 1. That is, the image at image$_1$ (or image$_2$) serves as the object for image$_3$. When viewing image$_3$, note that the rays reflect twice after leaving the object point at O.

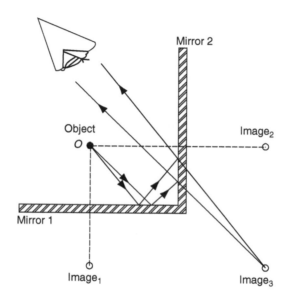

FIGURE 8.4 An object placed between two plane mirrors at right angles.

8.2.3 TRACING A LASER BEAM PASSING BETWEEN TWO PLANE MIRRORS AT AN ACUTE ANGLE

This is sometimes called double reflected light ray between two mirrors. Two mirrors make an acute angle, β, with each other, as shown in Figure 8.5. A light ray is incident on mirror 1 at an angle θ_1 to the normal. From the law of reflection, the reflected ray also makes an angle θ_1 with the normal.

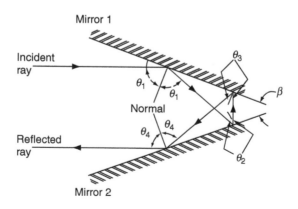

FIGURE 8.5 Laser beam path passes between two plane mirrors at an acute angle.

The reflected ray from mirror 1 is then incident on mirror 2, making an angle θ_2 with the normal. The reflected ray from mirror 2 also makes an angle θ_2 with the normal. Similarly, the reflected ray from mirror 2 is incident on mirror 1 with an angle θ_3 with the normal. The final reflected ray from mirror 1 onto mirror 2 makes angles of incidence and reflection θ_4 with the normal.

From the triangle made by the first reflected ray and the two mirrors, one can see the first reflected ray from mirror 1 and mirror 2 in relation to the acute angle between the mirrors. Figure 8.5 also shows that the first incident ray on mirror 1 is parallel to the last reflected ray from mirror 2. The first incident and last reflected rays are in opposite directions.

8.2.4 TRACING A LASER BEAM PASSING BETWEEN TWO PLANE MIRRORS AT RIGHT ANGLES

Two mirrors make a right angle with each other, as shown in Figure 8.6. A light ray is incident on mirror 1 at an angle 45° to the normal. From the law of reflection, the reflected ray also makes an angle 45° with the normal. The reflected ray from mirror 1 is then incident on mirror 2. The incident ray makes an angle of 45° with the normal, and the reflected ray from mirror 2 also makes an angle of 45° with the normal. From the triangle made by the first reflected ray and the two mirrors, it is possible to see the first reflected ray from mirror 1 and mirror 2 in relation to the right angle between the mirrors. Figure 8.6 also shows that the incident ray on mirror 1 is parallel to the reflected ray from mirror 2. The first incident and last reflected rays are in opposite directions.

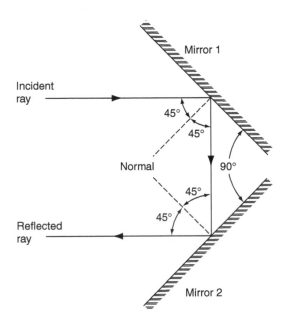

FIGURE 8.6 Laser beam path passes between two plane mirrors at right angles.

8.2.5 TRACING A LASER BEAM PASSING BETWEEN TWO PLANE MIRRORS AT AN OBTUSE ANGLE

Two mirrors make an obtuse angle β with each other, as shown in Figure 8.7. A light ray is incident on mirror 1 at an angle θ_1 to the normal. From the law of reflection, the reflected ray also makes an angle θ_1 with the normal, the reflected ray from mirror 1 is incident on mirror 2. The incident ray makes an angle θ_2 with the normal and the reflected ray from mirror 2 also makes an angle θ_2 with the normal. From the triangle made by the first reflected ray and the two mirrors, you can see the first reflected ray from mirror 1 and mirror 2 in relation to the obtuse angle between the mirrors. Figure 8.7 also shows that the incident ray on mirror 1 is not parallel to the reflected ray from mirror 2. The first incident and last

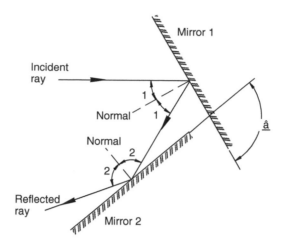

FIGURE 8.7 Laser beam path passes between two plane mirrors at an obtuse angle.

reflected rays are in opposite direction at an angle with each other. It is possible to direct the last reflected ray to any angle by rotating mirror 1 and/or mirror 2, i.e., changing the angle β.

8.2.6 Tracing a Laser Beam Passing between Three Plane Mirrors at Different Angles

Three mirrors make different angles with each other, as shown in Figure 8.8. A light ray is incident on mirror 1 at an angle θ_1 to the normal. From the law of reflection, the reflected ray also makes an angle θ_1 with the normal. The reflected ray from mirror 1 is then incident on mirror 2, making an

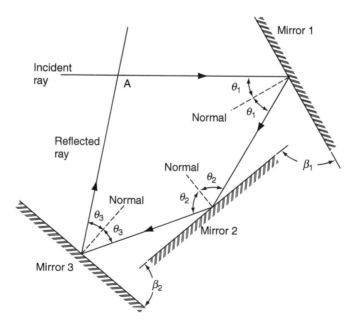

FIGURE 8.8 Laser beam path passes between three plane mirrors at different angles.

angle θ_2 with the normal. The reflected ray from mirror 2 also makes an angle θ_2 with the normal. Similarly, the reflected ray from mirror 2 is incident on mirror 3, with an angle θ_3 with the normal. The reflected ray from mirror 3 makes an angle θ_3 with the normal.

From the triangle made by the first reflected ray and the three mirrors, one can see the first reflected ray from mirror 1, mirror 2, and mirror 3 in relation to the angles between the mirrors. Figure 8.8 also shows that the incident ray on mirror 1 is not parallel to the reflected ray from mirror 3. Depending on the angles between the mirrors, the last reflected ray could be in any direction at an angle with the first incident ray. The last reflected and first incident rays can intercept each other at a point A. It is possible to direct the last reflected ray to any angle by rotating mirror 1, mirror 2, or mirror 3.

8.2.7 Tracing a Laser Beam Incident on a Rotating Mirror

Figure 8.9(a) shows a single ray of light incident normally upon a plane mirror. The angle of incidence is zero to the mirror; thus, the angle of reflection is also zero, according to the law of reflection (the angle of incidence is equal to the angle of reflection). If the mirror is now rotated through an angle, α, without changing the direction of the incident light ray, the angle of incidence also becomes equal to α. Therefore, the angle of reflection NBC becomes α as well, as shown in Figure 8.9(b). Thus, if the mirror rotates through an angle α, the separation of the incident and reflected rays changes by 2α. Since the incident ray remains fixed throughout, it is equally true that the rotation of the mirror through an angle α causes a movement of the reflected ray through an angle of 2α.

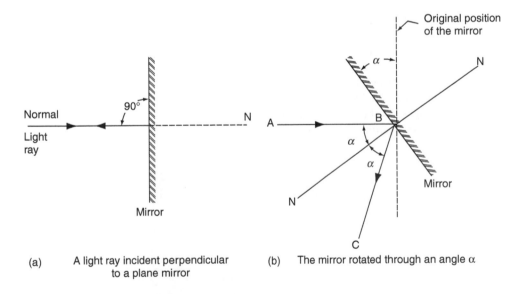

(a) A light ray incident perpendicular to a plane mirror

(b) The mirror rotated through an angle α

FIGURE 8.9 The rotating mirror.

8.3 EXPERIMENTAL WORK

The theory of this experiment is based on the first law of light, which is called the law of reflection. A light ray incident on a reflecting surface is described by an angle of incidence (θ_i). This angle is measured relative to a line normal (perpendicular) to the reflecting surface, as shown in Figure 8.1. Similarly, the reflected ray is described by an angle of reflection (θ_{refl}). The incident ray, the normal line, and the reflected ray lie in the plane of incidence. The relationship between these angles is given by the law of reflection: the angle of incidence is equal to the angle of reflection.

This experiment is a study of the formation of images when an object is placed in front of a plane mirror. The student will study the image formation by reflection. Mirrors are optical components that work on the basis of image formation by reflection. Such optical components are commonly used in manufacturing optical devices, instruments, and systems. In the theory of this experiment, the ray approximation assumes that light travels in straight lines. This corresponds to the field of geometric optical components. This experiment discusses the manner in which optical components, such as mirrors, form images. The plane mirrors are the simplest optical components to be considered in detail.

The purpose of this lab is to observe the image(s) and laser beam path when a candle or a laser beam source is placed in front of plane mirror arrangements for the following cases:

a. A candle placed in front of a fixed plane mirror.
b. A candle placed between two plane mirrors at right angles.
c. A laser beam passing between two plane mirrors at an acute angle.
d. A laser beam passing between two plane mirrors at right angles.
e. A laser beam passing between two plane mirrors at an obtuse angle.
f. A laser beam passing between three plane mirrors at different angles.
g. A laser beam incident on a rotating plane mirror.

8.3.1 Technique and Apparatus

Appendix A presents the details of the devices, components, tools, and parts.

1. 2×2 ft. optical breadboard.
2. HeNe laser light source and power supply.
3. Laser mount assembly.
4. Candles.
5. Hardware assembly (clamps, posts, screw kits, screwdriver kits, sundry positioners, etc.).
6. Mirrors and mirror holders, as shown in Figure 8.10.
7. Rotation stages, as shown in Figure 8.10.
8. Target card and cardholder, as shown in Figure 8.11.

FIGURE 8.10 A plane mirror mounted on a rotation stage.

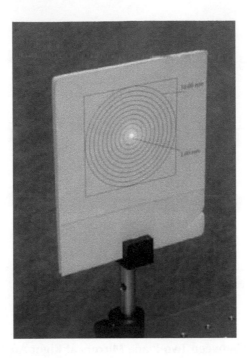

FIGURE 8.11 A target card and cardholder.

9. Protractor.
10. Ruler.

8.3.2 Procedure

Follow the laboratory procedures and instructions given by the instructor.

8.3.3 Safety Procedure

Follow all safety procedures and regulations regarding the use of optical components and instruments, light source devices, and optical cleaning chemicals.

8.3.4 Apparatus Setup

8.3.4.1 A Candle Placed in Front of a Fixed Plane Mirror

1. Figure 8.12 shows the experimental apparatus setup.
2. Mount a plane mirror on the breadboard so that the mirror lines up with the holes.
3. Place the candle in front of the plane mirror at a reasonable distance.
4. Measure the distance between the candle and the mirror by counting the holes on the breadboard. Fill out Table 8.1.
5. Measure the distance between the mirror and the image by counting the holes on the breadboard. Fill out Table 8.1.
6. Illustrate the locations of the candle, mirror, and image in a diagram.

FIGURE 8.12 A candle placed in front of a plane mirror.

8.3.4.2 A Candle Placed between Two Plane Mirrors at Right Angles

1. Figure 8.13 shows the experimental apparatus setup.
2. Mount two plane mirrors on the breadboard. Position the two mirrors to make a right angle with each other.
3. Place the candle between the mirrors.
4. Move the candle between the two mirrors and observe the formation of images.
5. Measure the distance between the candle and the mirrors by counting the holes on the breadboard. Fill out Table 8.2.
6. Measure the distance between the mirrors and the images by counting the holes on the breadboard. Fill out Table 8.2.
7. Count the number of images that are produced between the mirrors.
8. Illustrate the locations of the candle, mirrors, and images in a diagram.

FIGURE 8.13 A candle placed between two plane mirrors at right angles.

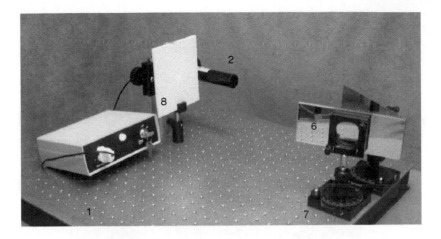

FIGURE 8.14 A laser beam passing between two plane mirrors at an acute angle.

8.3.4.3 A Laser Beam Passing between Two Plane Mirrors at an Acute Angle

1. Figure 8.14 shows the experimental apparatus setup.
2. Bolt the laser short rod to the breadboard.
3. Bolt the laser mount to the clamp using bolts from the screw kit.
4. Put the clamp on the short rod.
5. Place the HeNe laser into the laser mount and tighten the screw. Turn on the laser device. Follow the operation and safety procedures of the laser device in use.
6. Check the laser alignment with the line of bolt holes on the breadboard and adjust when necessary.

FIGURE 8.15 The path of the laser beam.

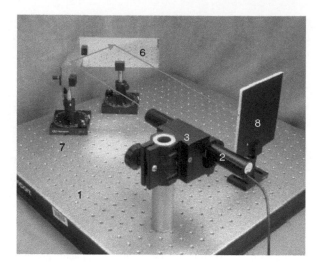

FIGURE 8.16 A laser beam passing between two plane mirrors at right angles.

7. Mount two plane mirrors on the breadboard.
8. Align the laser beam to be incident on the first mirror. Align the second mirror to make an acute angle with the first, as shown in Figure 8.15.
9. Observe the laser beam path between the two mirrors and intercept the laser beam with the target card, as shown in Default 8.14 and 8.15.
10. Measure the angles of the laser beam path relative to the mirrors. Fill out Table 8.3.
11. Illustrate the locations of the mirrors and the laser beam path in a diagram.

8.3.4.4 A Laser Beam Passing between Two Plane Mirrors at Right Angles

Repeat the procedure explained in Case (c) of this experiment. However, in Case (d), make a right angle between the two plane mirrors. Figure 8.16 shows the experimental apparatus setup and the laser beam path. Fill out Table 8.4.

8.3.4.5 A Laser Beam Passing between Two Plane Mirrors at an Obtuse Angle

Repeat the procedure explained in Case (c) of this experiment. However, in Case (e), make an obtuse angle between the two plane mirrors. Figure 8.17 shows the experimental apparatus setup and the laser beam path. Fill out Table 8.5.

8.3.4.6 A Beam Source Passing between Three Plane Mirrors at Different Angles

Repeat the procedure explained in Case (c) of this experiment. However, in Case (f), choose different angles between the three plane mirrors. Figure 8.18 shows the experimental apparatus setup and laser beam path. Fill out Table 8.6.

8.3.4.7 A Laser Beam Incident on a Rotating Mirror

1. Figure 8.19 shows the experimental apparatus setup.
2. Repeat steps 2 to 8 from Case (c).
3. Mount a plane mirror on a rotation stage and place it on the breadboard.
4. Set the rotation stage to zero degree or align the mirror to give you a back reflection into the laser source, as shown in Figure 8.19(a).

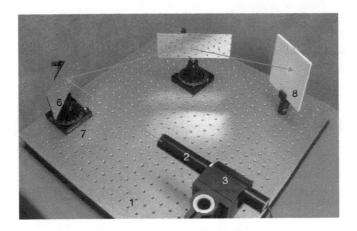

FIGURE 8.17 A laser beam passing between two plane mirrors at an obtuse angle.

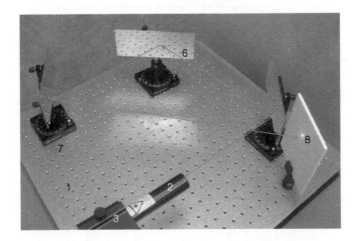

FIGURE 8.18 A laser beam passing between three plane mirrors at different angles.

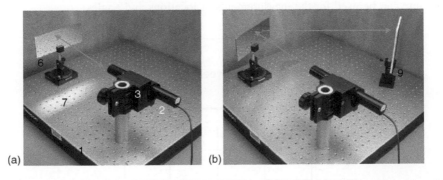

FIGURE 8.19 A laser beam incident on a rotating mirror.

5. Turn the rotation stage by an angle α (several degrees, say 20°), as shown in Figure 8.19(b).
6. Try to capture the reflected laser beam on the target card.
7. Measure the angle between the incident beam and the reflected beam; this is 2α. Fill out Table 8.7.
8. Illustrate the locations of the laser source, mirror, and the target card in a diagram.

8.3.5 DATA COLLECTION

8.3.5.1 A Candle Placed in Front of a Fixed Plane Mirror

Measure the locations of the candle, mirror, and image, and fill out Table 8.1.

TABLE 8.1
A Candle Placed in Front of a Plane Mirror

Location of		
Candle	Mirror	Image

8.3.5.2 A Candle Placed between Two Plane Mirrors at Right Angles

Measure the locations of the candle, mirrors, and images, and fill out Table 8.2.

TABLE 8.2
A Candle Placed between Two Plane Mirrors at Right Angles

Location of					
Candle	Mirror 1	Mirror 2	Image 1	Image 2	Image 3

8.3.5.3 A Laser Beam Passing between Two Plane Mirrors at an Acute Angle

Measure the locations of the laser source, mirrors, and target card, and fill out Table 8.3. A reference line can be used for the measurements of locations and angles.

TABLE 8.3
Laser Beam Passing between Two Plane Mirrors at an Acute Angle

Location of				Angle of	
Laser Source	Mirror 1	Mirror 2	Target Card	Mirror 1	Mirror 2

8.3.5.4 A Laser Beam Passing between Two Plane Mirrors at Right Angles

Measure the locations of the laser source, mirrors, and target card, and fill out Table 8.4. A reference line can be used for the measurements of locations and angles.

TABLE 8.4
Laser Beam Passing between Two Plane Mirrors at Right Angles

Location of				Angle of	
Laser Source	Mirror 1	Mirror 2	Target Card	Mirror 1	Mirror 2
				45°	45°

8.3.5.5 A Laser Beam Passing between Two Plane Mirrors at an Obtuse Angle

Measure the locations of the laser source, mirrors, and target card, and fill out Table 8.5. A reference line can be used for the measurements of locations and angles.

TABLE 8.5
Laser Beam Passing between Two Plane Mirrors at an Obtuse Angle

Location of				Angle of	
Laser Source	Mirror 1	Mirror 2	Target Card	Mirror 1	Mirror 2

8.3.5.6 A Laser Beam Passing between Three Plane Mirrors at Different Angles

Measure the locations of the laser source, mirrors, and target card, and fill out Table 8.6. A reference line can be used for the measurements of locations and angles.

TABLE 8.6
Laser Beam Passing between Three Plane Mirrors at Different Angles

Location of					Angle of		
Laser Source	Mirror 1	Mirror 2	Mirror 3	Target Card	Mirror 1	Mirror 2	Mirror 3

8.3.5.7 A Laser Beam Incident on a Rotating Mirror

Measure the locations of the laser source, mirror, and target card, and fill out Table 8.3.6. A reference line can be used for the measurements of locations and angles.

TABLE 8.7
A Laser Beam Incident on a Rotating Mirror

Location of			Angle of		
Laser Source	Mirror	Target Card	Mirror	Incident Ray	reflected Ray

8.3.6 CALCULATIONS AND ANALYSIS

8.3.6.1 A Candle Placed in Front of a Fixed Plane Mirror

Illustrate the locations of the candle, mirror, and image in a diagram; include your measurements.

8.3.6.2 A Candle Placed between Two Plane Mirrors at Right Angles

Illustrate the locations of the candle, mirror, and image in a diagram; include your measurements.

8.3.6.3 A Laser Beam Passing between Two Plane Mirrors at an Acute Angle

Illustrate the locations of the laser source, mirrors, target card, and laser beam path in a diagram; include your measurements.

8.3.6.4 A Laser Beam Passing between Two Plane Mirrors at Right Angles

Illustrate the locations of the laser source, mirror, target card, and laser beam path in a diagram; include your measurements.

8.3.6.5 A Laser Beam Passing between Two Plane Mirrors at an Obtuse Angle

Illustrate the locations of the laser source, mirrors, target card, and laser beam path in a diagram; include your measurements.

8.3.6.6 A Laser Beam Passing between Three Plane Mirrors at Different Angles

Illustrate the locations of the laser source, mirrors, target card, and laser beam path in a diagram; include your measurements.

8.3.6.7 A Laser Beam Incident on a Rotating Mirror

1. Illustrate the locations of the laser beam, mirror, target card, and laser beam path in a diagram; include your measurements.
2. Show that the reflected ray rotates 2α when the mirror rotates an angle α.

8.3.7 Results and Discussions

8.3.7.1 A Candle Placed in Front of a Fixed Plane Mirror

1. Discuss your observation of the formation of the image in a diagram.
2. Confirm that the distance of the object from the mirror is equal to the distance of the image formed behind the mirror.

8.3.7.2 A Candle Placed between Two Plane Mirrors at Right Angles

1. Discuss your observation of the formation of the images.
2. Confirm that the distance of the object from the mirror is equal to the distance of the image formed behind the mirrors.
3. Confirm the number of images formed by the two plane mirrors at a right angle with each other.

8.3.7.3 A Laser Beam Passing between Two Plane Mirrors at an Acute Angle

1. Discuss your observation of the laser beam path.
2. Confirm the law of reflection of the laser beam when incident on a mirror.

For Cases (d) through (g), repeat the steps as explained in Case (c).

8.3.8 Conclusion

Summarize the important observations and findings obtained in this lab experiment.

8.3.9 Suggestions for Future Lab Work

List any suggestions for improvements using different experimental equipment, procedures, and techniques for any future lab work. These suggestions should be theoretically justified and technically feasible.

8.4 LIST OF REFERENCES

List any references that were used in the report. Use one format in writing the references. Never mix reference formats in a report.

8.5 APPENDICES

List all of the materials and information that are too detailed to be included in the body of the report.

FURTHER READING

Beiser, A., *Physics,* 5th ed., Addison-Wesley Publishing Company, Reading, MA, 1991.
Cutnell, J. D. and Johnson, K. W., *Physics* 5th ed., John Wiley and Sons, Inc., New York, 2001a.
Cutnell, J. D. and Johnson, K. W., *Student Study Guide—Physics*, 5th ed., John Wiley and Sons, Inc., New York, 2001b.

Davis, C. C., *Lasers and Electro-Optics: Fundamentals and Engineering*, Cambridge University Press, Cambridge, 1996.

Edmund Industrial Optics, *Optics and Optical Instruments Catalog*, Edmund Industrial Optics, Barrington, NJ, 2004.

Ghatak, A. K., *An Introduction to Modern Optics*, McGraw-Hill Book Company, New York, 1972.

Giancoli, D. C., *Physics,* 5th ed., Prentice Hall, Engle Wood, Cliffs, NJ, 1998.

Halliday, D., Resnick, R., and Walker, J., *Fundamentals of Physics*, 6th ed., John Wiley & Sons, Inc, USA, 1997.

Heath, R. W., Macaughton, R. R., and Martindale, D. G., *Fundamentals of Physics*, D.C. Heath Canada, Ltd., Canada, 1979.

Hecht, E., *Optics,* 4th ed., Addison-Wesley Longman, Inc, USA, 2002.

Jenkins, F. W. and White, H. E., *Fundamentals of Optics*, McGraw Hill, New York, 1957.

Keuffel & Esser Co., *Physics*, Keuffel & Esser Audiovisual Educator—Approved Diazo Transparency Masters, Audiovisual Division, Keuffel & Esser Co., U.S.A., 1989.

Lambda Research Optics, Inc., *Catalog 2004*, Lambda Research Optics, Costa Mesa, CA, 2004.

Lerner, R. G. and Trigg, G. L., *Encyclopedia of Physics*, 2nd ed., VCH Publishers, Inc, New York, NY, 1991.

McDermott, L. C., *Introduction to Physics*, Preliminary Edition, Prentice Hall, Inc., Engle Wood, Cliffs, NJ, 1988.

Melles, G., *The Practical Application of Light*, Melles Griot Catalog, Irvine, CA, 2001.

Naess, R. O., *Optics for Technology Students*, Prentice Hall, Englewood, Cliffs, NJ, 2001.

Nichols, D. H., *Physics for Technology with Applications in Industrial Control Electronics*, Prentice Hall, Englewood, Cliffs, NJ, 2002.

Nolan, P. J., *Fundamentals of College Physics*, Wm. C. Brown Publishers, Dubuque, Iowa, 1993.

Pedrotti, F. L. and Pedrotti, L. S., *Introduction to Optics*, 2nd ed., Prentice Hall, Inc, Engle Wood, Cliffs, NJ, 1993.

Serway, R. A. and Jewett, J. W., *Physics for Scientists and Engineers* with Modern Physics, 6th ed., volume 2, Thomson Books/Cole, U.S.A., 2004.

Smith, W. J., *Modern Optical Engineering*, McGraw-Hill Book Co., New York, 1966.

Warren, M. L., *Introduction to Physics*, W.H. Freeman and Company, San Francisco, 1979.

White, H. E., *Modern College Physics*, 6th ed., Van Nostrand Reinhold Company, New York, 1972.

Williams, J. E. et al., *Teacher's Edition—Modern Physics*, Holt, Rinehart and Winston, Inc, New York, 1968.

Woods, N., *Instruction's Manual to Beiser Physics*, 5th ed., Addison-Wesley Publishing Company, Reading, MA, 1991.

9 Spherical Mirrors

9.1 INTRODUCTION

Like plane mirrors, spherical mirrors are a common type of mirror used in many applications. Spherical mirrors are similar to plane mirrors in image formation, but instead of being made from a flat piece of glass, spherical mirrors have the shape of a section from the surface of a hollow sphere. If the inside surface of the mirror is reflective, it is called a concave mirror. If the outside surface of the mirror is reflective, it is called a convex mirror. The concave and convex mirrors are sometimes also called converging and diverging mirrors, respectively. Spherical mirrors are used in many applications, such as simple reflections, collimating, light convergence and divergence, and reflecting telescopes.

There are also curved mirrors with parabolic or elliptical shapes. These mirrors are used in many applications, including light reflectors in automobile headlights and solar ray concentrators. Spherical and curved mirrors are used mostly for concentrating or diverging light.

Three experimental lab cases presented in this chapter cover the formation of images of an object placed in front of a spherical mirror. The student will study the image formation using the ray-tracing method with experimental measurements and calculations. The cases are: image formation by concave mirrors; image formation by convex mirrors; and image formation by two spherical concave mirrors.

9.2 IMAGES FORMED BY SPHERICAL MIRRORS

The theory of this chapter is based on the first law of light, which is called the law of reflection. A light ray incident on a reflecting surface is described by an angle of incidence (θ_i) and an angle of reflection (θ_{refl}). The law of reflection has been presented in detail among the Laws of Light in Chapter 7.

Most curved mirrors used in practical optical applications are spherical. A spherical mirror is a portion of a reflecting sphere, which has a radius of curvature R and centre of curvature located at

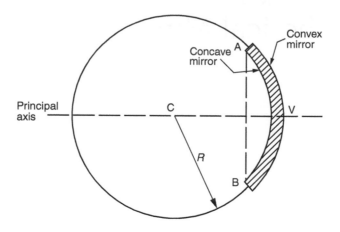

FIGURE 9.1 Definition of terms for spherical mirrors.

point C, as illustrated in Figure 9.1. The principal axis is an imaginary line that runs through C and the vertex V of the mirror. The opening AB, often used in solving optical problems, is called the linear aperture of the mirror.

There are two types of spherical mirrors: with the reflecting surface applied on the inner or on the outer surfaces. The reflecting surface can be created by polishing or reflective coating process. When the inner surface of the sphere is the reflecting surface, the mirror is called a concave mirror, as illustrated in Figure 9.2. When the outer surface of the sphere is the reflecting surface, the mirror is called a convex mirror, as illustrated in Figure 9.3.

Consider the reflection of parallel light rays from a spherical mirror. As illustrated in Figure 9.4, a light beam of parallel rays is incident on a concave mirror. Since the mirror is perpendicular to the principal axis of the mirror at the vertex V, the rays are reflected back and pass through a point F. This point is called the focal point of the mirror. The geometry of the reflection shows that the point F is located on the principal axis halfway between the centre of curvature C and the vertex V of the mirror. The distance from F to V is called the focal length f of the mirror. The focal length is defined

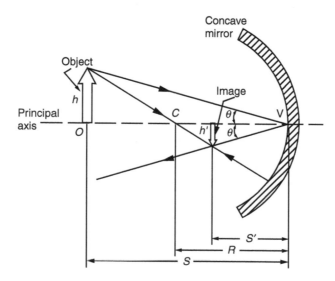

FIGURE 9.2 Formation of an image by a spherical concave mirror.

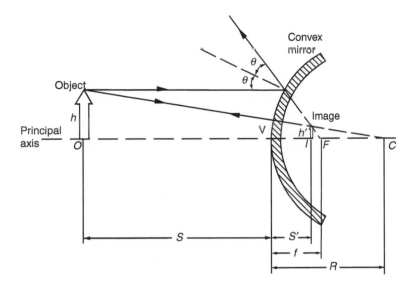

FIGURE 9.3 Formation of an image by a spherical convex mirror.

by Equation (9.1):

$$f = \frac{R}{2} \tag{9.1}$$

All light rays from a distant object, such as the sun, are parallel rays and will converge from the mirror at the focal point. For this reason, concave mirrors are frequently called converging mirrors. The focal point of a spherical mirror can be found experimentally by converging parallel light rays

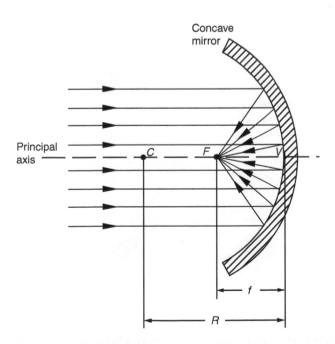

FIGURE 9.4 Light rays from a distant object reflect from a concave mirror through the focal point F.

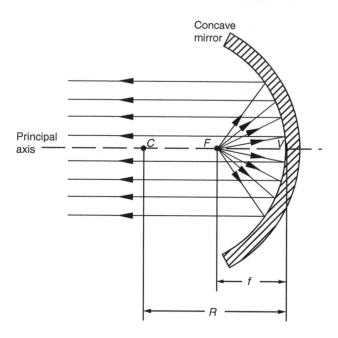

FIGURE 9.5 A light source at the focal point F of a concave mirror.

onto a target card. When sunlight is converged on a piece of paper at the focal point of the mirror, the paper can be burned by the intensity of the concentrated sunlight.

Since the light rays are reversible in direction, a source of light placed at the focal point of a concave mirror will form its image at infinity. In other words, the emerging light rays will be parallel to the principal axis of the mirror, as shown in Figure 9.5.

The same general principles can be applied for concave and convex mirrors. Light rays parallel to the principal axis, incident on the convex mirror surface, will diverge after reflecting off the mirror surface. The diverging reflected light rays appear to come from the focal point F located

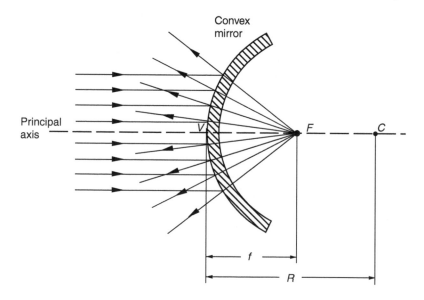

FIGURE 9.6 Light rays from a distant object reflect from a convex mirror and form the focal point F behind the mirror.

behind the mirror, but no light rays actually pass through it. Even though the focal point is virtual, the distance VF is called the focal length f, as shown in Figure 9.6. Since the actual light rays diverge when striking a convex mirror, convex mirrors are frequently called diverging mirrors. Equation (9.1) also applies in defining the focal length for a convex mirror.

9.2.1 CONCAVE MIRRORS

The geometrical optics method, sometimes called the ray-tracing method, is used for locating the images formed by spherical mirrors. This method traces the reflection of a few rays diverging from a point on an object O; this point must not be on the principal axis of the mirror. The point at which the reflected rays intersect will locate the image. Consider an object located on the principal axis away from the centre of curvature C of the concave mirror. Three rays originate from the point O on the top of the object. Figure 9.7 illustrates each of the three rays converging from the concave mirror.

Ray 1 is drawn from the top of the object from point O, parallel to the principal axis of the mirror, and then reflected back through the focal point F of a concave mirror.
Ray 2 is drawn from the top of the object from point O, through the focal point F of a concave mirror to the vertex V of the mirror, and then is reflected with the angle of incidence equal to the angle of reflection.
Ray 3 is drawn from the top of the object from point O, through the centre of curvature C, and then is reflected back along its original path.

The intersection of any two of these rays is enough to locate the image of an object. The third ray serves as a check on the image formation by the ray-tracing method. The image location obtained in the ray tracing method must always agree with the image location S' calculated from the spherical mirror equation. In Figure 9.7, the solid lines are used to identify real rays and real images.

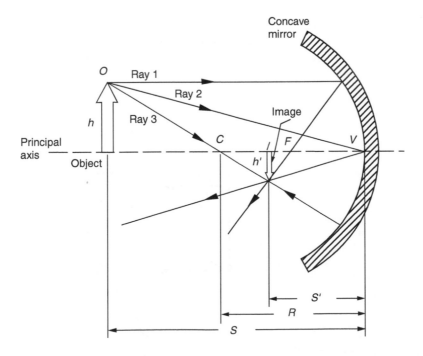

FIGURE 9.7 Formation of an image by a concave mirror using the ray tracing method.

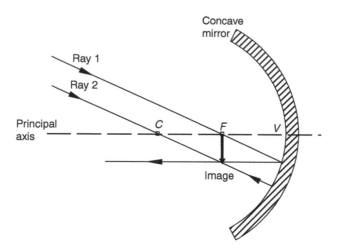

FIGURE 9.8 Image formed by a concave mirror for a distant object.

Consider the image formation by the ray-tracing method for an object which is placed in different positions in front of a concave mirror using two rays. In Figure 9.8, the parallel rays come from an object located at a distance far away from the centre of the concave mirror. The mirror forms an image at the focal point. The image is real, inverted, and smaller than the object.

In Figure 9.9, the object is located at the centre of curvature, C, of the concave mirror. The mirror forms an image at the centre of curvature. The image is real, inverted, and the same size as the object.

In Figure 9.10, the object is located at the focal point of F of the concave mirror. All reflected rays are parallel and will never intersect at any distance from the mirror. Therefore, no image will be formed in this situation.

In Figure 9.11, the object is located between the focal point, F, and the vertex, V, of the concave mirror. The rays form an image behind the mirror. The image can be located by extending the reflected rays to a point where they intersect behind the mirror. The image is virtual, erect, larger

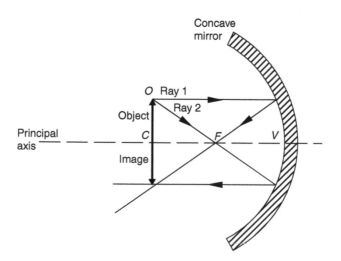

FIGURE 9.9 Image formed by a concave mirror for an object located at the centre of the curvature.

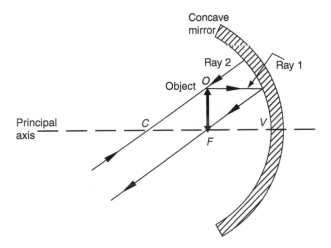

FIGURE 9.10 No image formed by a concave mirror for an object located at the focal point.

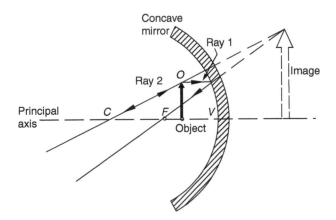

FIGURE 9.11 Image formed by a concave mirror for an object located inside the focal length.

than the object, and located behind the mirror. The magnification in this case follows the principle of spherical mirrors forming enlarged virtual images.

In summary, the characteristics of the image formed by a concave mirror depend on the location of the object. The images can be real or virtual, inverted or erect, and enlarged or reduced.

9.2.2 The Mirror Equation

The geometry shown in Figure 9.12 is used to calculate the image distance, S', knowing the object distance, S, and the radius of curvature, R, of a concave mirror. By convention, these distances are measured from the vertex point V of the concave mirror. Figure 9.12 shows two rays of light leaving from the point O on the top of the object. Ray 1 strikes the mirror at the vertex point V and reflects, as shown in the figure, obeying the law of reflection. Ray 1 is incident at an angle θ and reflected at an equal angle. Ray 2 passes through the centre of curvature C of the mirror, hits the mirror perpendicular to the surface, and reflects back on itself. The image is located where these two

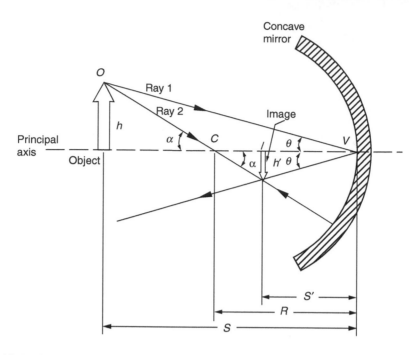

FIGURE 9.12 Deriving the mirror equation for an image formed by a concave mirror.

rays intersect. The image located at I is a real image, since actual light rays form it. The image is smaller than the object and is inverted.

From the largest triangle in Figure 9.12, one can write that $\tan \theta = h/s$, while the smallest triangle gives $\tan \theta = -h'/s'$. The negative sign signifies that the image is inverted. Thus, the magnification of the concave mirror can be written as:

$$M = \frac{h'}{h} = -\frac{S'}{S} \tag{9.2}$$

From the two triangles in Figure 9.12, one can also write the following relations:

$$\tan \alpha = \frac{h}{S - R} \quad \text{and} \quad \tan \alpha = -\frac{h'}{R - S'}$$

from which it is found that

$$\frac{h'}{h} = -\frac{R - S'}{S - R} \tag{9.3}$$

Combining Equation (9.2) and Equation (9.3), one has

$$\frac{S'}{S} = \frac{R - S'}{S - R} \tag{9.4}$$

Rearranging the terms using simple algebra, the following equation is obtained:

$$\frac{1}{S} + \frac{1}{S'} = \frac{2}{R} \tag{9.5}$$

Equation (9.5) is often written in terms of the focal length, f, of the mirror, instead of the radius of curvature. Again, the focal length of a mirror is given by Equation (9.1).

The mirror equation can be rewritten in terms of the focal length, by combining Equation (9.1) and Equation (9.5), to obtain:

$$\frac{1}{S} + \frac{1}{S'} = \frac{1}{f} \tag{9.6}$$

Equation (9.5) and Equation (9.6) are the two forms of the equation known as the spherical mirror equation. This equation is applicable to any spherical mirror, concave or convex.

9.2.3 Convex Mirrors

Light is reflected from the silver coating of the outer convex surface of a spherical mirror. This type of mirror is sometimes called a diverging mirror. The rays from a point on a real object diverge after reflection as though they were coming from some point behind the convex mirror. Figure 9.13 illustrates an image formation by a convex mirror. In this figure, the image is virtual, since it lies behind the mirror at the location where the reflected rays appear to originate.

The ray-tracing method for the formation of images by spherical convex mirrors is applied and illustrated in Figure 9.13. The point at which the reflected rays intersect will locate the image. Consider an object located on the principal axis away from the centre of curvature C of the convex mirror. Three rays originate from the point O on the top of the object.

Ray 1 is drawn from the top of the object from point O, parallel to the principal axis of the mirror, and then reflected back to the same side. If the ray is extended, it seems to come from the focal point F of the convex mirror.

Ray 2 is drawn from the top of the object from point O, proceeds toward the focal point F of the convex mirror, and then is reflected parallel to the principal axis.

Ray 3 is drawn from the top of the object from point O, proceeds toward the centre of curvature C, and then is reflected back along its original path.

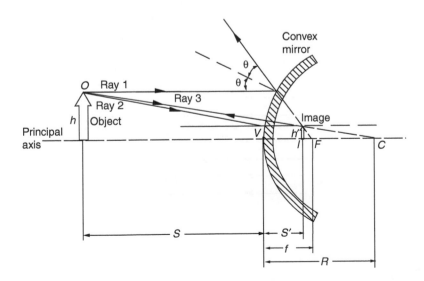

FIGURE 9.13 Ray tracing for a convex mirror.

The intersection of any two of these rays is enough to locate the image of an object. The third ray serves as a check on the image formation by the ray-tracing method.

The image location obtained in the ray-tracing method must always agree with the image location S' calculated from the spherical mirror equation. In Figure 9.13, the dashed lines are used to identify virtual rays and virtual images. Equation (9.6), the mirror equation, can be used to calculate the image location.

9.2.4 SIGN CONVENTION FOR SPHERICAL MIRRORS

S is *positive* if the object is in front of the mirror (real object).
S is *negative* if the object is in back of the mirror (virtual object).
S' is *positive* if the image is in front of the mirror (real image).
S' is *negative* if the image is in back of the mirror (virtual image).
Both f and R are positive if the centre of curvature is in front of the mirror (concave mirror).
Both f and R are negative if the centre of curvature is in back of the mirror (convex mirror).
If M is positive, the image is erect.
If M is negative, the image is inverted.

This convention applies only to the numerical values substituted into the mirror equations, Equation (9.5) and Equation (9.6). The quantities S, S', and f should maintain their signs unchanged until the substitution is made.

9.3 SPHERICAL ABERRATION

Spherical mirrors form sharp images as long as their apertures are small compared to their focal lengths. When large spherical mirrors are used, some of the rays from objects strike near the outer edges (away from the principal axis) and are focused to different points on the principal axis, producing a blurred image. This focusing defect is known as spherical aberration. The spherical aberration is illustrated in Figure 9.14. This defect can be corrected by using a parabolic mirror. Theoretically, parallel light rays incident on a parabolic mirror section will focus the rays at a single

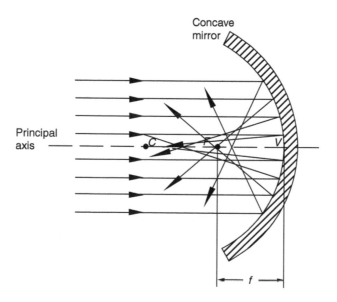

FIGURE 9.14 Spherical aberration occurring in a spherical mirror.

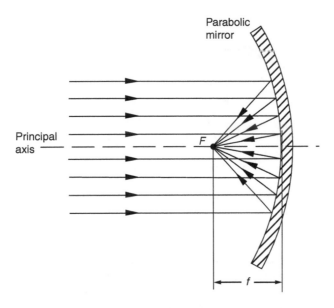

FIGURE 9.15 Parabolic mirror focuses all parallel light rays to the focal point.

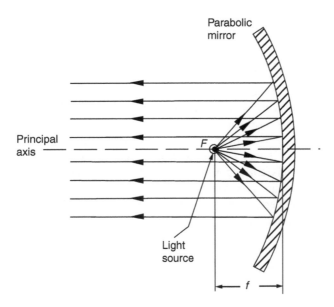

FIGURE 9.16 A light source at the focal point of a parabolic mirror.

point on the principal axis, as illustrated in Figure 9.15. Similarly, a small light source located at the focal point of a parabolic mirror is used in many optical devices, such as spotlights and searchlights. The light rays emitted from a light source are parallel to the principal axis of a parabolic mirror, as illustrated in Figure 9.16.

9.3.1 FORMATION OF IMAGE BY TWO SPHERICAL CONCAVE MIRRORS

Consider the image formation between two matching spherical concave mirrors, as illustrated in Figure 9.17. The two mirrors have back-silvered spherical glass surfaces, and each mirror has a hole

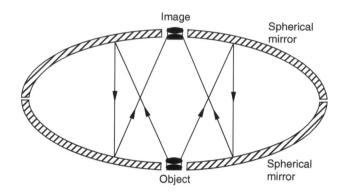

FIGURE 9.17 Forming a real three-dimensional image in the mirage.

at the centre of its spherical surface. The two mirrors are placed one on top of the other, with the silvered surfaces facing each other, and framed in a wooden box. An object is located at the hole of the lower mirror. Light rays from the object are first incident on the upper mirror, and then reflected onto the lower mirror. Next, the rays are all reflected from the lower mirror and intersect with each other to form the image in the hole of the upper mirror. The image appears as a real three-dimensional image in the mirage.

9.4 EXPERIMENTAL WORK

This experiment is a study of the formation of images of an object placed in front of a spherical mirror. The student will study the image formation using the ray-tracing method. Spherical mirrors are optical components that work on the basis of image formation by reflection. Such optical components are commonly used in manufacturing optical devices and instruments, such as reflecting telescopes. In the theory of this experiment, the ray approximation assumes that light travels in straight lines; this corresponds to the field of geometric optics. The theory of this experiment discusses the manner in which spherical mirrors form images. Spherical mirrors are similar to plane mirrors in image formation. The following cases deal with the image formation for different spherical mirror arrangements.

    a. Formation of images by concave mirrors.
    b. Formation of images by convex mirrors.
    c. Formation of image by two spherical concave mirrors.

9.4.1 TECHNIQUE AND APPARATUS

Appendix A presents the details of the devices, components, tools, and parts.

    1. Hardware assembly (clamps, posts, positioners, etc.).
    2. Spherical concave and convex mirrors, as shown in Figure 9.18.
    3. Spherical mirror holders.
    4. Two spherical concave mirrors, as shown in Figure 9.19.
    5. Optical rail.
    6. Candle.
    7. Black/white card and cardholder.
    8. Ruler.

FIGURE 9.18 Spherical concave and convex mirrors.

FIGURE 9.19 Spherical concave mirrors.

9.4.2 PROCEDURE

Follow the laboratory procedures and instructions given by the instructor.

9.4.3 SAFETY PROCEDURE

Follow all safety procedures and regulations regarding the use of light source, optical components, and optical cleaning chemicals.

9.4.4 APPARATUS SETUP

9.4.4.1 Formation of Images by Concave Mirrors

1. Figure 9.20 shows the experimental apparatus setup.
2. Place an optical rail on the table.

FIGURE 9.20 Formation of images by a concave mirror.

3. Mount a concave mirror on the table so that the centre of the mirror lines up perpendicular to the optical rail.
4. Place the candle on the optical rail in front of the concave mirror at a reasonable distance outside the centre of curvature of the concave mirror.
5. Measure the distance between the candle and the vertex of the mirror by counting the divisions on the optical rail. Fill out Table 9.1.
6. Measure the height of the candle and the height of the image. Fill out Table 9.1.
7. Choose three other positions for the candle in front of the concave mirror for this case. Repeat steps 5 and 6. Fill out Table 9.1.
8. Illustrate the locations of the candle, mirror, and image in diagrams using the ray-tracing method.

9.4.4.2 Formation of Images by Convex Mirrors

1. Figure 9.21 shows the experimental apparatus setup.
2. Place an optical rail on the table.
3. Mount a convex mirror on the table so that the centre of the mirror lines up perpendicular to the optical rail.
4. Place the candle on the optical rail in front of the convex mirror at a reasonable distance.
5. Measure the distance between the candle and the vertex of the mirror by counting the divisions on the optical rail. Fill out Table 9.2.
6. Measure the height of the candle and the height of the image. Fill out Table 9.2.
7. Choose three other positions for the candle in front of the convex mirror for this case. Repeat steps 5 and 6. Fill out Table 9.2.
8. Illustrate the locations of the candle, mirror, and image in diagrams using the ray-tracing method.

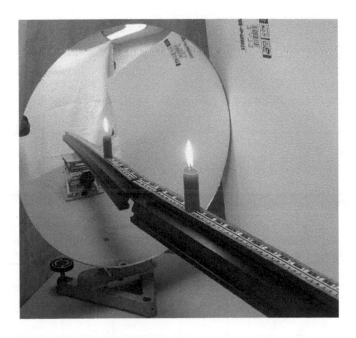

FIGURE 9.21 Formation of images by a convex mirror.

9.4.4.3 Formation of Image by Two Spherical Concave Mirrors

1. Mount two spherical mirrors, one on top of the other, on the table, as shown in Figure 9.22.
2. Place a paper clip in the centre hole of the lower spherical mirror.
3. Observe the image formation of the paper clip in the centre hole of the upper spherical mirror.
4. Illustrate the locations of the paper clip, mirrors, and image in a diagram using the ray-tracing method.

FIGURE 9.22 Formation of image by two spherical concave mirrors.

9.4.5 Data Collection

9.4.5.1 Formation of Images by Concave Mirrors

1. Measure the distance of the candle from the mirror and the heights of the candle and the image. Fill out Table 9.1.
2. Repeat step 1 for three additional candle positions. Fill out Table 9.1.

TABLE 9.1
A Candle Placed in Front of a Concave Mirror

Candle	Measured			Calculated		
Position	Candle Distance S (unit)	Candle Height h (unit)	Image Height h' (unit)	Magnification M	Image Distance S' (unit)	Mirror Focal Length f (unit)
1						
2						
3						
4						

9.4.5.2 Formation of Images by Convex Mirrors

1. Measure the distance of the candle from the mirror and the heights of the candle and the image. Fill out Table 9.2.
2. Repeat step 1 for three additional candle positions. Fill out Table 9.2.

TABLE 9.2
A Candle Placed in Front of a Convex Mirror

Candle	Measured			Calculated		
Position	Candle Distance S (unit)	Candle Height h (unit)	Image Height h' (unit)	Magnification M	Image Distance S' (unit)	Mirror Focal Length f (unit)
1						
2						
3						
4						

9.4.5.3 Formation of Image by Two Spherical Concave Mirrors

No data collection is required in this section.

9.4.6 Calculations and Analysis

9.4.6.1 Formation of Images by Concave Mirrors

1. For four candle positions, calculate the magnification of the mirror and the image distance using Equation (9.2), and the focal length of the mirror using Equation (9.6). Fill out Table 9.1.
2. For four candle positions, illustrate the locations of the candle, mirror, and image in diagrams, being sure to include dimensions.
3. Find the published value of the focal length of the mirror as provided by the manufacturer/supplier.

9.4.6.2 Formation of Images by Convex Mirrors

1. For four candle positions, calculate the magnification of the mirror and the image distance using Equation (9.2) and the focal length of the mirror using Equation (9.6). Fill out Table 9.2.
2. For four candle positions, illustrate the locations of the candle, mirror, and image in diagrams, being sure to include dimensions.
3. Find the published value of the focal length of the mirror as provided by the manufacturer/supplier.

9.4.6.3 Formation of Image by Two Spherical Concave Mirrors

Illustrate the locations of the paper clip, mirrors, and image in a diagram.

9.4.7 Results and Discussions

9.4.7.1 Formation of Images by Concave Mirrors

1. Discuss your measurements and calculations of the formation of the images in diagrams.
2. Verify the calculated focal length of the mirror against the value published by the manufacturer/supplier.

9.4.7.2 Formation of Images by Convex Mirrors

1. Discuss your measurements and calculations of the formation of the images in diagrams.
2. Verify the calculated focal length of the mirror against the value published by the manufacturer/supplier.

9.4.7.3 Formation of Image by Two Spherical Concave Mirrors

Illustrate the locations of the paper clip, mirrors, and image in a diagram.

9.4.8 Conclusion

Summarize the important observations and findings obtained in this lab experiment.

9.4.9 SUGGESTIONS FOR FUTURE LAB WORK

List any suggestions for improvements using different experimental equipment, procedures, and techniques for any future lab work. These suggestions should be theoretically justified and technically feasible.

9.5 LIST OF REFERENCES

List any references that were used in the report. Use one format in writing the references. Never mix reference formats in a report.

9.6 APPENDIX

List all of the materials and information that are too detailed to be included in the body of the report.

FURTHER READING

Beiser, A., *Physics*, 5th ed., Addison-Wesley Publishing Company, Reading, MA, 1991.

Cutnell, J. D. and Johnson, K. W., *Physics*, 5th ed., Wiley, New York, 2001.

Cutnell, J. D. and Johnson, K. W., *Student Study Guide—Physics*, 5th ed., Wiley, New York, 2001.

Edmund Industrial Optics, *Optics and Optical Instruments Catalog*, Edmund Industrial Optics, Barrington, NJ, 2004.

Ewald, W. P., Young, W. A., and Roberts, R. H., *Practical Optics*, Makers of Pittsford, Pittsford, Rochester, New York, 1982.

Ewen, D. et al., *Physics for Career Education*, 4th ed., Prentice Hall, Englewood Cliffs, NJ, 1996.

Ghatak, A. K., *An Introduction to Modern Optics*, McGraw-Hill Book Company, New York, 1972.

Giancoli, D. C., *Physics*, 5th ed., Prentice Hall, Englewood Cliffs, NJ, 1998.

Halliday, D., Resnick, R., and Walker, J., *Fundamentals of Physics*, 6th ed., Wiley, New York, 1997.

Heath, R. W., Macnaughton, R. R., and Martindale, D. G., *Fundamentals of Physics*, D.C. Heath Canada Ltd., Canada, 1979.

Hecht, E., *Optics*, 4th ed., Addison-Wesley Longman, Inc, Reading, MA, 2002.

Jenkins, F. W. and White, E. H., *Fundamentals of Optics*, McGraw-Hill, New York, 1957.

Keuffel & Esser Co., Physics, Keuffel & Esser Audiovisual Educator-Approved Diazo Transparency Masters, Audiovisual Division, Keuffel & Esser Co., U.S.A., 1989.

Lambda Research Optics, Inc., *Catalog 2004*, Lambda Research Optics, Costa Mesa, CA, 2004.

Lehrman, R. L., *Physics—The Easy Way*, 3rd ed., Barron's Educational Series, Inc., Hauppauge, NY, 1998.

Lerner, R. G. and Trigg, G. L., *Encyclopedia of Physics*, 2nd ed., VCH Publishers, Inc., New York, 1991.

Malacara, D., *Geometrical and Instrumental Optics*, Academic Press, Boston, MA, 1988.

McDermott, L. C. et al., *Introduction to Physics*, Preliminary Edition, Prentice Hall, Inc., Englewood Cliffs, NJ, 1988.

Naess, R. O., *Optics for Technology Students*, Prentice Hall, Englewood Cliffs, NJ, 2001.

Newport Corporation, *Optics and Mechanics Section*, Newport Resources 1999/2000 Catalog, Newport Corporation, Irvine, CA, 1999.

Nolan, P. J., *Fundamentals of College Physics*, Wm. C. Brown Publishers, Dubuque, IA, 1993.

Pedrotti, F. L. and Pedrotti, L. S., *Introduction to Optics*, 2nd ed., Prentice Hall, Inc., Englewood Cliffs, NJ, 1993.

Sears, F. W., Zemansky, M. W., and Young, H. D., *University Physics—Part II*, 6th ed., Addison-Wesley Publishing Company, Reading, MA, 1998.

Serway, R. A., *Physics for Scientists and Engineers*, 3rd ed., Saunders Golden Sunburst Series, Philadelphia, PA, 1990.

Smith, W. J., *Modern Optical Engineering*, McGraw-Hill Book Co., New York, 1966.

Urone, P. P., *College Physics*, Brooks/Cole Publishing Company, Pacific Grove, CA, 1998.

Walker, J. S., *Physics*, Prentice Hall, Englewood Cliffs, NJ, 2002.

Warren, M. L., *Introduction to Physics*, W.H. Freeman and Company, San Francisco, CA, 1979.

White, H. E., *Modern College Physics*, 6th ed., Van Nostrand Reinhold Company, New York, 1972.

Williams, J., Metcalfe, H., Trinklein, F., and Lefler, R., *Teacher's Edition—Modern Physics*, Holt, Rinehart and Winston, Inc., New York, 1968.

Wilson, J. D., *Physics—A Practical and Conceptual Approach*, Saunders College Publishing, Philadelphia, PA, 1989.

Wilson, J. D. and Buffa, A. J., *College Physics*, 5th ed., Prentice Hall, Englewood Cliffs, NJ, 2000.

10 Lenses

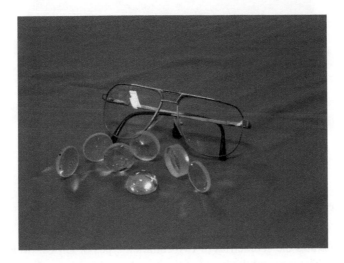

10.1 INTRODUCTION

A typical lens is made of glass or plastic. Lenses are classified in two categories: thin lenses and thick lenses. Each category has its applications and equations for calculations. Lenses have two refraction surfaces; each surface is a segment of either a sphere or a plane. Lenses are commonly used to form images by the refraction of light. Light refraction in lenses is one application of the theory of the second law of light: the law of refraction. Lenses are used in building optic/optical fibre devices and instruments, such as cameras, microscopes, slide projectors, and fibre optical switches. The graphical method for locating images formed by mirrors will be used in this chapter as well. The experimental work in this chapter will enable students to practise tracing image formation by a lens or lens combination.

10.2 TYPES OF LENSES

The following sections introduce the types of lenses which are used in building optic and optical fibre devices and instruments.

10.2.1 CONVERGING AND DIVERGING LENSES

Figure 10.1 and Figure 10.2 show some common shapes of lenses. Typical spherical lenses have two surfaces defined by two spheres. The surfaces can be convex, concave, or planar. Lenses are divided into two types: (a) converging and (b) diverging. Converging lenses have positive focal lengths and are thickest at the middle. Common shapes of converging lenses are (1) biconvex, (2) convex–concave, and (3) plano-convex. Diverging lenses have negative focal lengths and are thickest at the edges. Common shapes of diverging lenses are (1) biconcave, (2) convex–concave, and (3) plano-concave.

FIGURE 10.1 Converging and diverging lenses.

Figure 10.3 illustrates the geometries of two common types of lenses: (a) a biconvex lens and (b) a biconcave lens. The type and thickness of the lens depends on the radius of curvatures R_1 and R_2, and the distance between the centres of curvature.

The principle of operation for a lens forming an image is explained by the second law of light, the law of refraction. When light rays pass through a lens, they are bent or deviated from their

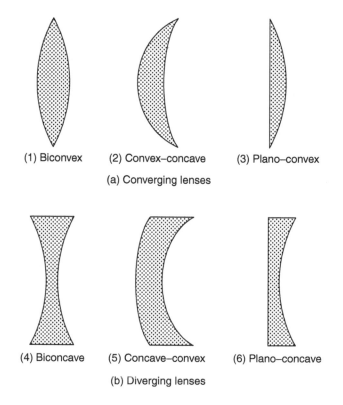

FIGURE 10.2 Various shapes of lenses.

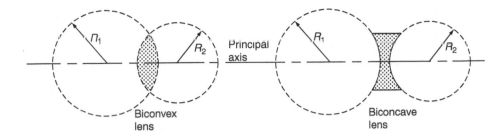

FIGURE 10.3 Two common types of lenses.

original paths, according to the law of refraction. The theory of light refraction through an optical medium is presented in Chapter 7, The Laws of Light. Refraction by a prism is addressed in Chapter 11, Prisms.

To study light refraction in a biconvex lens, two prisms can be placed base to base to approximate the convex lens operation, as shown in Figure 10.4(a). Parallel light rays that pass through the prisms are deviated so that the various rays intersect. They do not intersect or focus at a single point. However, if the surfaces of the prisms are curved rather than flat, then it becomes a converging lens,

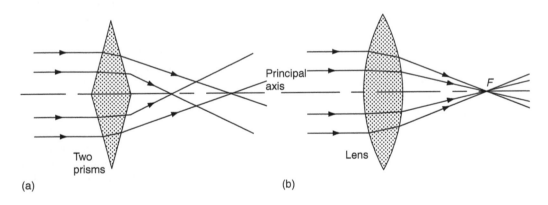

FIGURE 10.4 The principle of the converging lens.

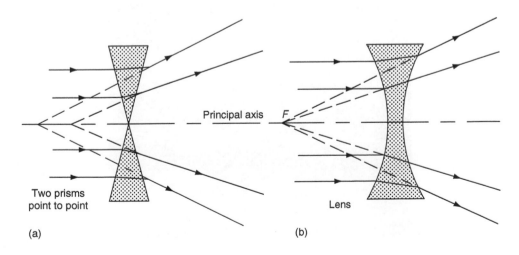

FIGURE 10.5 The principle of the diverging lens.

as shown in Figure 10.4(b). The converging lens brings incoming parallel rays of light to a single point F, called the focal point, at the principal axis. Because the refracted light rays pass through F, the focal point is real. This type of lens is often called a converging or convex lens.

Similarly, a biconcave lens can be approximated by two prisms with their apexes together, as shown in Figure 10.5(a). Parallel light rays that pass through the prisms are spread outward, but these diverging rays cannot be projected back to a single point. However, if the surfaces of the prisms are curved rather than flat, then it becomes a diverging lens, as shown in Figure 10.5(b). The diverging rays appear to originate from a single point on the incident side of the lens. The focal point F is not real; it is virtual because the rays do not actually pass through F. This type of lens is often called a diverging, or concave lens.

10.2.2 GRIN LENSES

GRIN is the acronym of the GRadient INdex lens. GRIN lenses have a cylindrical shape, as shown in Figure 10.6. One end is polished at an angle of 2, 6, 8, or 12 degrees, while the other end is polished at an angle of 2 or 90 degrees.

If the lens has one end polished, it is called a single angle lens; likewise, if both ends are polished, the lens is called a double angle lens. The choice of the angle for the polished face of the GRIN lens is dependent on the application. GRIN lenses are coated with anti-reflection (AR) coatings to reduce Fresnel losses occurring between two parallel optical surfaces. AR coatings also help protect the lens surfaces from humidity, chemical reaction, and other types of physical damage. The GRIN lens is constructed from multiple concentric layers, somewhat like the annular rings of a tree. Each concentric layer of glass, arranged radially from the central axis of the lens, has a lower index of refraction than the previous layer, as shown in Figure 10.7.

Figure 10.7 also shows the structure of a GRIN lens, and illustrates the paths of the light rays. Light travels faster in materials with a lower index of refraction. The further the light ray is from the lens centre, the greater its speed. The light traveling near the centre of the lens has the slowest average velocity. As a result, all rays tend to reach the end of the lens at the same time. Each layer within the lens refracts the light differently. Instead of being sharply reflected as in a step-index fibre, the light is now bent or continually refracted in a sinusoidal pattern. Those rays that follow the longest path by traveling near the outside of the lens have a faster average velocity. Collimated light entering from one end is focused on the focal point of the lens on the other end.

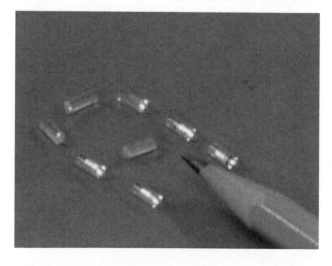

FIGURE 10.6 Actual GRIN lenses.

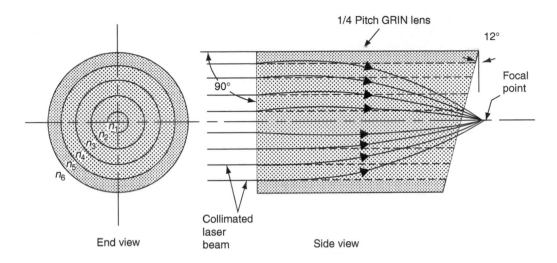

FIGURE 10.7 GRIN len's structure showing light rays.

Figure 10.8 illustrates a comparison between a GRIN lens and a convex lens. The GRIN lens equation is defined as:

$$P = \frac{Z\sqrt{A}}{2\pi}.$$ (10.1)

where

P the pitch
Z the lens length (mm); and
$\sqrt{A}$ the index gradient constant.

GRIN lenses are often specified in terms of their pitch lengths and operating wavelengths. Figure 10.9 illustrates GRIN lens of different pitch lengths. Figure 10.9(a) shows a GRIN lens with a length of more than one pitch. Where the length of a GRIN lens is very long and covered by cladding and jacket, a GRIN fibre optic cable is created. The principle of GRIN fibre optic cable will be discussed in detail later in this book. Figure 10.9(b) shows a one pitch GRIN lens. The image is erect and the same size as the object. In (c), which demonstrates a GRIN lens with a length of 1/2 pitch, the image is inverted and the same size as the object. And in (d), showing a GRIN lens

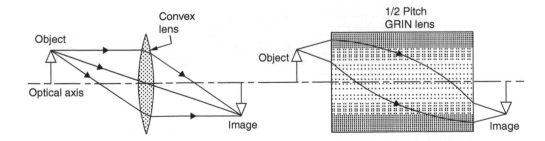

FIGURE 10.8 Comparison between a convex lens and a GRIN lens.

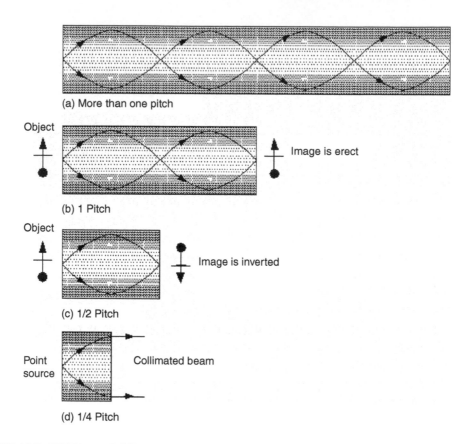

FIGURE 10.9 GRIN lens of different pitch lengths.

with a length of 1/4 pitch, a point light source from one end is converted into a collimated beam exiting from the other end.

Coupling light into a fibre cable can be accomplished by using a quarter-pitch GRIN lens, which focuses the collimated beam from the laser source onto a small spot on the core of the fibre cable. Figure 10.10 is a schematic diagram showing the GRIN lens between the laser source and the fibre cable. The input power increases by inserting a GRIN lens. When the GRIN lens focuses the

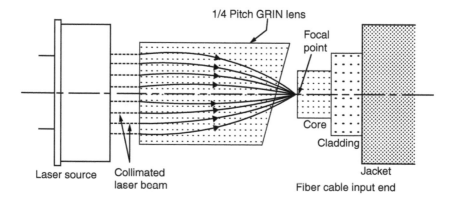

FIGURE 10.10 GRIN lens between a laser source and fibre cable input end.

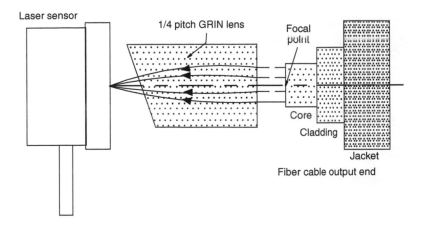

FIGURE 10.11 GRIN lens between a laser sensor and fibre cable output end.

power of the collimated laser beam onto the core of the fibre cable, the fibre input power is optimized.

Similarly, inserting a GRIN lens at the output of the fibre cable increases the output power, as shown in Figure 10.11.

GRIN lenses are used to focus and collimate light within a variety of fibre optical components. Passive component manufacturers use GRIN lenses in WDMs, optical switches, and attenuators. Active component manufacturers use GRIN lenses for fibre-to-detector and laser-to-fibre coupling.

GRIN lenses are useful for imaging and optical fibre communications systems because:

1. they are shaped for specific optical applications;
2. a completely flat piece of radial-GRIN material can act as a lens;
3. they can be used to image an object at one end and a detector at the other;
4. they are easy to use in both the design and alignment of optical fibre devices;
5. they can be used to produce micro-optical fibre devices; and
6. they are smaller than conventional lenses.

Radial graded-index (GRIN) lenses are useful in building fibre optical devices that are used in imaging and telecommunications systems. Made using a thermal diffusion technique, the new GRIN lenses are likely to be cheaper and of higher optical quality than existing GRIN lenses. This fabrication method offers flexibility in size and refractive index variation that is difficult to achieve with other methods.

10.2.3 BALL LENSES

Ball lenses, as shown in Figure 10.12(a), are used in manufacturing fibre optic devices for modern communication systems. Ball lenses couple fibres from micro-optic components to light sources, from laser diodes to fibre cables, and from fibre cables to detectors. They also collimate light from fibre cables in thin-film filters for dense wavelength division multiplexing. In a ball lens, transmitted light changes direction at the curved boundaries, traversing the interior in straight lines, as shown in Figure 10.12(b). An incoming parallel beam converges at the focal point on the other side of the ball lens. The effective focal length (EFL) depends upon the refractive index of the glass and the lens diameter. Placing a point light source at the focal point, F, of the lens results in collimated light.

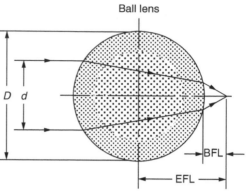

(a) Ball lens.   (b) Schematic diagram of a ball lens.

FIGURE 10.12 Ball lens (a) ball lens (b) schematic diagram of a ball lens.

Ball lenses benefit from cost and size advantages over gradient index, convex, and concave lenses; however, misperceptions concerning the relative ease of use of GRIN lenses have often led communications users to choose these over ball lenses. Due to recent improvements in ball lens antireflective coating technology, this preference for GRIN lenses has diminished. In some cases, ball lenses may even yield higher coupling efficiencies in smaller packages.

As shown in Figure 10.12, the EFL of a ball lens depends on two variables: the ball lens diameter (D) and the material index of refraction of the ball (n). The EFL is measured from the centre of the lens. The back focal length (BFL) is calculated as:

$$BFL = EFL - D/2 \tag{10.2}$$

$$EFL = nD/4(n-1) \tag{10.3}$$

The numerical aperture (NA) of a ball lens is dependent on the focal length of the ball and on the input beam diameter (d). The NA is defined in the following equation:

$$NA = 2d(n-1)/nD \tag{10.4}$$

When coupling light efficiently from a laser into a fibre optic cable, the choice of the ball is dependent on the NA of the fibre cable and the diameter of the laser beam. The diameter of the laser beam is used to determine the NA of the ball lens, as given by Equation 10.4. The NA of the

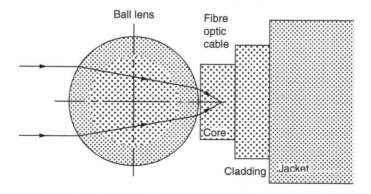

FIGURE 10.13 Ball lens coupling into a fibre optic cable.

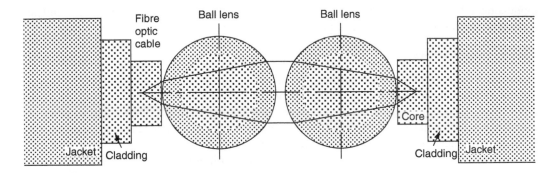

FIGURE 10.14 Coupling two fibre optic cables.

ball lens must be less than or equal to the NA of the fibre cable in order to couple all of the light into the fibre cable. This concept will be discussed in detail in the fibre optic cable chapter. The ball lens is placed directly onto the fibre cable for optimum coupling, as shown in Figure 10.13.

To couple light from one fibre cable to another fibre cable of similar NA, two identical ball lenses are used. The two lenses are placed in contact with the fibre cables, as shown in Figure 10.14. This figure also shows that light goes from one fibre cable to another in both directions.

Half-ball (hemispherical) lenses are available in the market. They are ideal for applications such as fibre communication, endoscopy, microscopy, optical pick-up devices, and laser interferometry.

10.2.4 FRESNEL LENSES

A Fresnel lens replaces the curved surface of a conventional lens with a series of concentric grooves molded into the surface of a thin, lightweight glass or plastic sheet, as shown in Figure 10.15. The grooves act as individual refracting surfaces, like tiny prisms, when viewed in cross section, bending parallel rays in a very close approximation to a common focal length. Because the lens is thin, very little light is lost by absorption compared to conventional lenses. Fresnel lenses are a compromise between efficiency and image quality. High groove density allows higher quality images (as needed in projection), while low groove density yields better efficiency (as needed in light gathering

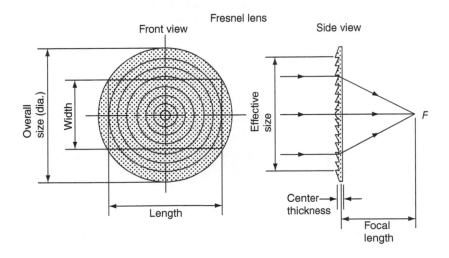

FIGURE 10.15 Fresnel lens.

applications). Fresnel lenses are most often used in light gathering applications, such as condenser systems or emitter/detector setups. Fresnel lenses can also be used as magnifiers or projection lenses.

10.2.5 LIQUID LENSES

Researchers have developed a unique lens system with a variable focus, which is able to function without mechanical moving parts. This development can be utilized in a wide range of optical applications, such as digital cameras, endoscopes, and security systems. The system simulates the human eye, which adjusts its focal length by changing its shape. The new lens overcomes the fixed focus disadvantages of many more economical image systems.

The liquid lens consists of two immiscible liquids with different indexes of refraction. Between these liquid layers are both an electrically conductive aqueous solution and an electrically nonconductive oil. This liquid lens is contained in a short tube with transparent end caps. The internal surface of the tube wall and one of the end caps are covered with a hydrophobic layer which forces the aqueous solution to take on a hemispherical shape at the opposite end of the tube.

The shape of the lens is controlled by applying an electric field across the hydrophobic coating. This causes the coating to become less hydrophobic as the surface tension changes. As a result, the aqueous solution starts to wet the side walls of the tube, and so changes the meniscus between the two liquids as well as the curve radius and focal length of the lens. In this way, the surface of the original convex lens can be made completely planar or concave by applying an electric field.

The liquid lens measures a few millimetres in diameter and length. The focusing range is from a few centimetres to infinite. The focal point switches quickly over the complete range. The lens is reliable, has very low loss, is shock-resistant, and functions over a wide temperature range.

10.3 GRAPHICAL METHOD TO LOCATE AN IMAGE FORMED BY CONVERGING AND DIVERGING LENSES

Graphical methods using ray diagrams are convenient for determining the location of an image formed by a lens. Figure 10.16 illustrates the method of locating the image of an object placed in front of a convex (converging) lens. Consider an object placed at location O and at distance S from the centre of the converging lens of focal length f. For a converging lens, the focal length is positive and is located on the right side of the lens (the side opposite to the object). To locate the image, the three rays are drawn from the top of the object as follow:

1. *Ray 1* is drawn from the top of the object parallel to the principal axis. After the ray is refracted through the lens, this ray passes through the focal point of the lens.

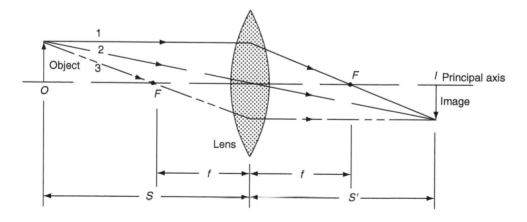

FIGURE 10.16 Graphical diagram to locate an image for a convex lens.

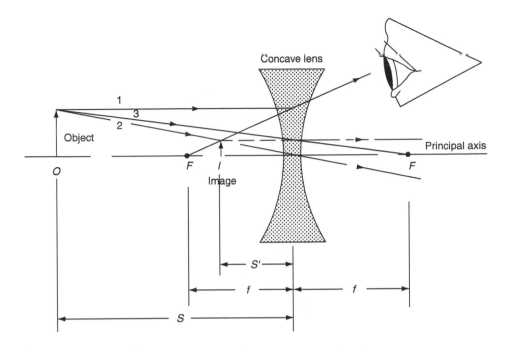

FIGURE 10.17 Graphical diagram to locate an image for a concave lens.

2. *Ray 2* is drawn from the top of the object passing through the centre of the lens. This ray continues in a straight line and intersects with ray 1 at a point at the top of the image. This intersection point of rays 1 and 2 shows the location of the image on the principal axis at point I at a distance S' away from the lens centre. Though the two rays can locate the image, it is better to confirm the image location using the third ray.

3. *Ray 3* is drawn from the top of the object passing through the focal point of the lens located on the left side (object side) of the lens. Ray 3 refracts from the lens parallel to the principal axis and intersects with rays 1 and 2 at the same point.

Figure 10.17 illustrates the method of locating the image of an object placed in front of a concave (diverging) lens. Consider an object placed at location O and at distance S from the centre of the diverging lens of focal length f. For a diverging lens, the focal length is negative and thus the focal point is on the left side of the lens (the object side).

A similar construction is used to locate the image from a diverging lens. In contrast, the deviated ray 1 appears to originate from the focal point and ray 3 heads towards the other focal point. The point of intersection of any two rays in the graphical diagram can be used to locate the image. Ray 3 serves as a check on the construction procedure of the image location. For diverging lenses, the rays appear to originate from a virtual image, as shown in dashed lines. This image is imaginary, located on the same side as the object, and is reduced in size.

10.4 IMAGE FORMATION BY CONVERGING LENSES

Figure 10.18 shows the cases of image formation by an object placed at different locations in front of a converging lens.

1. *Case (a)* occurs when the object is placed between the focal point F and the centre of the lens. The image is erect, enlarged, and virtual. The image is located on the object side of

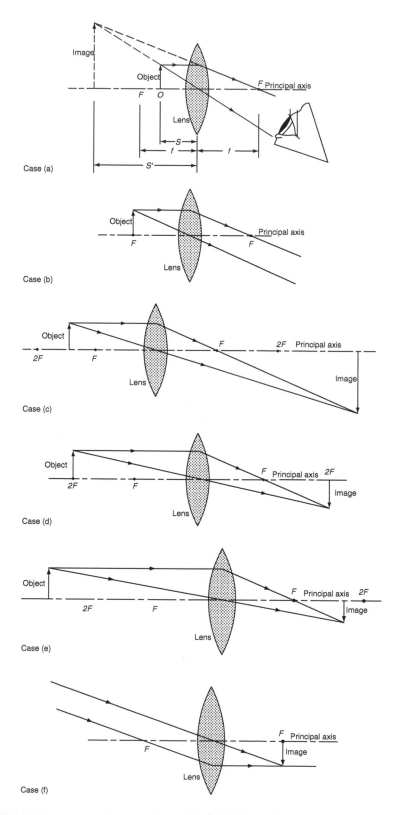

FIGURE 10.18 Images formation by a converging lens. Cases (a)–(f).

the lens because the refracted rays diverge as though coming from a point behind the lens. A magnifying glass is a simple example of this case.

2. *Case (b)* occurs when the object is located at the focal point *F* of the lens. All refracted rays are parallel and will never intersect at any distance from the lens. Therefore, no image will be formed; it seems the image is at an infinite distance from the lens. Light coming from a lighthouse is a common example.

3. *Case (c)* occurs when the object is placed beyond the focal point *F* of the lens. The image is inverted, real, and larger than the object. The image is located at a distance beyond 2*f*. This case illustrates the principle of image formation in an optical projector.

4. *Case (d)* occurs when the object is placed at a distance of 2*f*. The image is formed on the other side of the lens at location 2*f*. The image is real, inverted, and the same size as the object. This case is present in the operation of an office photocopier.

5. *Case (e)* occurs when the object is placed beyond 2*f*. The image is formed at a location between *f* and 2*f*. The image is real, inverted, and smaller than the object. This case shows the principle of the camera.

6. *Case (f)* occurs when the object is placed at infinity. The image is formed at the focal point of the lens. The image is real, inverted, and smaller than the object. This case presents the principle of the camera viewing a distant object.

In summary, a variety of image sizes, orientation, and locations are produced by a converging lens. In contrast, the image of a real object formed by a diverging lens is always virtual, erect, and smaller than the object, as shown in Figure 10.17. The further the object is located from the lens, the smaller the image.

10.5 THE LENS EQUATION

Consider defining the focal length *f* of a lens as a function of the object distance *S* and image distance *S'* from the lens centre, as shown in Figure 10.19. The lens equation can be derived geometrically as illustrated in Figure 10.19. The two triangles *ABO* and *CDO* are similar, which means that the corresponding sides of those triangles are proportional. Thus:

$$\frac{CD}{AB} = \frac{CO}{AO} = \frac{S'}{S}. \tag{10.5}$$

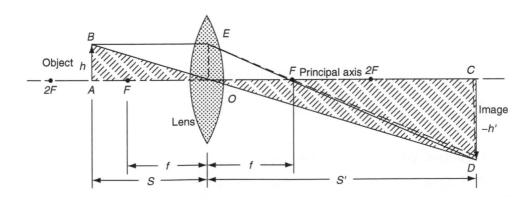

FIGURE 10.19 Ray diagram for deriving the lens equation.

The triangles OEF and CDF are also similar, so:

$$\frac{CD}{OE} = \frac{CF}{OF} = \frac{S'-f}{f} = \frac{S'}{f} - 1 \tag{10.6}$$

Since the light ray BE is parallel to the principal axis,

$$OE = AB \tag{10.7}$$

Therefore:

$$\frac{CD}{OE} = \frac{CD}{AB} \tag{10.8}$$

Which means the right side of Equation (10.5) and Equation (10.6) are equal, as shown in Equation 10.9:

$$\frac{S'}{f} - 1 = \frac{S'}{S} \tag{10.9}$$

Dividing each term in Equation (10.9) by S' gives:

$$\frac{1}{f} - \frac{1}{S'} = \frac{1}{S} \tag{10.10}$$

Rearranging terms gives the lens equation:

$$\frac{1}{S} + \frac{1}{S'} = \frac{1}{f} \tag{10.11}$$

Equation (10.11) is the basic equation for thin lenses. Thin lenses are defined as lenses whose thickness can be ignored. In the following sections, thin lenses are explained in detail.

The sign convention for thin lenses is given below:

S is *positive* if the object is in front of the lens (real object).
S is *negative* if the object is behind the lens (virtual object).
S' is *positive* if the image is behind the lens (real image).
S' is *negative* if the image is in front of the lens (virtual image).
f is *positive* if the focal point of the incident surface is in front of the lens (convex lens).
f is *negative* if the focal point of the incident surface is in back of the lens (concave lens).

This convention applies only to the numerical values substituted into the thin lens equation. The quantities S, S', and f should maintain their signs unchanged until the substitution is made into the lens equation.

10.6 MAGNIFICATION OF A THIN LENS

Figure 10.19 shows a converging lens that forms the image CD of the object AB. The object height is $AB = h$ and the image height $CD = -h'$ is negative because the image is inverted. The triangles ABO and CDO are similar, and their corresponding sides are proportional. Hence:

$$\frac{CD}{AB} = \frac{CO}{AO} \tag{10.12}$$

Since $CD = -h'$, $AB = h$, $CO = S'$, and $AO = S$, then Equation (10.12) becomes the lateral magnification M of the lens:

$$M = \frac{h'}{h} = -\frac{S'}{S} \tag{10.13}$$

In other words, the lateral magnification M of the thin lens is defined as the ratio of the image height h' to the object height h or also defined as the image distance S' to the object distance S, as given in Equation 10.14:

$$M = \frac{\text{Image height}}{\text{Object height}} = -\frac{\text{Image distance}}{\text{Object distance}} \tag{10.14}$$

When M is positive, the image is erect and on the same side of the lens as the object. When M is negative, the image is inverted and on the side of the lens opposite the object.

10.7 THE LENSMAKER'S EQUATION

Figure 10.20 shows a converging lens that has a focal length f, index of refraction n, and two radii with surface curvatures R_1 and R_2. The thin lens is a lens whose thickness is small compared with R_1 and R_2. The Lensmaker's equation, which is the relation between the focal length, the index of refraction, and the surface curvatures, is given by:

$$\frac{1}{f} = (n-1)\left(\frac{1}{R_1} - \frac{1}{R_2}\right) \tag{10.15}$$

Both f and R are *positive* if the centre of curvature of the incident surface is in front of the lens (convex lens).

Both f and R are *negative* if the centre of curvature of the incident surface is in back of the lens (concave lens).

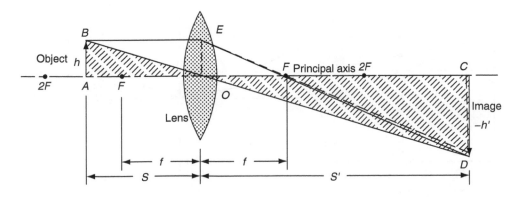

FIGURE 10.20 Biconvex lens.

10.8 COMBINATION OF THIN LENSES

Two lenses can be combined to form an image in optical devices and instruments, such as microscopes and telescopes. The graphical method can be used to locate the image, as explained previously. The first image of the first lens can be located by the graphical method. It can be calculated applying Equation 10.11 as if the second lens was not present. The light then approaches the second lens as if it had originated from the image formed by the first lens. Therefore, the image of the first lens is treated as the object to the second lens. The image of the second lens is the final image of the combination of two lenses. Draw and apply Equation 10.11 to the second lens as if the first lens were not present. The same procedure can be applied to any optical device or instrument with three or more lenses. The total magnification of the combination of lenses equals the product of the magnification of the separate lenses. Applications of lens combinations are presented in Chapter 14, covering optical instruments for viewing applications.

Suppose two thin lenses of focal lengths f_1 and f_2 are placed at a distance L between each other, as shown in Figure 10.21. If S_1 is the distance of the object from lens 1, then apply Equation 10.16 to find the location of the image S_1' formed by the first lens. The first image formed by the first lens becomes the object for the second lens. Apply the thin lens Equation 10.17 to find the location of the second image S_2' formed by the second lens. Now the distance between the two lenses can be calculated. Remember to consider the types of lens combinations when locating the images. The lenses involved in these combinations could be convex–convex, convex–concave, or any other combination. Lens combinations can form real or virtual images.

$$\frac{1}{S_1} + \frac{1}{S'_1} = \frac{1}{f_1} \tag{10.16}$$

$$\frac{1}{S_2} + \frac{1}{S'_2} = \frac{1}{f_2} \tag{10.17}$$

The distance L between the lenses can be calculated by Equation 10.17:

$$L = S'_1 + S_2 \tag{10.18}$$

The distances S'_1 and S_2 could be real or virtual, i.e., positive and negative. The total magnification for the combination of the two lenses can be found by multiplying the magnification of first and second lenses, which are calculated using Equation 10.14. The total magnification M_{total} for the

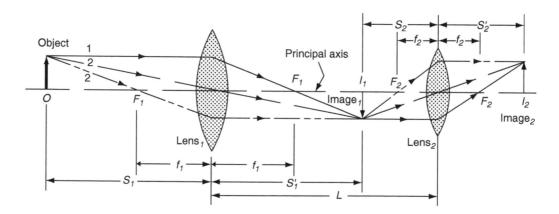

FIGURE 10.21 Combination of two types of lenses.

combinations of lenses is given by:

$$M_{\text{total}} = M_1 M_2. \qquad (10.19)$$

The conventional signs for M_1 and M_2 carry through the product to indicate, from the sign of M_{total}, whether the final image is erect or inverted.

10.9 LENS ABERRATIONS

Aberration is a deviation from the ideal focusing of light by a lens. As a result, the image is not sharply focused. Some of the problems may lie with the geometry of the lens in use. Other aberrations may occur because different wavelengths of light are not refracted equally at the same time. The latter problem is called dispersion of light. This is clearly visible when white light passes through a prism. White light dispersion in a prism is discussed in detail in Chapter 11, Prisms. Lenses, like spherical mirrors, also have aberrations. Here are some common types of aberrations.

10.9.1 SPHERICAL ABERRATION

As explained in lenses chapter, the spherical mirrors, aberration occurring in spherical mirrors is the same phenomena that occurs in lenses. Spherical aberration results because the light rays which are far from the principal axis focus at different focal points than the rays which are passing near the axis. Figure 10.22 illustrates spherical aberration for parallel rays passing through a converging lens. Rays near the middle of the lens are imaged farther from the lens than rays near the edges. Therefore, there is no single focal point for a spherical lens.

Spherical aberration can be minimized by using an aperture to reduce the effective area of the lens, so that only light rays near the axis are transmitted. This happens in most cameras equipped with an adjustable aperture to control the light intensity, and when possible, reduce spherical aberration. To compensate for the light loss of a smaller aperture, a longer exposure time is used. Another way to minimize the effect of aberration is by combinations of converging and diverging lenses. The aberration of one lens can be compensated by the optical properties of another lens.

10.9.2 CHROMATIC ABERRATION

As described in the previous chapters, the index of refraction of a material varies slightly with different wavelengths; therefore, different wavelengths of light refracted by a lens focus at different points, giving rise to chromatic aberration. When white light passes through a lens, the transmitted rays of different wavelengths (colours) do not refract at the same angle. As a result, there is no

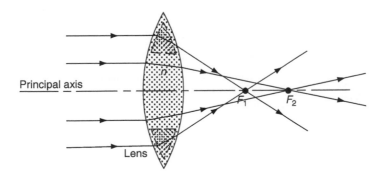

FIGURE 10.22 Spherical lens aberration.

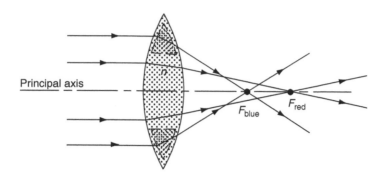

FIGURE 10.23 Chromatic lens aberration.

common focal point for all the wavelengths. Figure 10.23 shows that the images of different colours are produced at different locations on the principal axis. The chromatic aberration for a diverging lens is the opposite of that for a converging lens. The red colour focuses before the blue colour. Chromatic aberration can be greatly reduced by the use of a combination of converging and diverging lenses made from two different types of glass with differing indexes of refraction. The lenses are chosen so that the dispersion produced by one lens is compensated for by opposite dispersion produced by the other lens.

10.9.3 ASTIGMATISM

Astigmatism is the imaging of a point off-set from the axis as two perpendicular lines in different planes. This effect is shown in Figure 10.24. In astigmatism, the horizontal rays from a point object converge at a certain distance from the lens to a line called the primary image. The primary image is perpendicular to the plane defined by the optic axis and the object point. At a somewhat different distance from the lens, the vertical rays converge to a second line, called the secondary image. The secondary image is parallel to the vertical plane. The circle of least confusion (greatest convergence) appears between these two positions.

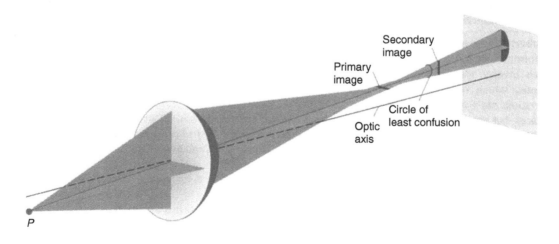

FIGURE 10.24 Astigmatism of a lens.

The location of the circle of least confusion depends on both the object point's transverse distance from the axis, as well as its longitudinal distance from the lens. As a result, object points lying in a plane are generally not imaged in a plane, but are imaged in some curved surface. This effect is called curvature of field.

Finally, the image of a straight line that does not pass through the optic axis may be curved. As a result, the image of a square with the axis through its centre may resemble a barrel (sides bent outward) or a pincushion (sides bent inward). This effect, called distortion, is not related to the lack of sharpness of the image, but results from a change in lateral magnification with distance from the axis. In high-quality optical instruments, astigmatism can be minimized by using properly designed, nonspherical surfaces or specific lens combinations.

10.10 LENS POLISHING TECHNOLOGY

Many lens polishing techniques are available in the industry. The technique chosen depends on the lens performance requirements and the cost. A new fabrication technology called magnetorheological finishing has been developed, which provides some advantages over traditional techniques. The main advantage of this technology is that it offers a way to produce high precision asphere lenses in moderate volume.

The spherical shape of any lens surface causes focus error, which is called a spherical aberration, as explained above. The manufacturers of the lenses use multiple surfaces to overcome spherical aberration. The problem lies with the traditional techniques used for optics fabrication shown in Figure 10.25. To produce spherical surfaces, the lens blanks are polished between convex and concave surfaces of the intended radius of curvature, using an abrasive slurry.

The semi-random movement of a convex spherical surface against a concave spherical surface tends to produce a spherical wear pattern. Originally, the lens blanks are not a perfect spherical shape. But over time, the wear from the polishing process causes them to take on a spherical shape. With process control, this method can yield an extremely high degree of precision. For example, lenses with $\lambda/4$ or better surface accuracy at 632.8 nm (HeNe laser) have a surface that is spherical to within a fraction of micrometre.

To produce aspheric surfaces, the movement of the polishing tool must be constrained. For lenses in which the aspheric profile departs from a sphere by only a small amount, typical lens

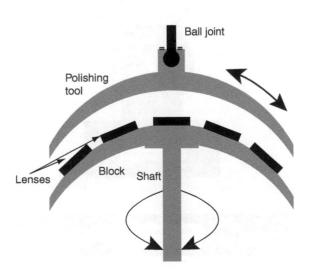

FIGURE 10.25 Traditional polishing technique.

polishing starts with the closest-fit sphere using standard techniques. The final aspheric shape is then achieved with a polishing tool that is much smaller than the optic (a sub-aperture tool) on another polishing machine. Controlling the amount of time that the tool spends working a particular zone of the lens surface can produce a specific aspheric profile.

If the aspheric shape departs significantly from spherical, a computer numerical control machine would typically be used to generate the aspheric shape directly into the lens blank. Alternatively, a special grinding tool can be used to transform the closest-fit sphere into a close approximation of the final desired lens shape. Once the blank closely resembles the desired asphere, the goal is to polish the blank without removing a significant amount of material. This is accomplished using a modified form of a traditional polishing apparatus along with sub-aperture polishing.

Unfortunately, because there are many environmental and operator dependent variables in this type of polishing, successful operation of this process requires a high degree of operator skill. Given the mechanical limitations of the machinery commonly used, it is hard to achieve surface precision near $\lambda/20$. Process times for most common lens polishing methods are measured in hours, and even in days or weeks in some cases. Thus, high-quality aspheres produced in this way can be prohibitively expensive compared with spherical lenses.

Magnetorheological finishing technology was developed to overcome the limitations of traditional lens polishing techniques. It is a controlled way of producing high precision aspheres that can be used reliably and in high-volume production. It also potentially requires less operator know-how than traditional approaches, reduces process times significantly, and can yield surface precision to $\lambda/40$.

Figure 10.26 shows the basic elements of the magnetorheological system including a pump that continuously supplies a small amount of magnetorheological fluid from a reservoir. This fluid is sprayed onto a rotating wheel that draws out the fluid into a thin ribbon. The work piece (lens) to be

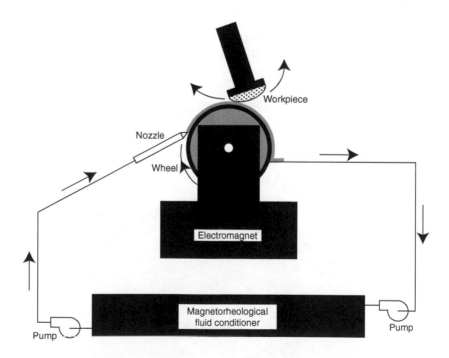

FIGURE 10.26 Magnetorheological system.

polished is partially immersed in this moving ribbon of fluid. A specially designed electromagnet beneath the polishing wheel produces a strong local magnetic field gradient. When the moving fluid enters this magnetic field, it stiffens substantially in a matter of milliseconds; it then returns to its fluid state as it leaves the field. This zone of stiffened fluid is the polishing tool. After the fluid passes the lens, it is removed by suction and pumped back into the reservoir.

A polishing session begins by running a test routine that characterizes the exact polishing shape and properties of the polishing fluid. To polish a given lens, the initial surface error (the difference between the present and desired shape) of the lens is supplied to the system software. The software generates a set of instructions for the computer numerical control machine, which then starts polishing the lens. During the polishing process, the machine rotates the lens on the spindle, as shown in Figure 10.27. At the same time, the machine slowly tilts the lens so that different zones in the lens are exposed to the fluid. During this process, the machine maintains a constant gap and surface normal for the lens in the fluid. The computer program controls the speeds of these movements to remove only the necessary amount of material. In other words, the system will spend more time polishing a zone with a high spot than it will spend polishing a low spot. The machine continuously monitors and controls the properties of the polishing fluid, such as temperature, viscosity, and evaporation rate, to maintain process invariance.

After the lens is completely polished, it has the correct specifications and is entirely transparent. The lens material is very sensitive and can scratch easily. Therefore, a harder coating must be applied to the surfaces of the lens. There are many types of coating materials. One coating is an acrylic monomer solution that is spin-coated onto the lens. The coating is then cured under a UV lamp for a specific time. This coating helps protect the lens from scratches and wear.

Most of the lenses and other optical components are coated with at least one antireflective (AR) layer. AR coating is used to reduce back reflection and to filter undesirable light waves. An antireflection coating material is a composite thin film. There are between four to six layers that have several purposes. Some layers help bond the other layers to the lens while other layers provide the antireflection properties. Each layer of the composite coating depends on the properties of the surrounding layers in order to achieve the desired specifications. AR coatings made of multi-layer

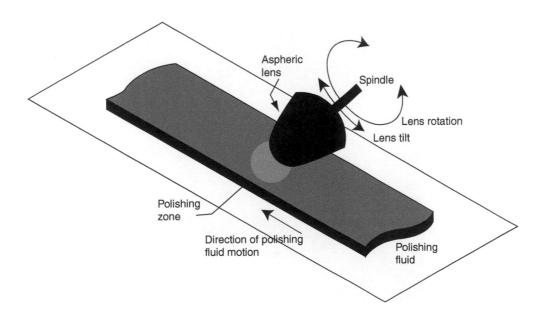

FIGURE 10.27 Polishing process.

thin film metallic oxide can increase the light transmittance to nearly 100%. Typically, titania and zircona are used because they have a high index of refraction. Silica is usually used with material that has a low index of refraction. The thickness of the layers determines the optical performance of the coating, thus the specifications need to be precisely calculated and carefully controlled.

10.11 EXPERIMENTAL WORK

The theory of this experiment is based on the second law of light, which is called the law of refraction. Students will practise tracing image formation by lenses. Lenses are optical components that work on the basis of image formation by refraction. Such optical components are commonly used in manufacturing optical devices, instruments, and systems. The purpose of this experiment is to observe the image formed when an object is placed in front of a lens or combination of two lenses. The lens applications are shown in the following cases:

a. Image formed by a lens.
b. Image formed by a combination of two lenses.

10.11.1 TECHNIQUE AND APPARATUS

Appendix A presents the details of the devices, components, tools, and parts.

1. Two thin convex lenses, as shown in Figure 10.1, Figure 10.29, and Figure 10.31.
2. Lens holders and positioners, as shown in Figure 10.29 and Figure 10.31.
3. Hardware assembly (clamps, posts, screw kits, screwdriver kits, sundry positioners, etc.).
4. Optical rail.
5. Light source/object and positioner, as shown in Figure 10.28.
6. Black/white card and cardholder.
7. Ruler.

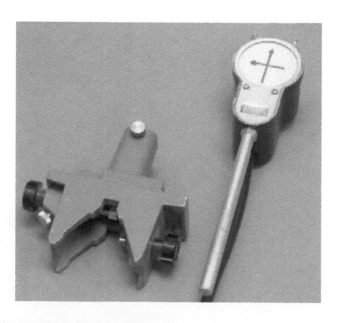

FIGURE 10.28 Light source/object and positioner.

10.11.2 Procedure

Follow the laboratory procedures and instructions given by the professor and/or instructor.

10.11.3 Safety Procedure

Follow all safety procedures and regulations regarding the use of optical components, light source devices, and optical cleaning chemicals.

10.11.4 Apparatus Set-Up

10.11.4.1 Image Formed by a Lens

1. Figure 10.29 shows the experimental apparatus set-up.
2. Place an optical rail on the table.
3. Mount a light source on one end of the optical rail.
4. Mount a convex lens and lens holder on the optical rail.
5. Mount the black/white card on the optical rail.
6. Connect the light source to the power supply.
7. Turn off the lights of the lab.
8. Move the lens back and forth until you capture a focused image on the card.
9. Make a fine adjustment to the lens position to get a clear and focused image, as shown in Figure 10.30. Describe the characteristics of the image.
10. Measure the object and image distances. Record in Table 10.1.
11. Measure the height of the object and the height of the image. Record in Table 10.1.
12. Turn on the lights of the lab.
13. Illustrate the locations of the light source/object, lens, and image, in a diagram using the ray tracing method.

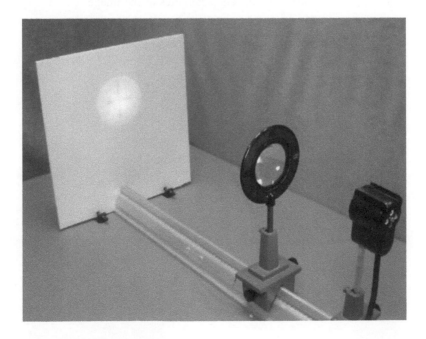

FIGURE 10.29 Image formation by a lens.

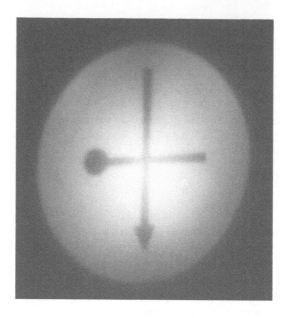

FIGURE 10.30 Image formation by a convex lens.

10.11.4.2 Image Formed by a Combination of Two Lenses

1. Figure 10.31 shows the experimental set-up.
2. Repeat the steps in Case (a) to find the image formed by the lens.
3. Mount another lens between the first lens and the card.
4. Find the image formed by lens 2 as explained in Case (a).
5. Make a fine adjustment to the position of the lenses to get a clear and focused image, as shown in Figure 10.32. Describe the characteristics of the image.

FIGURE 10.31 Image formation by a combination of two lenses set-up.

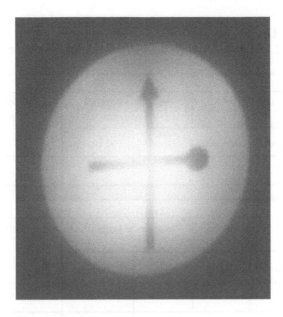

FIGURE 10.32 Image formation by a combination of two lenses.

6. Measure the distances between the light source, lens 1, lens 2, and the card, to determine the object and image distances. Record in Table 10.2.
7. Turn on the lights of the lab.
8. Illustrate the locations of the light source/object, lens 1, lens 2, and image, in a diagram using the ray tracing method.

10.11.5 DATA COLLECTION

10.11.5.1 Image Formed by a Lens

Measure the object and imaged distances. Record in Table 10.1.

TABLE 1
10.1 Image Formed by a Lens

Distance of (unit)		Lens Focal Length f (unit)		Magnification
Object S	Image S'	Given	Calculated	M

10.11.5.2 Image Formed by a Combination of Two Lenses

Measure the distances between the light source, lens 1, lens 2, and the card, to determine the object and image distances. Record in Table 10.2.

TABLE 2

10.2 Image Formed by a Combination of Two Lenses

		Lens 1		
Distance of (unit)		**Lens Focal Length** f (unit)		**Magnification**
Object S	**Image** S'	**Given**	**Calculated**	M

		Lens 2		
Distance of (unit)		**Lens Focal Length** f (unit)		**Magnification**
Object S	**Image** S'	**Given**	**Calculated**	M
Distance between the lenses L				
Total magnification M_{total}				

10.11.6 CALCULATIONS AND ANALYSIS

10.11.6.1 Image Formed by a Lens

1. Calculate the focal length of the lens, using Equation 10.11. Calculate the lateral magnification M of the lens, using Equation 10.13. Record in Table 10.1.
2. Find the published value of the focal length of the lens, as provided by the manufacturer/supplier. Record in Table 10.1.
3. Illustrate the locations of the light source/object, lens, and image in a diagram; include the dimensions.

10.11.6.2 Image Formed by a Combination of Two Lenses

1. Calculate the focal length of lenses 1 and 2, using Equation 10.16 and Equation 10.17, respectively. Record in Table 10.2.
2. Calculate the distance L between the lenses, using Equation 10.18. Record in Table 10.2.
3. Calculate the lateral magnification M of lens 1 and lens 2, using Equation 10.14. Record in Table 10.2.
4. Calculate the total magnification M_{total}, using Equation 10.19. Record in Table 10.2.
5. Find the published value of the focal length for each of the lenses, as provided by the manufacturer/supplier. Record in Table 10.2.
6. Illustrate the locations of the light source/object, lens 1, lens 2, and image, in a diagram; include the dimensions.

10.11.7 RESULTS AND DISCUSSIONS

10.11.7.1 Image Formed by a Lens

1. Discuss your measurements and calculations of the formation of the image by a thin lens.
2. Discuss the characteristics of the image.

3. Verify the calculated focal length of the lens against the value published by the manufacturer/supplier.

10.11.7.2 Image Formed by a Combination of Two Lenses

1. Discuss the measurements and calculations of the formation of the image by combinations of two thin lenses.
2. Discuss the characteristics of the image
3. Verify the calculated focal length of the lenses against the values published by the manufacturer/supplier.

10.11.8 CONCLUSION

Summarize the important observations and findings obtained in this lab experiment.

10.11.9 SUGGESTIONS FOR FUTURE LAB WORK

List any suggestions for improvements using different experimental equipment, procedures, and techniques for any future lab work. These suggestions should be theoretically justified and technically feasible.

10.12 LIST OF REFERENCES

List any references that were used in the report. Use one format in writing the references. Never mix reference formats in a report.

10.13 APPENDICES

List all of the materials and information that are too detailed to be included in the body of the report.

FURTHER READING

Beiser, Arthur, *Physics*, 5th ed., Addison-Wesley, Reading, MA, 1991.

Blaker, J. Warren and Peter, Schaeffer, *Optics an Introduction for Technicians and Technologists*, Prentice Hall, Inc., Englewood Cliffs, NJ, 2000.

Born, M. and Wolf, E., *Elements of the Theory of Diffraction Principles of Optics: Electromagnetic Theory of Propagation, Interference, and Diffraction of Light*, 7th Ed., Cambridge University Press, Cambridge, pp. 370–458, 1999.

Born, M. and Wolf, E., *Rigorous Diffraction Theory Principles of Optics: Electromagnetic Theory of Propagation, Interference, and Diffraction of Light*, 7th Ed., Cambridge University Press, Cambridge, pp. 556–592, 1999.

Clark, Timothy and Wanser, Keith, Ball vs gradient index lenses, *Photonics Spectra*, February, 94–96, 2001.

Cox, Arthur, *Photographic Optics*, 15th ed., Focal Press, London, New York, 1974.

Cutnell, John D. and Johnson, Kenneth W., *Physics*, 5th Ed., Wiley, New York, 2001.

Cutnell, John D. and Johnson, Kenneth W., *Student Study Guide-Physics*, 5th Ed., Wiley, New York, 2001.

DiCon Fibre Optics, Inc. *GRIN Lenses Catalog*, Rev. D. Dicon Fibre Optics, Inc., Richmond, CA, U.S.A., 2002.

Drisoll, Walter G. and Vaughan, William, *Handbook of Optics*, McGraw Hill Book Company, New York, 1978.

Drollette, Dan, Liquid lenses make better beams,*Photonics Spectra*, pp. 25–26, August, U.S.A., 2001.

Edmund Industrial Optics, *Optics and Optical instruments Catalog*, Edmund Industrial Optics, New Jersey, 2004.

EPLAB, *EPLAB Catalogue 2002*. The Eppley Laboratory, Inc., 12 Sheffield Avenue, P.O. Box 419, Newport, Rhode Island 02840, U.S.A.

Ewald, Warren J. et al., *Optics: The Matrix Theory*, Marcel Dekker, New York, 1972.

Ewald, William P. et al., *Practical Optics*, 1982.

Francon, M., *Optical Interferometry*, Academic Press, New York, 1966.

Ghatak, Ajoy K., *An Introduction to Modern Optics*, McGraw-Hill, New York, 1972.

Giancoli, Douglas C., *Physics*, 5th Ed., Prentice Hall, Englewood Cliffs, NJ, 1998.

Halliday, R., Resnick, D., and Walker, J., *Fundamental of Physics*, 6th Ed., Wiley, New York, 1997.

Heath, R. W., Macnaughton, R. R., and Martindale, D. G., *Fundamentals of Physics*, D.C. Heath Canada, Ltd, Canada, 1979.

Hecht, Eugene, *Optics*, 4th Ed., Addison-Wesley Longman, Inc., New York, 2002.

Hewitt, Paul G., *Conceptual Physics*, 8th Ed., Addison-Wesley, Inc., Reading, MA, 1998.

Jenkins, F. W. and White, H. E., *Fundamentals of Optics*, McGraw Hill, New York, 1957.

Jones, Edwin and Richard, Childers, *Contemporary College Physics*, McGraw-Hill Higher Education, New York, 2001.

Kennedy, T. P., Understanding Ball Lenses. *Optics and Optical Instruments Catalog*, Edmund Industrial Optics Co., U.S.A., 2004.

Keuffel & Esser Co., *Physics*, Keuffel & Esser Audiovisual Educator—Approved Diazo Transparency Masters, Audiovisual Division, Keuffel & Esser Co., U.S.A., 1989.

Lambda Research Optics, Inc., *Catalog 2004*, Lambda Research Optics, Inc., California, 2004.

Lehrman, Robert L., *Physics-The Easy Way*, 3rd Ed., Barron's Educational Series, Inc., Hauppage, NY, 1998.

Lerner, Rita G. and Trigg, George L., *Encyclopedia of Physics*, 2nd Ed., VCH Publishers, Inc., New York, 1991.

McDermott, L. C. and Shaffer, P. S., *Tutorials in Introductory Physics*, Preliminary Edition, Prentice Hall Series in Educational Innovation, Upper Saddle River, New Jersey, U.S.A., 1988.

Melles Griot, *The Practical Application of Light Melles Griot Catalog, 2001*, Melles Griot, Rochester, NY, 2001.

Newport Corporation. Optics and mechanics. *Newport 1999/2000 catalog*, Newport Corporation.

Nichols, Daniel H, *Physics for Technology with Applications in Industrial Control Electronics*, Prentice Hall, Englewood Cliffs, NJ, 2002.

Nolan, P. J., *Fundamentals of College Physics*, Wm. C. Brown Communications, Inc., Dubuque, Iowa, U.S.A., 1993.

Ocean Optics, Inc. *Product Catalog, 2003*, Florida.

Pedrotti, Frank L. and Pedrotti, Leno S., *Introduction to Optics*, 2nd Ed., Prentice Hall, Inc., Englewood Cliffs, NJ, 1993.

Robinson, Paul, *Laboratory Manual to Accompany Conceptual Physics*, 8th Ed., Addison-Wesley, Inc., Reading, MA, 1998.

Romine, Gregory S., *Applied Physics Concepts into Practice*, Prentice Hall, Inc., Englewood Cliffs, NJ, 2001.

Sears, F. W., Zemansky, M. W., and Young, H. D., *University Physics- Part II*, 6th Ed., Addison-Wesley Publishing Company, Reading, MA, U.S.A., 1998.

Serway, R. A., and Jewett, J. W., *Physics for Scientists and Engineers*, with Modern Physics, 6th Ed., Volume 2, Thomson Books/Cole, U.S.A., 2004.

Sciencetech, *Designers and Manufacturers of Scientific Instruments Catalog*, Sciencetech, London, Ont., Canada, 2003.

Smith, W. J., *Modern Optical Engineering*, McGraw-Hill Book Co., New York, 1966.

Sterling, Donald J. Jr. *Technician's Guide to Fibre Optics*, 2nd Ed., Delmar Publishers Inc., New York, 1993.

Tippens, P. E., *Physics*, 6th Ed., Glencoe McGraw-Hill, Westerville, OH, U.S.A., 2001.

Tom, Miller, Aspherics come of age, *Photonics Spectra*, February, 76–81, 2004.

Urone, Paul Peter, *College Physics*, Brooks/Cole Publishing Company, Belmont, CA, 1998.

Walker, James S., *Physics*, Prentice Hall, Englewood Cliffs, NJ, 2002.

White, Harvey E., *Modern College Physics*, 6th Ed., Van Nostrand Reinhold Company, New York, 1972.

Wilson, Jerry D., *Physics—A Practical and Conceptual Approach Saunders Golden Sunburst Series*, Saunders College Publishing, London, 1989.

Wilson, Jerry D. and Buffa, Anthony J., *College Physics*, 5th Ed., Prentice Hall, Inc., Englewood Cliffs, NJ, 2000.

11 Prisms

11.1 INTRODUCTION

The purpose of this chapter is to explain the basic principles that govern light passing through a prism or a combination of prisms. Types of prisms and image formation are also presented. The concept of light passing through prisms is very important to photonics and has significant use in image formation and building optical devices. Particular emphasis will be given to calculating the index of refraction of a prism, producing a rainbow of colours, and mixing a rainbow of colours using a glass rod or tube. Also in this chapter, along with the theoretical presentation, five experimental cases demonstrate the principles of light passing through prisms.

11.2 PRISMS

Prisms are blocks of optical material with flat polished faces arranged at precisely controlled angles, as shown in the figure above. In many situations it is necessary to direct a beam of light entering from one side and exiting from the other side of the prism. Light passing through prisms is governed by the laws of light.

Prisms are widely used in building optical devices, such as a prism spectrometer, which is commonly used to study the wavelengths emitted by a light source. Prisms are also used in building optical fibre devices, such as an opt-mechanical switch, which deflects or deviates an optical signal through a telecommunication system. Prisms can invert or rotate an image, deviate a light beam, disperse light into its component wavelengths, and separate states of polarization. The orientation of a prism (with respect to the incident light beam) determines its effect on the beam. Prisms can be designed for specific applications. The most popular prisms are right angle prisms, Brewster's angle dispersing prisms, Penta prisms, solid glass retro-reflectors, equilateral dispersing prisms, littrow dispersion prisms, wedge prisms, roof prisms, and Dove prisms. Some of these commonly used prisms are discussed in detail in this chapter.

The light beam can be arranged to exit the prism either parallel or horizontal to the input light and could also exit on the same or the opposite side of the input light. Depending on the desired characteristics of the image, the distance between the input and output beams can be determined using a combination of two prisms. This is usually done with different types of prisms.

11.3 PRISM TYPES

There are many types of prisms that are used in different applications. They are commonly used in building optic/optical fibre devices. In the following sections, the types of prisms and applications are presented in detail.

11.3.1 RIGHT ANGLE PRISMS

Right angle prisms are generally used to achieve a 90° bend in a light path and to change the orientation of an image. Depending on the prism orientation, images viewed through the prism will be inverted while maintaining correct left-to-right orientation, as shown in Figure 11.1(a). If the prism is rotated by 90°, images viewed through the prism will be erect, but they will be reversed left-to-right. Prisms can also be used in combination for image/beam displacement. A right angle prism may be used as a front surface mirror. Right angle prisms are also used to reverse the light beam, as shown in Figure 11.1(b). The input and output light beams are on the same side of the prism. It is possible to achieve a fixed distance between the input and the output in a right angle prism.

Figure 11.2 illustrates the arrangement of a combination of two right angle prisms. The light beam is incident on the first prism and exits from the second prism. The light beam exiting from the second prism is parallel to the incident light beam at the first prism. In this arrangement, the

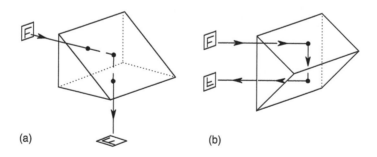

FIGURE 11.1 (a) Right angle prism as a mirror. (b) The direction of the light beam is reversed.

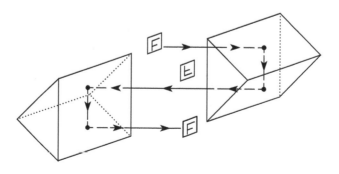

FIGURE 11.2 Light beam passing through two prisms.

distance between the incident and the exit beams can be varied. The common application of a two-prism combination is use in a submarine periscope.

11.3.2 EQUILATERAL PRISMS

Equilateral prisms, also called dispersing prisms, have three equal 60° angles, as shown in Figure 11.3. They are used for wavelength separation applications. A light beam is twice refracted while passing through the prism with a deviation angle (δ). Deviation is a function of the index of refraction and the wavelength of the light beam. These prisms are used within a laser cavity to compensate for dispersion of fixed laser cavity optics, such as a Ti:Sapphire crystal, or external to the cavity to manipulate the pulse characteristics.

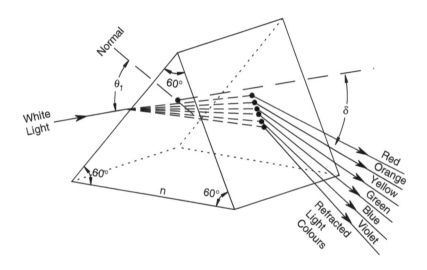

FIGURE 11.3 White light dispersing through an equilateral prism.

11.3.3 DOVE PRISMS

Dove prisms are commonly used as image rotators, as shown in Figure 11.4. As the prism is rotated, the image will rotate at twice the angular rate of the prism. Dove prisms can also be used with parallel or collimated light.

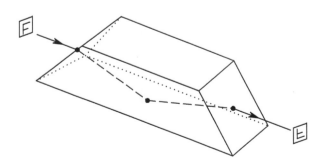

FIGURE 11.4 The Dove prism.

11.3.4 Roof Prisms

The roof or Amici prism is essentially a truncated right-angle prism whose hypotenuse has been replaced by a 90° total internal reflection roof, as shown in Figure 11.5. The prism deviates, or deflects, the image through an angle of 90°. Roof prisms are commonly used to split the image down the middle of the prism and interchange the right and left portions. They are often used in building simple telescope systems to correct for the image reversal introduced by the telescope lenses.

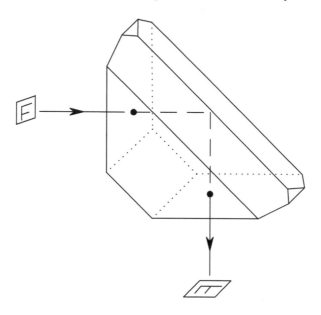

FIGURE 11.5 The Roof prism.

11.3.5 Penta Prisms

The Penta prism deviates the beam by precisely 90° without affecting the orientation of the image (neither inverting nor reversing the image), as shown in Figure 11.6. Penta prisms will

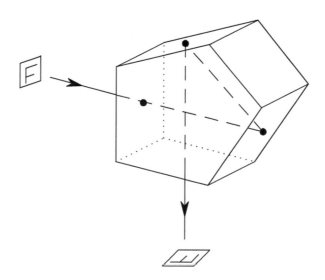

FIGURE 11.6 The Penta prism.

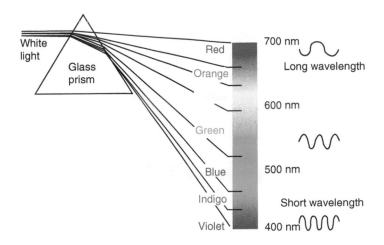

FIGURE 6.1 Dispersion of white light into the visible spectrum by a prism.

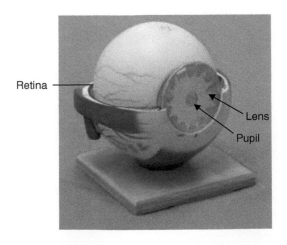

FIGURE 6.2 The major parts of the human eye.

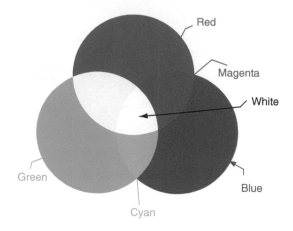

FIGURE 6.4 Additive colour mixing.

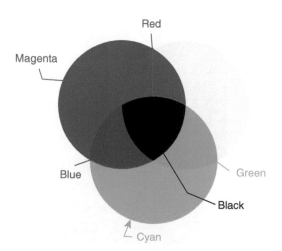

FIGURE 6.6 Subtractive colour mixing.

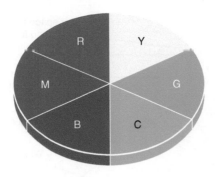

FIGURE 6.7 The Relationship between the additive and subtractive primary's complement.

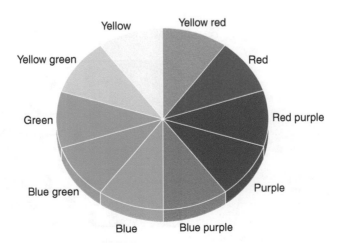

FIGURE 6.8 The colour wheel (the three primary additive colours sandwiched between variations of coloured light).

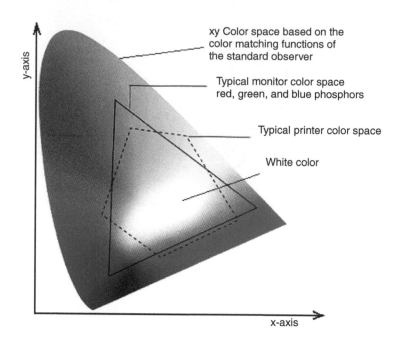

FIGURE 6.10 The C. I. E. Chromaticity diagram.

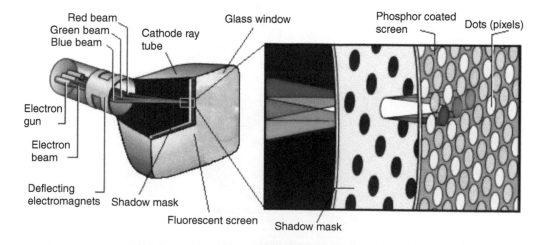

FIGURE 6.11 The additive theory is used in colour television.

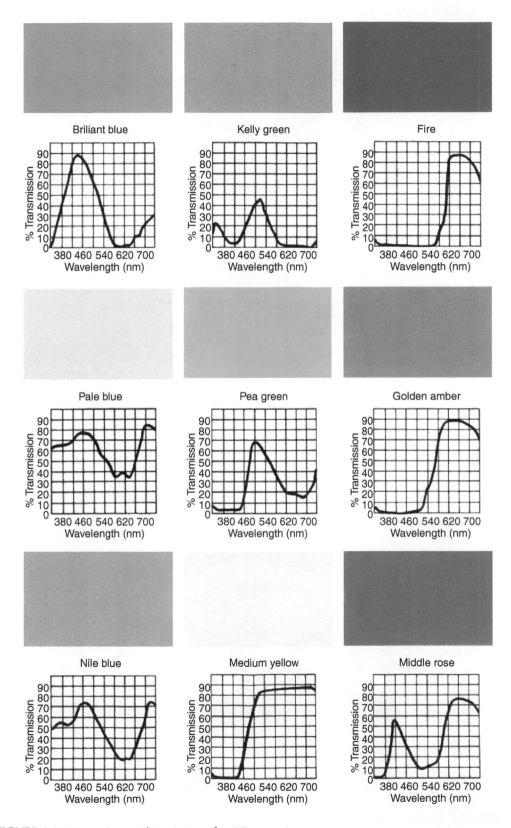

FIGURE 6.12 Spectral transmittance curve for different colours.

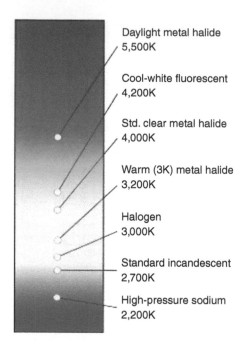

Daylight metal halide
5,500K

Cool-white fluorescent
4,200K

Std. clear metal halide
4,000K

Warm (3K) metal halide
3,200K

Halogen
3,000K

Standard incandescent
2,700K

High-pressure sodium
2,200K

FIGURE 6.13 Colour temperature chart.

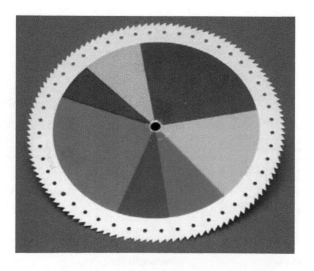

FIGURE 6.14 Newton's colour wheel.

FIGURE 6.19 Subtractive colour filter set.

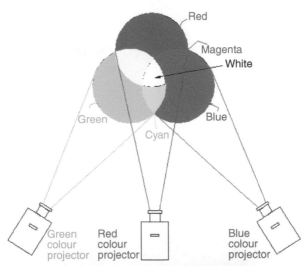

(a) Three slide projectors project lights onto the white cardboard.

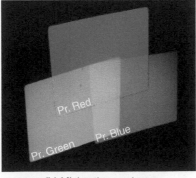

(b) Mixing three colours

FIGURE 6.22 Additive colour mixing. (a) Three slide projectors project lights onto the white cardboard. (b) Mixing three colours.

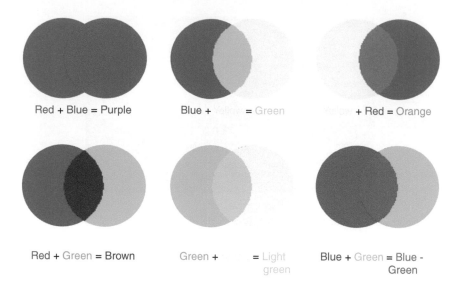

Red + Blue = Purple   Blue + Yellow = Green   Yellow + Red = Orange

Red + Green = Brown   Green + Yellow = Light green   Blue + Green = Blue - Green

FIGURE 6.23 Two colour additive mixing.

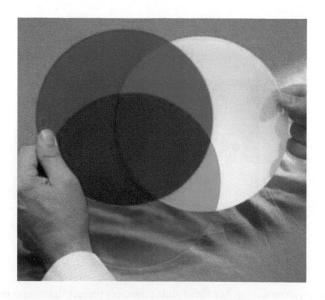

FIGURE 6.24 Subtractive colour mixing.

FIGURE 6.25 Newton's colour wheel apparatus set-up.

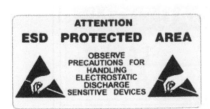

FIGURE 17.8 Electrostatic discharge warning symbols and signs.

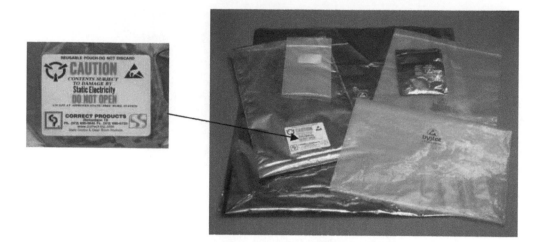

FIGURE 17.10 ESD bags.

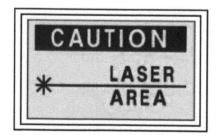

FIGURE 17.13 Laser warning labels.

direct a beam through the same angle regardless of the prism's orientation to the beam. They are used in applications requiring an exact 90° deviation without having to orient the prism precisely. They are also often used as end reflectors in small range finders. The design of a Penta prism also makes it inherently more stable than a system consisting of two mirrors.

11.3.6 DOUBLE PORRO PRISMS

The double Porro prism, as shown in Figure 11.7, consists of two right-angle prisms. The Porro prism is made with rounded corners to reduce weight and size. They are relatively easy to manufacture. A small slot is often cut in the hypotenuse face of the prism to obstruct rays that are internally reflected at glancing angles. Since there are four reflections between the two Porro prisms, the exiting image will be right-handed. For example, two matching Porro prisms are used in binoculars to produce erect final images. At the same time, they permit the distance between the object-viewing lenses to be greater than the normal eye-to-eye distance, thereby enhancing the stereoscopic effect produced by ordinary binocular vision.

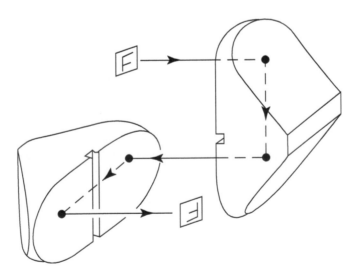

FIGURE 11.7 The double Porro prism.

11.3.7 LEMAN–SPRINGER PRISMS

Figure 11.8 shows that the Leman–Springer prism has a 90° roof like the roof prism. The input beam is displaced without being deviated, but the emerging image is right-handed and rotated through 180°. This prism can be used to correct image orientation in telescope systems.

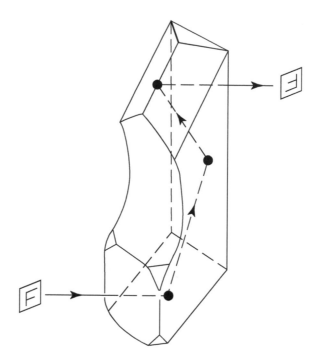

FIGURE 11.8 The Leman–Springer prism.

11.3.8 WEDGE PRISMS

Wedge prisms can be used individually to deviate a laser beam by a set angle (θ_d), as shown in Figure 11.9. Also, two wedge prisms can be combined together as an anamorphic pair. A single prism's ability to deviate the angle of an incident beam is measured in dioptres. One dioptre is defined as the angular deviation of the beam of 1 cm at a distance of one metre from the wedge prism. Wedge prisms are used as beamsteering elements in optical systems.

The apex angle of a wedge prism necessary to produce a given minimum deviation angle (θ_d) or deflection is determined by the wedge angle (θ_w):

$$\theta_w = \arctan\left[\frac{\sin \theta_d}{(n - \cos \theta_d)}\right] \qquad (11.1)$$

where n is the index of refraction of a wedge prism.

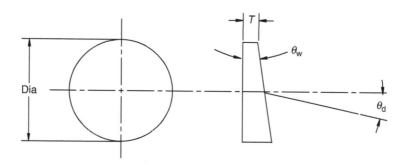

FIGURE 11.9 A Wedge prism.

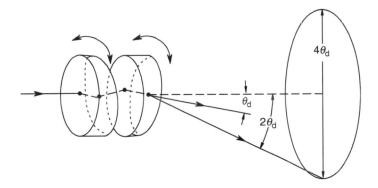

FIGURE 11.10 Wedge prisms used in beamsteering applications.

Figure 11.10 shows a pair of matching wedge prisms, which can steer a beam anywhere within a circle described by the full angle $(4\theta_d)$, where θ_d is the deviation angle from a single prism as defined in Equation 11.2. The equation to determine the deviation angle (θ_d) for the same input direction but other wavelengths is:

$$\theta_d = \text{arc } \sin(n \sin \theta_w) - \theta_w \tag{11.2}$$

This beam steering is accomplished by rotating the two wedge prisms independently of each other. Beam steering is typically used to scan a beam to different locations in imaging applications. By combining two wedge prisms of equal power in near contact, and by independently rotating them about an axis parallel to the normals of their adjacent faces, a ray can be steered in any direction within the narrow cone. The deviation will change with the input angle.

11.3.9 PRISMS WITH SPECIAL APPLICATIONS

Prisms can be combined to produce achromatic overall behaviour. The achromatic prism has a net dispersion of zero for two given wavelengths, even though the deviation is not zero. On the other hand, the direct vision prism accomplishes zero deviation for a particular wavelength while providing chromatic dispersion. Figure 11.11 illustrates special combinations that form these two prism types. The arrangement of prisms, as illustrated in Figure 11.11(a), is combined so that one prism cancels the dispersion of the other. These can also be reversed so that the dispersion is additive, providing double dispersion. Figure 11.11(b) shows a direct vision prism when light with a wavelength of λ passes through. These prisms can be used in laser beam alignment.

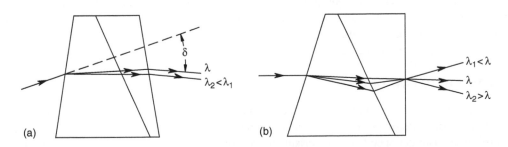

(a)

(b)

FIGURE 11.11 Nondispersive and nondeviating prisms. (a) Achromatic prism. (b) Direct vision prism for wavelength λ.

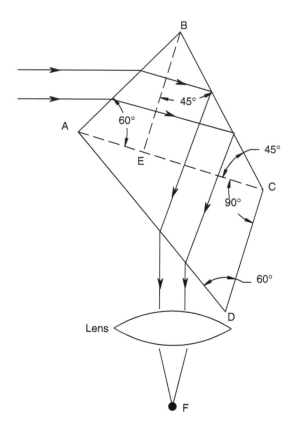

FIGURE 11.12 Pellin–Broca prism of constant deviation.

A prism design useful in spectrometers is one that produces a constant deviation for all wavelengths as they are observed or detected. One example is the Pellin–Broca prism illustrated in Figure 11.12. A collimated beam of light enters the prism at face AB and departs at face AD, making an angle of 90° with the incident direction. The dashed lines are merely added to assist in analyzing the operation of the prism. Of the incident wavelengths, only one will refract at the precise angle of deviation, as shown, with the light rays parallel to the prism base AE. At face BC total internal reflection occurs to direct the light beam into the prism section ACD. Since the prism section BEC acts only as a mirror, the beam passes through sections ABE and ACD, which together constitute a prism of 60° apex angle. The beams are focused at F, the focal point of lens.

An observing telescope may be rigidly mounted, as the prism is rotated on its prism table (or about an axis normal to the page). As the prism rotates, various wavelengths in the incident beam deviate and follow the path indicated and focus at F.

11.3.10 Other Types of Prisms

There are many other types of prisms that are used in building various optical devices and systems. Many reflecting prisms perform specific functions. Non-standard prism designs can be made according to customer specifications. Prisms made from birefringent crystals are useful for producing a highly polarized light wave or polarization splitting of light, such as the Wollaston prism. This prism will be addressed in detail in Chapter 15.

11.4 PRISMS IN DIFFERENT COMBINATIONS

Prism combinations are commonly used in optical instruments for purposes such as inversion of an image in prism binoculars. Laser light can be passed through two or more prisms aligned one beside the other, as shown in Figure 11.13. Some prism combinations are described above in the prism types. The laser light can be directed to a point by moving one prism in the prism combination. Consider the orientation of the final image when designing any prism combination.

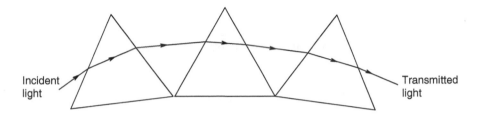

FIGURE 11.13 Laser beam passing through prism combinations.

11.5 LIGHT BEAM PASSING THROUGH A PRISM

Refraction of light passing through a prism, as shown in Figure 11.14, is understood using light rays and Snell's law. This figure shows how a single light ray that is incident on the prism from the left emerges bent away from its original direction of travel. When the ray exits the prism, the emerged ray is bent by an angle of δ, called the angle of deviation (δ). The angle of deviation will change as the angle (θ_1) of the incident ray changes. The magnitude of the deviation angle also depends on the apex angle (A) of the prism, the index of refraction of the prism material, and the wavelength of the incident light. The angle of deviation reaches its minimum value when the light passing through the prism is symmetrical; i.e., when $\theta_1 = \theta_2$, as shown in Figure 11.14. Then, the angle is called the minimum deviation angle (δ_m).

The minimum angle of deviation (δ_m) is related to the apex angle of the prism (A) and the index of refraction (n) of the prism. The index of refraction of a prism in Equation 11.3 is calculated by applying Snell's law and using simple geometry.

$$n = \frac{\sin\left(\frac{A+\delta_m}{2}\right)}{\sin\left(\frac{A}{2}\right)} \tag{11.3}$$

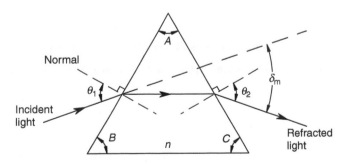

FIGURE 11.14 Light ray incident on the prism surface.

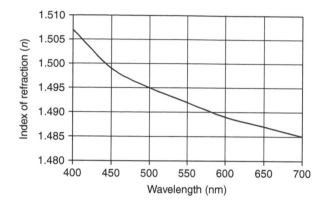

FIGURE 11.15 Index of refraction as a function of wavelength for a transparent plastic material.

White light is a uniform mixture of all visible light. Natural light, like sunlight, is considered to be white light and contains the entire visible spectrum.

Chromatic dispersion is the spreading of white light into its spectrum of wavelengths. This type of dispersion occurs when the refraction process deviates the light. The angle of deviation depends on the wavelength of the light, since the index of refraction of any optical material is a function of wavelength.

The light of shorter wavelengths travels with slightly smaller wave velocities than that of longer wavelengths. This means that the prism's index of refraction is not constant across the visible spectrum, but decreases continuously as the wavelengths increase from violet to red light. Figure 11.15 shows the relation between the light wavelength and the index of refraction of a transparent plastic material, which has the typical dispersive behaviour of decreasing the index of refraction in conjunction with increasing the wavelength.

Water, glass, transparent plastics, and quartz are all optically dispersive materials. Table 11.1 lists the indices of refraction for such materials. Again, these indices are dependent on the wavelength of light.

The index of refraction is greater for violet light than for red light, thus violet light deviates through a greater angle than does red light, as shown in Figure 11.16.

TABLE 11.1
Index of Refraction for Different Materials at Various Wavelengths

Material	Red λ=660	Orange λ=610	Yellow λ=580	Green λ=550	Blue λ=470	Violet λ=410
Water	1.331	1.332	1.333	1.335	1.338	1.342
Diamond	2.410	2.415	2.417	2.426	2.444	2.458
Glass (crown)	1.512	1.514	1.518	1.519	1.524	1.530
Glass (flint)	1.662	1.665	1.667	1.674	1.684	1.698
Polystyrene	1.488	1.490	1.492	1.493	1.499	1.506
Quartz (fused)	1.455	1.456	1.458	1.459	1.462	1.468

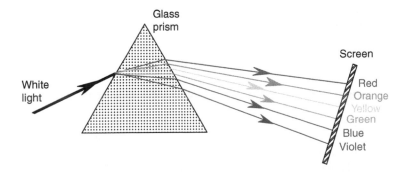

FIGURE 11.16 White light is dispersed by a prism.

11.6 FACTORS GOVERNING DISPERSION OF LIGHT BY A PRISM

Dispersion of white light by a prism depends on the following factors:

1. The optical properties of the prism (index of refraction n)
2. The angle of incidence (θ) of the light ray at the prism surface
3. The apex angle (A) of the prism
4. The wavelength of light (λ)

11.7 DISPERSION OF WHITE LIGHT BY A PRISM

Isaac Newton's first scientific paper, published in 1672, described his experiments with light and colour. Newton passed a beam of sunlight through a prism, spreading the light into a spectrum of colours. A second prism turned the opposite way re-combined the spectrum of colours back into a narrow beam of white light. Newton found that when white light is passed through a prism, a spectrum of colours emerges as shown in Figure 11.16. This separation of white light into its constituent spectral wavelengths is called chromatic dispersion. In real life, dispersion can be observed in such examples as rainbows, the surface of CD's, and an oil layer on water.

700 nm	The visible light spectrum	400 nm
Red		Violet
long λ		short λ

FIGURE 11.17 The visible spectrum of white light.

TABLE 11.2
Wavelength of the Colours in the Visible Spectrum

Colour	Wavelength Range λ (nm)
Red	630–700
Orange	590–630
Yellow	570–590
Green	500–570
Blue	450–500
Violet	400–450

Colours are associated with different wavelengths of light, as shown in Figure 11.17. It is possible to assign an approximate wavelength range to each colour of the spectrum, as given in Table 11.2.

11.8 MIXING SPECTRUM COLOURS USING A GLASS ROD AND TUBE

Colour mixing theory has been presented earlier in this book. A rainbow of colours can be mixed using a glass rod or tube, as shown in Figure 11.18. A narrow beam of white light from a light source passes through a lens and enters a glass prism where it is dispersed into the complete visible spectrum. If a mirror is placed after the prism, all colours (after reflection) can be directed onto a translucent glass rod or tube. A card held in front of the mirror can control the colours that are permitted to mix by the rod or tube. For example, if the violet, blue, and green colours are blocked, then the remaining colours of red, orange, and yellow combine, and the rod appears orange.

The visible spectrum can be divided into three equal parts, as shown in Figure 11.19. These colours (red, green, and blue) are the additive primaries and appear as three large circular areas in Figure 11.19.

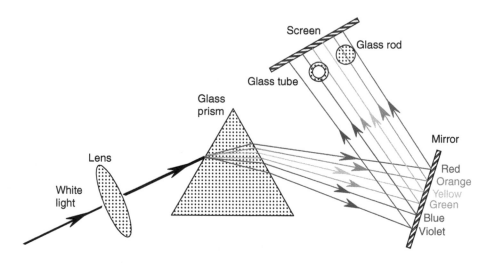

FIGURE 11.18 Experimental arrangement for mixing spectrum colours using a translucent rod.

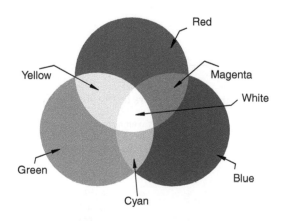

FIGURE 11.19 Additive colour mixing.

Consider mixing the primary colours of light (two at a time) and observing their resultant colour mixture, which is called a secondary colour. When primary red and primary green mix, they produce yellow; red and blue produce magenta; and green and blue produce cyan, which is a light blue–green. When red light and orange light mix, the rod appears as bright red; when yellow and green mix, the rod appears bright green; and when blue and violet mix, the rod appears blue–violet.

11.9 EXPERIMENTAL WORK

This experiment is designed to demonstrate the theory of light by passing light through both a single prism and through prism combinations. It also demonstrates light dispersion into a spectrum by a prism. Since the index of refraction varies with the wavelength, the angles of refraction are wavelength dependent. When white light enters a prism, the colour rainbow (a sequence of red, orange, yellow, green, blue, and violet colours) exits the prism, because the refractive deviation increases steadily with decreasing wavelength. In this experiment the student will perform the following cases:

a. Observe a laser beam passing through a right angle prism and draw the laser beam path.
b. Observe a laser beam passing through a Dove prism and draw the laser beam path.
c. Observe a laser beam passing through a Porro prism and draw the laser beam path.
d. Observe a laser beam passing through a prism and draw the laser beam path.
e. Observe a laser beam passing through prisms in different arrangements and draw the laser beam path through the prism combinations.
f. Use a laser beam to calculate the index of refraction of a prism.
g. Use white light passing through a prism to observe a rainbow of colours. Measure the refractive deviation angle of the rainbow colours and calculate the index of refraction of the prism.
h. Mix the spectrum colours using a glass rod and tube.

11.9.1 TECHNIQUE AND APPARATUS

Appendix A presents the details of the devices, components, tools, and parts.

1. 2×2 ft. optical breadboard
2. HeNe laser source and power supply

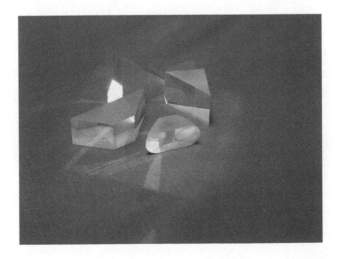

FIGURE 11.20 Prisms.

3. Laser mount assembly
4. Light source (white light)
5. Lab jack
6. Multi-axis translation stage
7. 360° rotational stage
8. Hardware assembly (clamps, posts, screw kits, screwdriver kit, positioners, post holder, laser holder/clamp, etc.)
9. Different types of prisms, as shown in Figure 11.20
10. Glass rod and tube, as shown in Figure 11.21
11. Lens and lens holder/positioner, as shown in Figure 11.22
12. Mirror and mirror holder, as shown in Figure 11.23
13. Black/white card and cardholder (large and small sizes)
14. Protractor
15. Ruler

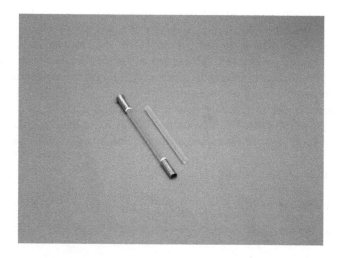

FIGURE 11.21 A glass rod and tube.

FIGURE 11.22 A lens and lens holder/positioner.

FIGURE 11.23 A mirror and mirror holder.

11.9.2 Procedure

Follow the laboratory procedures and instructions given by the professor and/or instructor.

11.9.3 Safety Procedure

Follow all safety procedures and regulations regarding the use of optical components, electrical and optical devices, and optical cleaning chemicals.

11.9.4 Apparatus Set-up

Case (a): Laser Beam Passing through a Right Angle Prism

1. Figure 11.24 shows the apparatus set-up.
2. Bolt the laser short rod to the breadboard.

FIGURE 11.24 Laser beam passing through a right angle prism.

3. Bolt the laser mount to the clamp using bolts from the screw kit.
4. Put the clamp on the short rod.
5. Place the HeNe laser into the laser mount and tighten the screw. Turn on the laser device. Follow the operation and safety procedures of the laser device in use.
6. Mount a 360° rotational stage to the breadboard so that the stage is approximately 10 cm in front of the laser source. Secure the stage to the breadboard.
7. Mount a right angle prism on top of the rotational stage platform and ensure that one side of the prism is facing the laser beam. The right angle prism should be orthogonal to the laser beam and parallel to the optical breadboard.
8. Arrange the location of the black/white card to capture the laser beam from the right angle prism.
9. Make the necessary adjustments so that the laser beam passes through the right angle prism, exits from the other side of the prism, and projects onto the white/black card.
10. Finely align the laser beam, using the rotational stage under the right angle prism, to direct the laser beam passing through the prism.
11. Observe and draw the laser beam path in a diagram.

Case (b): Laser Beam Passing through a Dove Prism

Figure 11.25 shows the apparatus set-up. Using a Dove prism, repeat the procedure explained in Case (a).

FIGURE 11.25 Laser beam passing through a Dove prism.

Case (c): Laser Beam Passing through a Porro Prism

Figure 11.26 shows the apparatus set-up. Using a Porro prism, repeat the procedure explained in Case (a).

FIGURE 11.26 Laser beam passing through a Porro prism.

Case (d): Laser Beam Passing through a Prism

Figure 11.27 shows the apparatus set-up. Using a prism repeat the procedure explained in Case (a).

FIGURE 11.27 A laser beam passing through a prism.

Case (e): Laser Beam Passing through Prism Combination

Figure 11.28 shows the apparatus set-up. Repeat the procedure as explained in Case (a) from step 2 through step 5. Add the following steps to complete the lab set-up for a laser beam passing through a combination of three prisms:

1. Mount three 360° rotational stages to the breadboard so that the first stage is approximately 10 cm in front of the light source. Mount the second and the third rotation stages offset as show in the Figure 11.28. Secure the stages to the breadboard.
2. Mount a prism on top of each rotational stage platform and ensure that one side of the prism is facing the laser beam. The prism should be orthogonal to the laser beam and parallel to the optical breadboard.

FIGURE 11.28 A laser beam passing through prism combination.

3. Arrange the location of the black/white card to capture the laser beam from the third prism.
4. Make the necessary adjustments so that the laser beam passes through the first prism, exits from the other side of the first prism, is incident on the second prism, and the transmitted light is incident on the third prism. Try to project the transmitted laser beam from the third prism onto the white/black card.
5. Finely align the laser beam, using rotational stages under the prisms, to direct the laser beam passing through the prisms.
6. Observe and draw the laser beam path in a diagram.

Case (f): Laser Beam Passing through a Prism to Calculate the Index of Refraction

Figure 11.27 also shows the apparatus set-up. Repeat the procedure explained in Case (a) from step 2 through step 11. Add the following steps to complete the lab set-up for a laser beam passing through a prism:

1. Use a protractor to measure the prism angles, the angle of incidence of the laser beam to the prism, the angle of deviation (δ), and the angle of refraction of the laser beam exiting the prism.
2. Record the measured angles in Table 11.3.

Case (g): Dispersion of White Light by a Prism

1. Figure 11.29 shows the apparatus set-up.
2. Mount the white light source on the lab jack to the optical breadboard.
3. Mount the lens and lens holder in front the white light beam to focus the light.
4. Use the lab jack to adjust the height of the light beam to the centre of the lens.
5. Mount a multi-axis translation stage on the optical breadboard.
6. Mount a prism on top of the multi-axis translation stage platform and ensure that one side of the prism is facing the light source. The prism should be orthogonal to the light source and parallel to the optical breadboard.
7. Turn off the laboratory light to be able to see the light exiting from the prism.

FIGURE 11.29 Apparatus set-up for light dispersion by a prism.

8. Make the necessary adjustments so that the light passes through the lens, prism, exits from the other side of the prism, and projects onto the white/black card.

9. Arrange the location of the black/white card to capture the rainbow colours exiting from the prism.

10. Finely align the light, using multi-axis translation stages under the prism, to maximize the amount of light passing through the prism. This will brighten the spectrum on the black/-white card. Make a very fine alignment to the stage to ensure that the light exiting the prism appears as a clear rainbow on the black/white card, as shown in Figure 11.30.

11. Note that the rainbow of colours will be different from one prism to other. This depends on the prisms optical properties (index of refraction), apex angles, and wavelength of the light source.

12. Use a protractor to measure the prism angles, the angle of incidence of the light source to the prism, the angle of deviation (δ), and the angle of refraction of each of the six colours exiting the prism.

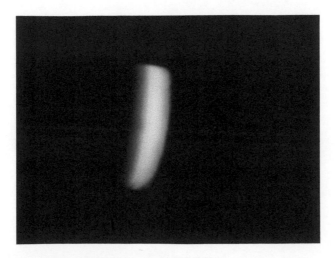

FIGURE 11.30 Rainbow colour exiting from the prism captured on the card.

13. Turn on the laboratory light.
14. Record the measured angles in Table 11.4.

Case (h): Mixing the Spectrum Colours Using a Glass Rod and Tube

Figure 11.31 shows the apparatus set-up. Repeat the procedure explained in Case (g) from step 1 through step 10. Add the following steps to mix the colours by using a glass rod and tube:

1. Mount a mirror in place of the black/white card.
2. Mount the black/white card to the optical breadboard to capture the rainbow colours reflected from the mirror, as shown in Figure 11.31.
3. Place a glass rod in front of the black/white card, as shown in Figure 11.32.
4. Place the glass rod between two adjacent colours of the rainbow starting from red towards violet.

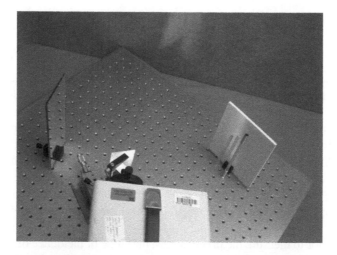

FIGURE 11.31 Apparatus set-up for light dispersion by a prism and mixing spectrum colours.

FIGURE 11.32 A glass rod and tube intercepting the rainbow colours.

5. Observe the mixing of spectrum colours exiting from the glass rod. When primary red and primary green mix at the glass rod, they produce yellow; red and blue produce magenta; and green and blue produce cyan.
6. Repeat step 3 through step 5 using a glass tube in front of the black/white card, as shown in Figure 11.32.
7. Report your observations when mixing spectrum colours.

11.9.5 Data Collection

11.9.5.1 Laser Beam Passing through a Right Angle Prism

No data collection is required for this case.

11.9.5.2 Laser Beam Passing through a Dove Prism

No data collection is required for this case.

11.9.5.3 Laser Beam Passing through a Porro Prism

No data collection is required for this case.

11.9.5.4 Laser Beam Passing through a Prism

No data collection is required for this case.

11.9.5.5 Laser Beam Passing through Prism Combination

No data collection is required for this case.

11.9.5.6 Laser Beam Passing through a Prism to Calculate the Index of Refraction

Measure the angles of the prism: the apex angle (A) and the base angles (B) and (C). Measure the angle of incidence of the laser beam incident on the prism, the angle of deviation (δ), and the refracted angle of the laser beam that leaves the prism. Record the measured angles in Table 11.3.

TABLE 11.3
Prism Angles, and Incidence and Refracted Angles of the Laser Beam

Prism Angles	Incidence Angle of Light	Angle of Deviation	Calculated Refractive Index
$A = (\quad^{o})$			
$B = (\quad^{o})$	$\theta_1 = (\quad^{o})$	$\delta = (\quad^{o})$	$n =$
$C = (\quad^{o})$			

11.9.5.7 Dispersion of White Light by a Prism

1. Measure the angles of the prism: the apex angle (A) and the base angles (B) and (C). Measure the angle of incidence of light incident on the prism, the angle of deviation (δ), and the refracted angles of the rainbow colours that leave the prism.
2. Record the measured angles in Table 11.4.

TABLE 11.4
Prism Angles, and Incidence and Refracted Angles of the Light

Prism Angles

$A = ($ $^{0})$

$B = ($ $^{0})$

$C = ($ $^{0})$

Incidence Angle of Light

$\theta_1 = ($ $^{0})$

Angle of Deviation

$\delta = ($ $^{0})$

Rainbow Colours	Rainbow Colours Refracted Angle from Prism (^{0})
Red	
Orange	
Yellow	
Green	
Blue	
Violet	

11.9.5.8 Mixing the Spectrum Colours Using a Glass Rod and Tube

No data collection is required for this case.

11.9.6 CALCULATIONS AND ANALYSIS

11.9.6.1 Laser Beam Passing through a Right Angle Prism

No calculations and analysis are required for this case.

11.9.6.2 Laser Beam Passing through a Dove Prism

No calculations and analysis are required for this case.

11.9.6.3 Laser Beam Passing through a Porro Prism

No calculations and analysis are required for this case.

11.9.6.4 Laser Beam Passing through a Prism

No calculations and analysis are required for this case.

11.9.6.5 Laser Beam Passing through Prism Combination

No calculations and analysis are required for this case.

11.9.6.6 Laser Beam Passing through a Prism to Calculate the Index of Refraction

1. Calculate the index of refraction (n) of the prism, using Equation 11.3.
2. Record the calculated value for the index of refraction of the prism in Table 11.3.
3. Find the published value of the index of refraction of the prism, as provided by the manufacturer.

11.9.6.7 Dispersion of White Light by a Prism

1. Calculate the index of refraction (n) of the prism for one colour, using Equation 11.3.
2. Record the calculated value for the index of refraction of the prism in Table 11.4.
3. Find the published value of the index of refraction of the prism, as provided by the manufacturer.

11.9.6.8 Mixing the Spectrum Colours Using a Glass Rod and Tube

No calculations and analysis are required for this case.

11.9.7 RESULTS AND DISCUSSIONS

11.9.7.1 Laser Beam Passing through a Right Angle Prism

Report your observations when a laser beam passes through a right angle prism. Compare the laser beam diagram with the theory.

11.9.7.2 Laser Beam Passing through a Dove Prism

Report your observations when a laser beam passes through a Dove prism. Compare the laser beam diagram with the theory.

11.9.7.3 Laser Beam Passing through a Porro Prism

Report your observations when a laser beam passes through a Porro prism. Compare the laser beam diagram with the theory.

11.9.7.4 Laser Beam Passing through a Prism

Report your observations when a laser beam passes through a prism. Compare the laser beam diagram with the theory.

11.9.7.5 Laser Beam Passing through Prism Combination

Report your observations when a laser beam passes through the combination of prisms. Compare the laser beam diagram with the theory.

11.9.7.6 Laser Beam Passing through a Prism to Calculate the Index of Refraction

1. Report the angle measurements for the prism.
2. Explain how a prism works and provide examples of optical devices that use prisms combined with other optical components.
3. Compare the calculated index of refraction with the published value of the prism as provided by the manufacturer. Explain the possible reasons for any differences.
4. List the most important factors to be considered when choosing the specifications of a prism.

11.9.7.7 Dispersion of White Light by a Prism

1. Report the colours of the rainbow.
2. Report the angle measurements for the prism.
3. Explain how a prism works and provide examples of optical devices that use prisms with other optical components.
4. Compare the calculated index of refraction with the published value of the prism as provided by the manufacturer. Explain the possible reasons for any differences.
5. List the most important factors to be considered when choosing the specifications of a prism.

11.9.7.8 Mixing the Spectrum Colours Using a Glass Rod and Tube

Report your observations when mixing spectrum colours using a glass rod and tube.

11.9.8 Conclusion

Summarize the important observations and findings obtained in this lab experiment.

11.9.9 Suggestions for Future Lab Work

List any suggestions for improvements using different experimental equipment, procedures, and techniques for any future lab work. These suggestions should be theoretically justified and technically feasible.

11.10 LIST OF REFERENCES

List any references that were used in the report. Use one format in writing the references. Never mix reference formats in a report.

11.11 APPENDICES

List all of the materials and information that are too detailed to be included in the body of the report.

FURTHER READING

Beiser, A., *Physics*, 5th ed., Addison-Wesley Publishing Company, Inc., USA, 1991.

Blaker, J. W. and Schaeffer, P., *Optics: An Introduction for Technicians and Technologists*, Prentice Hall, Inc., Upper Saddle River, NJ, 2000.

Born, M. and Wolf, E., Elements of the Theory of Diffraction, In *Principles of Optics: Electromagnetic Theory of Propagation, Interference, and Diffraction of Light*, 7th ed., Cambridge University Press, Cambridge, pp. 370–458, 1999.

Born, M. and Wolf, E., *Rigorous Diffraction Theory Principles of Optics: Electromagnetic Theory of Propagation, Interference, and Diffraction of Light*, 7th ed., Cambridge University Press, Cambridge pp. 556–592, 1999.

Cox, A., *Photographic Optics*, 15th ed., Focal Press, London, 1974.

Cutnell, J. D. and Johnson, K. W., *Physics*, 5th ed., Wiley, New York, 2001.

Cutnell, J. D. and Johnson, K. W., *Student Study Guide—Physics*, 5th ed., Wiley, New York, 2001.

Drisoll, W. G. and Vaughan, W., *Handbook of Optics*, McGraw Hill Book Company, New York, 1978.

Edmund Industrial Optics, *Optics and Optical Instruments Catalog*, Edmund Optics, Inc., New Jersey, 2004.

EPLAB, *EPLAB Catalogue 2002*, The Eppley Laboratory, Inc., 12 Sheffield Avenue, P.O. Box 419, Newport, Rhode Island 02840, USA, 2002.

Ewald, W. P., Young, W. A., and Roberts, R. H., *Practical Optics*, Makers of Pittsford, Rochester, New York, 1982.

Francon, M., *Optical Interferometry*, Academic Press, New York pp. 97–99, 1966.

Giancoli, D. C., *Physics*, 5th ed., Prentice Hall, Upper Saddle River, NJ, 1998.

Halliday, D., Resnick, R., and Walker, J., *Fundamental of Physics*, 6th ed., Wiley, New York, 1997.

Heath, R. W. et al., *Fundamentals of Physics*, Heath Canada Ltd., Toronto, DC, 1979.

Hecht, E., *Optics*, 4th ed., Addison-Wesley Longman, Inc., New York, 2002.

Jenkins, F. W. and White, H. E., *Fundamentals of Optics*, McGraw Hill, New York, 1957.

Jones, E. and Childers, R., *Contemporary College Physics*, McGraw-Hill Higher Education, New York, 2001.

Keuffel & Esser Co., *Physics Approved Diazo Transparency Masters*, Keuffel & Esser Audiovisual Educator, New York, 1989.

Lambda, *Catalog 2004*, Research Optics, Inc., California, 2004.

Lehrman, R. L., *Physics—The Easy Way*, 3rd ed., Barron's Educational Series, Inc., Hauppauge, NY, 1998.

Lerner, R. G. and George, L. T., *Encyclopedia of Physics*, 2nd ed., VCH Publishers, Inc., New York, NY, 1991.

McDermott, L. C. et al., *Introduction to Physics*, Preliminary Edition, Prentice Hall, Inc., Upper Saddle River, NJ, 1988.

Melles, G., *The Practical Application of Light Melles Griot Catalog*, Melles Griot, Rochester, NY, 2001.

Newport Corporation, Optics and Mechanics Section, the *Newport Resources 1999/2000 Catalog*, Newport Corporation, Irvine, CA, USA, 1999/2000.

Nolan, P. J., *Fundamentals of College Physics*, Wm. C. Brown Publishers, Dubuque, Iowa, 1993.

Ocean Optics, *Product Catalog*, Ocean Optics, Inc., Florida, 2003.

Pedrotti, F. L. and Pedrotti, L. S., *Introduction to Optics*, 2nd ed., Prentice Hall, Inc., Upper Saddle River, NJ, 1993.

Romine, G. S., *Applied Physics Concepts into Practice*, Prentice Hall, Inc., Upper Saddle River, NJ, 2001.

Sears, F. W., Zemansky, M. W., and Young, H. D., *University Physics—Part II*, 6th ed., Addison-Wesley Publishing Company, Massachusetts, 1998.

Serway, R. A. and Jewett, J. W., *Physics for Scientists and Engineers*, with Modern Physics, 6th ed., Vol. 2, Thomson Books/Cole, USA, 2004.

Sterling, D. J. Jr., *Technician's Guide to Fiber Optics*, 2nd ed., Delmar Publishers Inc., Albany, NY, 1993.

Tippens, P. E., *Physics*, 6th ed., Glencoe McGraw-Hill, Westerville, OH, U.S.A., 2001.

Smith, W. J., *Modern Optical Engineering*, McGraw-Hill Book Co., New York, 1966.

Urone, P. P., *College Physics*, Brooks/Cole Publishing Company, Florence, KY, 1998.

Walker, J. S., *Physics*, Prentice Hall, Englewood Cliffs, NJ, 2002.

White, H. E., *Modern College Physics*, 6th ed., Van Nostrand Reinhold Company, New York, 1972.

Wilson, J. D., *Physics—A Practical And Conceptual Approach Saunders Golden Sunburst Series*, Saunders College Publishing, Philadelphia, PA, 1989.

Wilson, J. D. and Buffa, A. J., *College Physics*, 5th ed., Prentice Hall, Inc., Upper Saddle River, NJ, 2000.

12 Beamsplitters

12.1 INTRODUCTION

A beamsplitter is a common optical component that partially transmits and partially reflects an incident light beam; this splitting usually occurs in unequal proportions. This chapter explains the basic principles that govern light passing through a beamsplitter. Various types of beamsplitters and their applications are presented. In addition to the function of dividing light, beamsplitters can be employed to recombine two separate light beams or images into a single path. The concept of light passing through beamsplitters is applicable to image formation and building optic/optical devices. Three experimental cases presented in this chapter demonstrate the principles of the light passing through beamsplitters in optical fibre devices.

12.2 BEAMSPLITTERS

As explained in the prisms chapter, the beamsplitters are also common optical components. Beamsplitters are blocks of optical material with flat polished faces. Beamsplitters partially transmit and partially reflect an incident light beam, usually in unequal ratios. They can also be used to recombine two separate light beams into a single light beam. Beamsplitters are used where it is necessary to direct a beam of light entering from one side and exiting from the other sides of the beamsplitter. Light passing through beamsplitters is governed by the laws of light. Many types of beamsplitters are used to build optical devices. The most common types of beamsplitters split the incoming light beam into two components of a 50/50 beamsplitting ratio. Rectangular beamsplitters are used to manufacture polarization beamsplitter devices with operating wavelengths of 1480 and 1550 nm, which are then used in communications systems. Some beamsplitters are shown in the figure above.

The light beam output components can be arranged exiting the beamsplitters either parallel or perpendicular to the input light. Each output light component exits from one side of the beamsplitter. Depending on the characteristics of the input light and the desired output light components, a specific beamsplitter can be chosen to accomplish the design requirements.

12.3 BEAMSPLITTER TYPES

There are many types of beamsplitters that are used in different applications. Some are used with polarizing and non-polarizing light, or with visible and near infrared waves. In the following sections, some beamsplitters and their applications are presented in detail.

12.3.1 STANDARD CUBE BEAMSPLITTERS

Standard cube beamsplitters consist of matched pairs of right angle prisms cemented together along their hypotenuses, as shown in Figure 12.1. The hypotenuse of one prism has a partial reflective coating. A black dot on the bottom side of the prism is used to identify which prism has the partial reflectore. The incident beam must enter the prism containing the partial reflector first. This prism splits the ray incident at the normal into two orthogonal components on the surface of the prism hypotenuse. It splits the incident beam into a 50/50 beamsplitting ratio. One component is called transmission, while the other is called reflection. Transmission and reflection approach 50%, though the output is partially polarized.

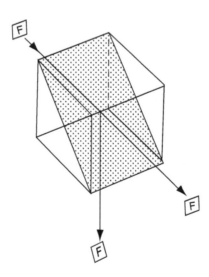

FIGURE 12.1 Standard cube beamsplitter.

The non-polarizing cube beamsplitters are constructed from a pair of precision, high-tolerance right angle prisms cemented together with a metallic-dielectric coating on the hypotenuse of one of the prisms. A broadband antireflective coating is applied to each face of the beamsplitter in order to produce maximum transmission efficiency.

Cube beamsplittes are available in two categories: polarizing and non-polarizing. If polarization sensitivity is critical in an application, then using the polarizing cube beamsplitters is recommended. Cube beamsplitters are available in three types: broadband hybrid, broadband dielectric, and laser-line non-polarizing.

12.3.2 POLARIZING CUBE BEAMSPLITTERS

Polarizing cube beamsplitters split randomly-polarized light into two orthogonal, linearly polarized components; S-polarized light is reflected at a 90° angle (perpendicular), while P-polarized light is transmitted (horizontal).

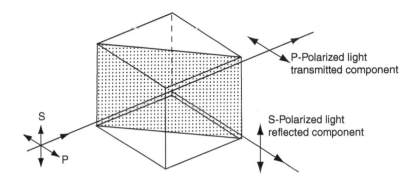

FIGURE 12.2 Polarizing cube beamsplitter.

Polarizing cube beamsplitters are constructed similarly to the standard cube beamsplitters, as shown in Figure 12.2. The beamsplitter consists of a pair of precision, high-tolerance right angle prisms cemented together with a dielectric coating on the hypotenuse of one of the prisms. A multi-layer antireflective coating is applied to each face of the beamsplitter in order to produce maximum transmission efficiency.

12.3.3 RECTANGULAR POLARIZING BEAMSPLITTERS

Rectangular polarizing beamsplitters perform the same function as cube beamsplitters. They consist of three prisms carefully cemented together along their hypotenuses, as shown in Figure 12.3. Rectangular polarization beamsplitters are used in optical devices where both of the output components are required to exit the side opposite from the input signal. This produces a lateral displacement between the two output components. Figure 12.3 shows a beamsplitter splitting the input signal into its two orthogonal 50/50 components. Rectangular beamsplitters are used in building polarizing beamsplitter devices for 1480 and 1550 nm wavelengths that are employed in communication systems.

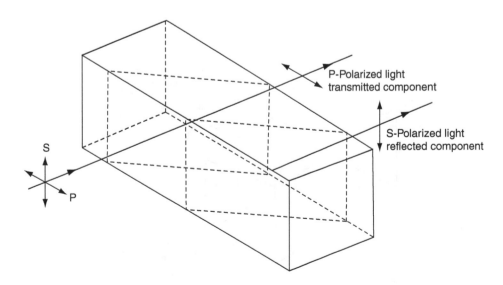

FIGURE 12.3 Rectangular polarizing beamsplitter.

12.3.4 LATERAL DISPLACEMENT POLARIZING BEAMSPLITTERS

As explained in rectangular beamsplitters, lateral displacement beamsplitters output two parallel beams separated by a distance that depends on the size chosen, as shown in Figure 12.4. These beamsplitters consist of a precision rhomboid prism cemented to a 1/8-wavelength right angle prism. The entrance and exit faces have an antireflective coating layer to increase efficiency. Polarizing and non-polarizing lateral displacement beamsplitters are available.

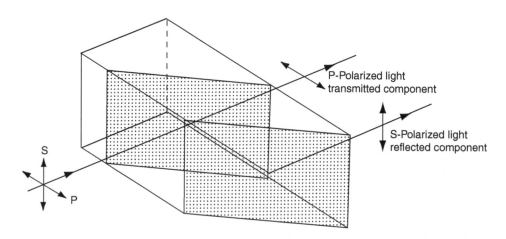

FIGURE 12.4 Lateral displacement polarizing beamsplitter.

12.3.5 GLAN THOMPSON POLARIZING BEAMSPLITTERS

The Glan Thompson polarizing beamsplitter is designed to permit the output of the S-polarized beam at 45° from the straight-through P-polarized beam, as shown in Figure 12.5. These are useful for utilizing both linear polarization states. They are commonly sold mounted in a rectangular metal cell.

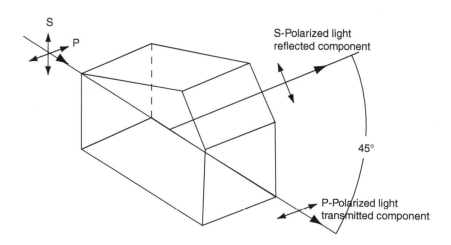

FIGURE 12.5 Glan Thompson polarizing beamsplitter.

12.3.6 POLKA-DOT BEAMSPLITTERS

Polka-dot beamsplitters offer a distinct advantage over standard dielectric beamsplitters; they have a constant 50/50 reflection-to-transmission ratio over a large spectral range. They are ideal for use with broadband, extended sources, such as tungsten, halogen, deuterium, and xenon lamps, and in monochromators, spectrophotometers, and other optical systems. The beamsplitter has an aluminum coating cell. The coated-to-uncoated surface areas resemble polka-dots. Input beams are split evenly: 50% of the incident light is reflected by the coating and 50% is transmitted through the clear glass, as shown in Figure 12.6. Since the polka-dot beamsplitters are not sensitive to the incident angle, wavelength, or polarization, they are ideal for splitting energy from a radiant light source.

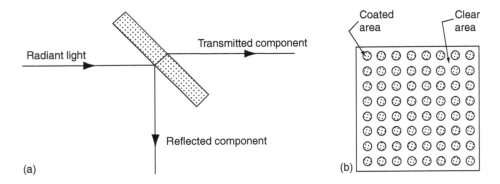

FIGURE 12.6 Polka-dot beamsplitter.

12.3.7 ELLIPTICAL PLATE BEAMSPLITTERS

The elliptical plate beamsplitter creates a circular aperture equal to the diameter of the minor axis, when oriented at 45°, as shown in Figure 12.7. They maximize beamsplitting efficiency while minimizing required mounting space. The elliptical plate beamsplitters are ideal for diffuse axial and in-line illumination. They operate in either visible light or near infrared light regions. They are a 50/50 beamsplitter coated with a high-efficiency multi-layer antireflection coating to reduce back reflections.

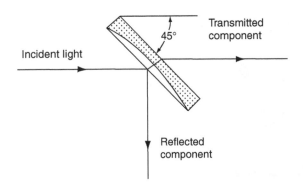

FIGURE 12.7 Elliptical plate beamsplitter.

12.3.8 MIRROR-TYPE BEAMSPLITTERS

A mirror-type (plate) beamsplitter is an optical window with a semi-transparent mirrored coating that breaks a light beam into two beams, as shown in Figure 12.8. This beamsplitter will reflect a portion of the incident light, absorb a relatively small portion, and transmit the remaining light.

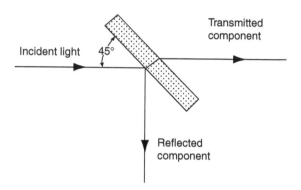

FIGURE 12.8 Mirror-type beamsplitter.

12.3.9 PELLICLE BEAMSPLITTERS

Pellicle beamsplitters are very thin, nitrocellulose membranes (pellicle) bonded to lapped aluminum frames. Figure 12.9 shows a comparison between a pellicle beamsplitter and a glass plate beamsplitter. Ghost images occur in the glass beamsplitter shown in Figure 12.9(b). In a pellicle beamsplitter, ghost images are eliminated by the thinness of the membrane, since the second surface's reflection superimposes on the first surface reflection. The uncoated pellicle reflects 8% and transmits 92% of light in the visible and near infrared regions. Pellicle beamsplitters can be coated with various thin films to change the reflectivity and transitivity, and thus the splitting ratio. Pellicles are usually mounted on a precision-lapped, hard-black, anodized aluminum alloy frame. Pellicle beamsplitters are not affected by mechanical shock or variations in temperature and humidity. Pellicle membranes are extremely delicate and can be easily punctured.

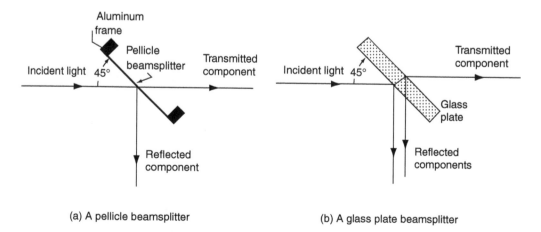

FIGURE 12.9 Comparison between a pellicle beamsplitter and a glass plate beamsplitter.

12.3.10 VISIBLE AND NEAR INFRARED REGION PLATE BEAMSPLITTERS

Plate beamsplitters are designed for a 45° angle of incidence and random polarization, as shown in Figure 12.10. Plate beamsplitters are available in 30% reflection/70% transmission, as well as the standard 50% reflection/50% transmission ratios, for both the visible (400–700 nm) and near infrared (700–1100 nm) wavelengths. These plate beamsplitters have the optimum combination of polished, high quality, optical-grade glass with minimum substrate thickness. The broadband dielectric coating can be chosen to cover spectral ranges to meet the application needs. Plate beamsplitters with low-absorption coatings allow maximum throughput to minimize ghost images and have much less light loss than metallic coatings. The back surfaces of these plate beamsplitters are multi-layer antireflection coated, which reduces back reflections to less than about 1% for each wavelength range. Typical applications include use in dual magnification imaging systems and in combining low-power laser beams.

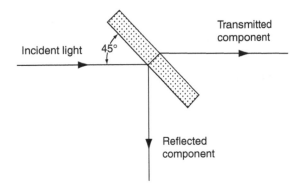

FIGURE 12.10 Plate beamsplitter.

12.3.11 QUARTZ BEAMSPLITTERS

Quartz beamsplitters are used to attenuate laser light. They attach directly to a power-meter probe and expose the probe to less than 10% of the laser light. The laser attenuators are ideal when pulsed laser light density levels approach damage limits of the power-meter probe coatings.

12.3.12 DICHROIC PLATE BEAMSPLLITERS

The dichroic manufacturing process gives the dichroic plate beamsplitters the steepest edges, and the flattest and the highest reflection and transmission bands. More complete transmission and reflection means that less stray light gets through the imaging system, yielding higher-contrast images and a better signal-to-noise ratio. Dichroic beamsplitters are critical components in fluorescence microscopy, genomics, proteomics, gel plate, and microplate readers, as well as in original equipment manufacturer instrumentation requiring beam separation, beam combination, or multispectral detection. These beamsplitters can also be used as highly efficient bandpass filters that cover several colours.

12.3.13 OTHER TYPES OF BEAMSPLITTERS

There are many other types of beamsplitters used in building optical devices and systems, such as low- and high-laser line polarizing cube beamsplitters, medium and high extinction broadband polarizing cube beamsplitters, circular variable metallic beamsplitters, variable beamsplitter/attenuators, and electronic variable beamsplitters. These beamsplitters are useful for producing polarized and non-polarized light. Other polarizing optical components, such as polarizing cubes and polarizing plate beamsplitters, and polarizing rectangular beamsplitters, will be explained in polarization of light in Chapter 15. Non-standard beamsplitter design can be made according to the customers' required specifications.

12.4 EXPERIMENTAL WORK

This experiment is designed to demonstrate the theory of light passing through beamsplitters. It also demonstrates light splitting into two components by a cube beamsplitter and a rectangular beamsplitter. In this experiment the student will perform the following cases:

a. Observe a laser beam passing through a cube beamsplitter and draw the laser beam components.
b. Observe a laser beam passing through a rectangular beamsplitter and draw the laser beam components.
c. Observe a laser beam passing through the Glan Thompson polarizing beamsplitter and draw the laser beam components.

12.4.1 Technique and Apparatus

Appendix A presents the details of the devices, components, tools, and parts.

1. 2×2 ft. optical breadboard
2. HeNe laser source and power supply
3. Laser light sensor
4. Laser light power meter
5. Laser mount assembly
6. Hardware assembly (clamps, posts, screw kits, screwdriver kits, sundry positioners, etc.)
7. Beamsplitter holder/positioner assembly
8. Cube beamsplitter, as shown in Figure 12.11
9. Rectangular beamsplitter, as shown in Figure 12.12
10. Glan Thompson polarizing beamsplitter, as shown in Figure 12.13

FIGURE 12.11 Cube beamsplitter.

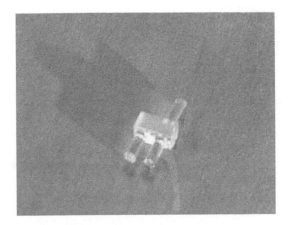

FIGURE 12.12 Rectangular beamsplitter.

FIGURE 12.13 Glan Thompson polarizing beamsplitter.

11. Black/white card and cardholder
12. Protractor
13. Ruler

12.4.2 PROCEDURE

Follow the laboratory procedures and instructions given by the professor and/or instructor.

12.4.3 SAFETY PROCEDURE

Follow all safety procedures and regulations regarding the use of optical components and instruments, light source devices, and optical cleaning chemicals.

12.4.4 APPARATUS SET-UP

12.4.4.1 Cube Beamsplitter

1. Figure 12.14 shows the apparatus set-up.
2. Bolt the laser short rod to the breadboard.
3. Bolt the laser mount to the clamp using bolts from the screw kit.
4. Put the clamp on the short rod.
5. Place the HeNe laser into the laser mount and tighten the screw. Turn on the laser device. Follow the operation and safety procedures of the laser device in use.
6. Check the laser alignment with the line of bolt holes and adjust when necessary.
7. Mount a beamsplitter holder/positioner to the breadboard.
8. Mount the cube beamsplitter on the beamsplitter holder/positioner, as shown in Figure 12.14.
9. Carefully align the laser beam so that the laser beam follows the direction of the normal to the centre of cube beamsplitter face.
10. Try to capture the laser beam exiting the cube beamsplitter on the black/white card, as shown in Figure 12.14.
11. Turn off the lights of the lab before taking measurements.
12. Measure the power of the laser beam at the input face and output face of the cube beamsplitter. Fill out Table 12.1 for Case (a).
13. Turn on the lights of the laboratory.

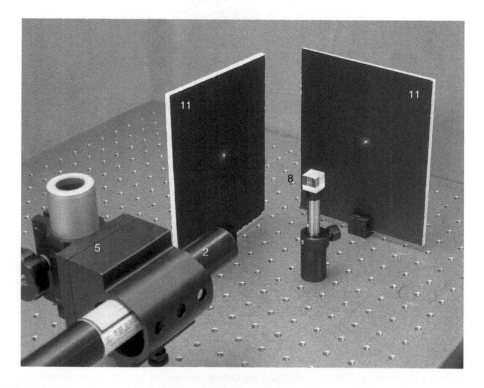

FIGURE 12.14 Laser beam passing through a cube beamsplitter.

12.4.4.2 Rectangular Beamsplitter

Figure 12.15 shows the apparatus set-up. Repeat the procedure from steps 1–7 in Case (a) using a rectangular beamsplitter; then add the following steps:

1. Mount a rectangular beamsplitter on the beamsplitter holder/positioner, as shown in Figure 12.15.
2. Carefully align the laser beam so that the laser beam impacts normally to the centre of the wide face of the rectangular beamsplitter.
3. Try to capture the laser beam exiting the rectangular beamsplitter on the black/white card, as shown in Figure 12.15. Two laser spots can be caught on the black/white card of the laser beam components exiting from the output face of the rectangular beamsplitter, as shown in Figure 12.16.
4. Turn off the lights of the laboratory before taking measurements.
5. Measure the laser beam power at the input face and output face of the rectangular beamsplitter. Fill out Table 12.1 for Case (b).
6. Turn on the lights of the laboratory.

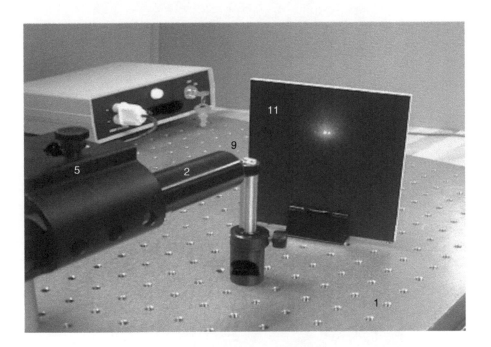

FIGURE 12.15 Laser beam passing through a rectangular beamsplitter.

12.4.4.3 Glan Thompson Polarizing Beamsplitter

Figure 12.17 shows the apparatus set-up. Repeat the procedure from step 1–7 in Case (a) using a Glen Thompson polarizing beamsplitter; then add the following steps:

1. Mount a Glan Thompson polarizing beamsplitter on the beamsplitter holder/positioner, as shown in Figure 12.17.
2. Carefully align the laser beam so that the laser beam impacts normally to the centre of the face Glan Thompson polarizing beamsplitter.

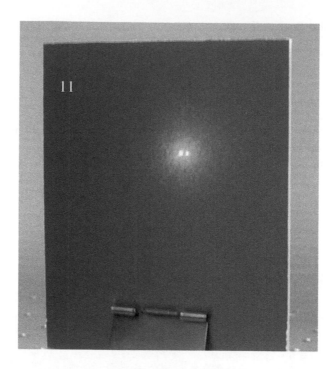

FIGURE 12.16 Laser beam outputs exiting the rectangular beamsplitter.

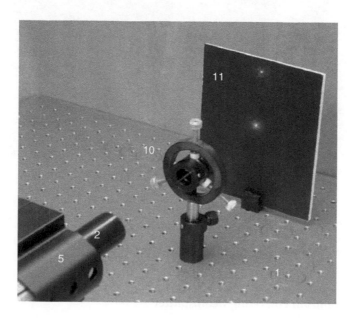

FIGURE 12.17 Laser beam passing through the Glan Thompson polarizing beamsplitter.

3. Try to capture the laser beam exiting from the face of the Glan Thompson polarizing beamsplitter on the black/white card, as shown in Figure 12.17. Two laser spots can be caught on the black/white card of the laser beam components exiting from the output faces of the Glan Thompson polarizing beamsplitter.

4. Turn off the lights of the laboratory before taking measurements.
5. Measure the laser power at the input face and output faces of the Glan Thompson polarizing beamsplitter. Fill out Table 12.1 for Case (c).
6. Turn on the lights of the laboratory.

12.4.5 DATA COLLECTION

12.4.5.1 Cube Beamsplitter

1. Measure the laser beam power at the input face and output faces of the cube beamsplitter.
2. Fill out Table 12.1 with the collected data for Case (a).

12.4.5.2 Rectangular Beamsplitter

1. Measure the laser beam power at the input face and output face of the rectangular beamsplitter.
2. Fill out Table 12.1 with the collected data for Case (b).

12.4.5.3 Glan Thompson Polarizing Beamsplitter

1. Measure the laser beam component power at the input face and output faces of the Glan Thompson polarizing beamsplitter.
2. Fill out Table 12.1 with the collected data for Case (c).

TABLE 12.1
Beamsplitters Data Collection

Case	Beamsplitter	Laser Power Input (unit)	P Component Power (unit)	S Component Power (unit)	Splitting Ratio
(a)	Polarizing cube beamsplitter				
(b)	Rectangular polarizing beamsplitter				
(c)	Glan Thompson Polarizing beamsplitter				

12.4.6 CALCULATIONS AND ANALYSIS

12.4.6.1 Cube Beamsplitter

1. Compare the laser beam power at the input face and output faces of the cube beamsplitter.
2. Calculate the splitting ratio of the cube beamsplitter. Fill out Table 12.1 with the collected data for Case (a).

12.4.6.2 Rectangular Beamsplitter

1. Compare the laser beam power at the input face and output face of the rectangular beamsplitter.
2. Calculate the laser beam splitting ratio of the rectangular beamsplitter. Fill out Table 12.1 with the collected data for Case (b).

12.4.6.3 Glan Thompson Polarizing Beamsplitter

1. Compare the laser beam power at the input face and output faces of the Glan Thompson polarizing beamsplitter.
2. Calculate the laser beam splitting ratio of the Glan Thompson polarizing beamsplitter. Fill out Table 12.1 with the collected data for Case (c).

12.4.7 RESULTS AND DISCUSSIONS

12.4.7.1 Cube Beamsplitter

1. Report the laser beam power measurements at the input face and output faces of the cube beamsplitter.
2. Verify the results with the technical specifications provided by the manufacturer.

12.4.7.2 Rectangular Beamsplitter

1. Report the laser beam power measurements at the input face and output face of the rectangular beamsplitter.
2. Verify the results with the technical specifications provided by the manufacturer.

12.4.7.3 Glan Thompson Polarizing Beamsplitter

1. Report the laser beam power measurements at the input face and output faces of the Glan Thompson polarizing beamsplitter.
2. Verify the results with the technical specifications provided by the manufacturer.

12.4.8 CONCLUSION

Summarize the important observations and findings obtained in this lab experiment.

12.4.9 SUGGESTIONS FOR FUTURE LAB WORK

List any suggestions for improvements using different experimental equipment, procedures, and techniques for any future lab work. These suggestions should be theoretically justified and technically feasible.

12.5 LIST OF REFERENCES

List any references that were used in the report. Use one format in writing the references. Never mix reference formats in a report.

12.6 APPENDICES

List all of the materials and information that are too detailed to be included in the body of the report.

FURTHER READING

Al-Azzawi, A. and Casey, R. P., *Fiber Optics Principles and Practices*, Algonquin Publishing Centre, Ontario, 2002.

Beiser, A., *Physics*, 5th ed., Addison-Wesley, Publishing Company, Inc., USA, 1991.

Blaker, J. W. and Schaeffer, P., *Optics: An Introduction for Technicians and Technologists*, Prentice Hall, Englewood, Cliffs, NJ, 2002.

Cox, A., *Photographic Optics*, 15th ed., Focal Press, Doncaster, UK, 1974.

Cutnell, J. D. and Johnson, K. W., *Physics*, 5th ed., John Wiley & Sons, New York, USA, 2001.

Cutnell, J. D. and Johnson, K. W., *Student Study Guide: Physics*, 5th ed., Wiley, New York, 2001.

Drisoll, W. G. and William, V., *Handbook of Optics*, McGraw Hill, New York, 1978.

Edmund Industrial Optics, *Optics and Optical Instruments Catalog 2004, Edmund Industrial*, New Jersey, USA, 2004.

Eppley Laboratory, Inc., *EPLAB Catalogue 2002*, Epplay Laboratory, Inc., Newport, Rhode Island, USA, 2002.

Ewald, W. P., Young, W. A., and Roberts, R. H., *Practical Optics*, Image Makers of Pittsford, USA, 1982.

Francon, M., Optical Interferometry,, Academic Press, New York, pp. 97–99, 1966.

Ghatak, A. K., *An Introduction to Modern Optics*, McGraw-Hill, New York, 1972.

Giancoli, D. C., *Physics*, 5th ed., Prentice Hall, Englewood, Cliffs, NJ, 1998.

Halliday, D., Resnick, R., and Walker, J., *Fundamentals of Physics*, 6th ed., Wiley, New York, 1997.

Heath, R. W., Macnaughton, R. R., and Martindale, D. G., *Fundamentals of Physics*, D.C. Heath Canada Ltd., Canada, 1979.

Hecht, E., *Optics*, 4th ed., Addison-Wesley Longman, Boston, MA, 2002.

Jenkins, F. W. and White, H. E., *Fundamentals of Optics*, McGraw Hill, New York, 1957.

Keuffel & Esser Co., *Physics*. Keuffel & Esser Audiovisual Educator—Approved Diazo Transparency Masters, Audiovisual Division, Keuffel & Esser Co., USA, 1989.

Lambda, Research Optics, Inc., *Catalog 2004*. California, USA, 2004.

Lehrman, R. L., *Physics-The Easy Way*, 3rd ed., Barron's Educational Series, Inc., USA, 1998.

Lerner, R. G. and Trigg, G. L., *Encyclopedia of Physics*, 2nd ed., VCH Publishers, Inc., New York, USA, 1991.

McDermott, L. C. et al., *Introduction to Physics*, Preliminary Edition, Prentice Hall, Englewood, Cliffs, NJ, 1988.

Melles Griot. The Practical application of light, *Melles Griot catalog Melles Griot, Inc., Rochester, NY*, 2001.

Newport Corporation, Optics and Mechanics. Section, the Newport Resources 1999/2000 Catalog, Newport Corporation, Irvine, CA, USA, 1999/2000.

Nolan, P. J., *Fundamentals of College Physics,* Wm. C. Brown Publishers, Inc., Dubugue IA, USA, 1993.

Ocean Optics, Inc., *Product Catalog 2003*. Ocean Optics, Inc., Florida, USA, 2003.

Pedrotti, F. L. and Pedrotti, L. S., *Introduction to Optics*, 2nd ed., Prentice Hall, Englewood, Cliffs, NJ, 1993.

Romine, G. S., *Applied Physics Concepts Into Practice*, Prentice Hall, Englewood, Cliffs, NJ, 2001.

Sears, F. W. et al., *University Physics—Part II*, 6th ed., Addison-Wesley, Wokingham, UK, 1998.

Smith, W. J., *Modern Optical Engineering*, McGraw-Hill, New York, 1966.

Serway, R. A., *Physics for Scientists and Engineers*, 3rd ed., Saunders College Publishing, London, 1990.

Sterling, D. J. Jr., *Technician's Guide to Fiber Optics*, 2nd ed., Delmar, Publishing Inc., Albany, New York, USA, 1993.

Tippens, P. E., *Physics*, 6th ed., Glencoe McGraw-Hill, Westerville, OH, USA, 2001, 1999.

Urone, P. P., *College Physics*, Brooks/Cole publishing Company, New York, 1998.

Walker, J. S., *Physics*, Prentice Hall, Upper Saddle River New Jersey, USA, 2002.

White, H. E., *Modern College Physics*, 6th ed., Van Nostrand Reinhold Company, New York, 1972.

Wilson, J. D., *Physics—A Practical and Conceptual Approach*, Saunders College Publishing, London, 1989.

Wilson, J. D. and Buffa, A. J., *College Physics*, 5th ed., Prentice Hall, Englewood, Cliffs, NJ, 2000.

13 Light Passing through Optical Components

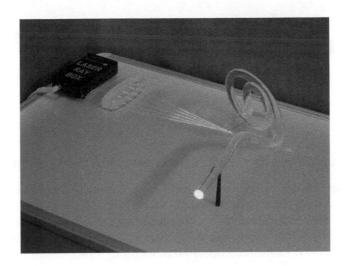

13.1 INTRODUCTION

This chapter demonstrates the behaviour of light passing through various optical components. Optical components can be lenses, mirrors, prisms, beamsplitters, glass rods and tubes, filters, polarizers, gratings, etc. These components are used in the manufacture of optical devices and systems for many applications, such as telecommunications, imaging, scanning, and microscopy. Theoretical principles of light propagation through optical components are explained in the previous chapters. This chapter presents a few cases involving different light sources passing through various optical components.

13.2 EXPERIMENTAL WORK

Students will practise aligning and observing the effects of several sources of light propagating through optical components in the following cases:

a. Laser, incandescent, or halogen light passing through various optical components from the laser optics kit
b. Laser light passing through various optical components from the ray optics laser set
c. Laser, incandescent, or halogen light passing through a glass rod and tube
d. Laser, incandescent, or halogen light passing through a spiral bar
e. Laser, incandescent, or halogen light passing through a fibre-optic cable bundle

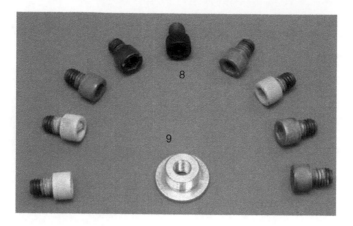

FIGURE 13.1 Laser optics component kit.

13.2.1 Light Passing through Optical Components from the Laser Optics Kit

This case shows that the light exiting each different optical component produces a different image. The image produced is also dependent on the light source. Three types of light sources can be used in this experiment: laser, incandescent light, and halogen light. A comparison can be made between the light sources and the light images produced from the optical components. Figure 13.1 shows the laser optics components from the laser optics kit as follows:

1. Dark-blue-coloured optical component creates a spider web effect.
2. Green-coloured optical component creates a star pattern.
3. Light-blue-coloured optical component produces bright and dark area of constructive and destructive interference of the light waves.
4. Orange-coloured optical component produces an offset Fresnel pattern.
5. Red-coloured optical component creates a diverging and dispersing laser beam pattern.
6. Violet-coloured optical component creates a laser scanner effect.
7. Yellow-coloured optical component projects a curved slice of light.
8. Gold-coloured optical component displays a swirling light show when rotated.
9. Black-coloured optical component polarizes the laser light beam.

13.2.2 Laser Light Passing through Optical Components from the Ray Optics Laser Set

The ray optics laser set contains a laser ray box, a variety of optical components, and templates to simulate actual optical devices. The optical components are lenses, reflecting surfaces, rods, and prisms. The ray optics laser set uses a laser ray box that projects five sharp parallel laser beams of 635 nm monochromatic light at 2 cm apart. The laser beams are easily visible in a normally lit room. In a moderately darkened room, it is easy to see secondary phenomena, such as front-surface and total internal reflections. The bottom surface of the laser ray box has a magnetized coating, so it can be mounted on the magnetic whiteboard when upright, as shown in Figure 13.7. The ray optics set includes thirteen large acrylic optical components with a magnetized coating on their back surfaces. The laser beams can pass through any optical components and make many combinations with two or more optical components. Additionally, the set includes six full-size template sheets, each 30×40 cm, printed on flexible white plastic with magnetized strips. These templates can be used to simulate a camera, the Galilean telescope, the Kepler telescope, the human eye and eyeglasses, spherical aberration of lenses and its correction, an angle scale for reflection and refraction, and any combination of optical components.

13.2.3 LIGHT PASSING THROUGH A GLASS ROD AND TUBE

Imagine a glass rod that is simply a core with a large diameter, as shown in Figure 13.3. Light enters the glass rod at an angle, as shown in Figure 13.8, that is perpendicular to the entrance face. This angle can be increased from perpendicular to its maximum value that is equal to the acceptance angle. Light will propagate in the glass rod by the principles of total internal reflection when the incidence angle of the light inside the rod is less than or equal to the critical angle. When the incident angle at the glass-rod face becomes greater than the acceptance angle, most of the light will reflect away. In this case, the light cannot enter the glass rod. Therefore, no light will propagate through the glass rod. Some optical materials absorb most of the light when a small amount of light enters. As a result, the optical material glows. This principle is applicable to the glass tube and it is even more evident when the glass rod and tube are joined together. The length of the optical material has a significant effect on the amount of light absorption. Similarly, when light propagating in a fibre-optic cable is subjected to many types of light losses. Equation 13.1 is the general loss equation that can be used to calculate light loss (in dB) in optical components. The common types of light losses in optical components, devices, and systems are represented as light attenuation in optical components and fibre-optic cables theory.

$$dB = -10 \log_{10} \frac{P_{out}}{P_{in}}, \tag{13.1}$$

where P_{out} and P_{in} are the power of light at the output and input, respectively.

13.2.4 LIGHT PASSING THROUGH A SPIRAL BAR

Similarly, as explained in Case (a), imagine a spiral bar that is simply a core with a large diameter, as shown in Figure 13.4. Light propagates from the entrance to the exit of the spiral bar by the principle of total internal reflection, as shown in Figure 13.9. When light passes through the spiral section of the bar, some of the light escapes to the outside when the incident angle is greater than the critical angle. Therefore, part of the light is lost to the outside of the bar. Light propagation in a spiral bar is similar to that in a fibre-optic cable with a large bend.

13.2.5 LIGHT PASSING THROUGH A FIBRE-OPTIC CABLE BUNDLE

The principles of this case are similar to Cases (c) and (d). Figure 13.10 shows the experimental set-up for this case.

13.2.6 TECHNIQUE AND APPARATUS

Appendix A presents the details of the devices, components, tools, and parts.

1. 2×2 ft. optical breadboard
2. HeNe laser source and power supply
3. Laser mount assembly
4. Laser light detector
5. Incandescent light source
6. Halogen light source
7. Light-source positioners

8. Laser optics component kit, as shown in Figure 13.1
9. Optical component mounting adapter, as shown in Figure 13.1
10. Optical component mount (holder/positioner assembly), as shown in Figure 13.2
11. Types of glass rods and tubes, as shown in Figure 13.3
12. Spiral bar, as shown in Figure 13.4
13. Ray optics laser set, as shown in Figure 13.7
14. Fibre-optic cable bundle, as shown in Figure 13.10
15. Hardware assembly (clamps, posts, screw kits, screwdriver kits, sundry positioners, etc.)
16. Black/white card and cardholder
17. Ruler

FIGURE 13.2 Optical component mount.

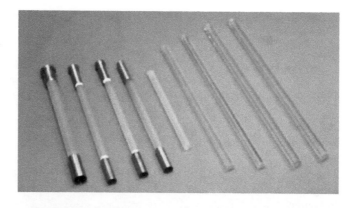

FIGURE 13.3 Types of glass rods and tubes.

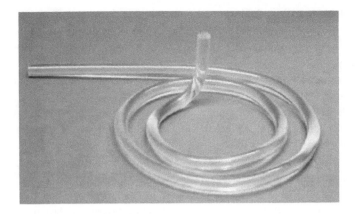

FIGURE 13.4 A spiral bar.

13.2.7 PROCEDURE

Follow the laboratory procedures and instructions given by the professor and/or instructor.

13.2.8 SAFETY PROCEDURE

Follow all safety procedures and regulations regarding the use of laser light, light source devices, optical components, and optical cleaning chemicals.

13.2.9 APPARATUS SET-UP

13.2.9.1 Light Passing through Optical Components from the Laser Optics Kit

1. Figure 13.5 shows the apparatus set-up.
2. Bolt the laser short rod to the breadboard.
3. Bolt the laser mount to the clamp using bolts from the screw kit.
4. Put the clamp on the short rod.

FIGURE 13.5 Laser light and coloured optical components apparatus set-up.

5. Place the HeNe laser into the laser mount and tighten the screw. Turn on the laser device. Follow the operation and safety procedures of the laser device in use.

6. Check the laser alignment with the line of bolt holes on the breadboard and adjust if necessary.

7. Place the optical component mount in front the laser source beam at a reasonable distance.

8. Align the laser source and the optical component mount so that the laser beam passes through the centre hole. Place the optical component adaptor into the optical component mount.

9. Inspect the dark-blue-coloured optical component under the microscope. Screw the component into the optical component mounting adapter.

10. Place the black/white card and cardholder at the output of the dark-blue-coloured component to display the image on the black/white board, as shown in Figure 13.6.

11. Turn off the laboratory light to be able to observe clear images.

12. Observe the output of the dark-blue-coloured component and report your observation in Table 13.1.

13. Unscrew the dark-blue-coloured optical component and replace it with the next coloured optical component. Report your observation in Table 13.1 for all colours.

14. Repeat steps 9–13 using incandescent light and halogen light sources with all coloured optical components. Report your observations in Table 13.1.

15. Turn on the lights of the laboratory.

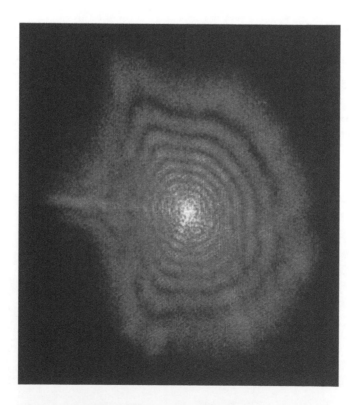

FIGURE 13.6 Blue-coloured component creates a spider-web effect.

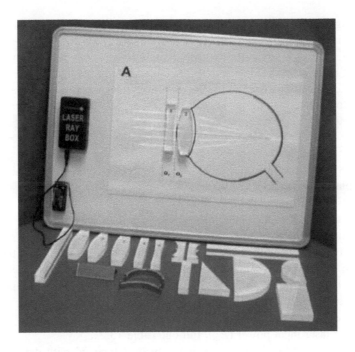

FIGURE 13.7 Ray optics laser set.

13.2.9.2 Laser Light Passing through Optical Components from the Ray Optics Laser Set

1. Figure 13.7 shows the apparatus set-up.
2. Place the whiteboard on a table using the built-in swiveling foot, which allows the whiteboard to stand vertically on the table.
3. Attach a paper to the whiteboard to trace the rays' paths.
4. Mount the laser ray box on the whiteboard. Connect the power cord of the laser ray box to the power supply.
5. Turn off the laboratory light to be able to observe clear laser rays reflecting, refracting, bending, and bouncing back and forth between the optical components.
6. Use the optical components and templates to simulate real optical devices.
7. Build three different optical devices, such as a camera, a Galilean telescope, a Kepler telescope, the human eye and eyeglass, and spherical aberration of lenses and its correction, using the setup template sheets.
8. Try to conduct three combinations between different optical components set-ups using two or more optical components.
9. Document your work in a laboratory report.

13.2.9.3 Light Passing through a Glass Rod and Tube

This experiment has three parts: measuring the light-beam output power after passing through a glass rod alone, a glass tube alone, and a glass rod and tube together. The output power of the light beam is measured when the light beam enters the glass rod and tube individually or together at different angles. Light-beam output power is measured by a light sensor that is located at a fixed distance from the end of the glass rod and tube. The output power of the light beam in these cases depends on the angle at which the glass rod, the glass tube, or the glass rod and tube are arranged. Most likely, there is an angle at which the output power is the greatest; at all other angles the output power will be less. If there is an angle at which the light beam is not passing through the glass rod and tube, the light beam is reflected back to the incoming side of the light beam. At a certain angle, the light beam is completely absorbed by the glass rod and tube. As the glass rod and tube are combined, the output power will decrease because the light is being absorbed more as the length of the glass rod and tube increases.

The following steps illustrate the experimental lab setup:

1. Figure 13.8 shows the apparatus setup using a laser light source. (Note: the glass rod and tube are placed together in this figure).
2. Bolt the laser short rod to the breadboard.
3. Bolt the laser mount to the clamp using bolts from the screw kit.
4. Put the clamp on the short rod.
5. Place the HeNe laser into the laser mount and tighten the screw. Turn on the laser device. Follow the operation and safety procedures of the laser device.
6. Check the laser alignment with the line of bolt holes and adjust if necessary.
7. Place the laser sensor in front of the laser source at a fixed distance for all cases.
8. Turn off the lights of the laboratory.
9. Measure the laser input power (P_{in}). Fill out Table 13.2.
10. Prepare a glass rod by cleaning it using a swab dampened with denatured ethanol. Follow the optical components' cleaning procedure.
11. Mount the glass rod into the slide holder/positioner.
12. Place the glass rod and glass-rod holder/positioner in front of the laser source at a fixed distance between the laser source and the laser sensor, as shown in Figure 13.8. Make certain that the glass rod is perpendicular to the laser beam. Keep the same distance for all cases.
13. Measure the laser output power (P_{out}) from the glass rod. Fill out Table 13.2.
14. Rotate the laser source 15 degrees from the normal. Rearrange the laser beam positioner to direct the laser beam at the centre of the glass rod. Measure the laser output power from the glass rod. Fill out Table 13.2.

FIGURE 13.8 Laser light loss in a glass rod and/or glass tube.

15. Repeat step 14 and rotate the laser source 15 degrees each time. Measure the laser output power. Fill out Table 13.2.
16. Repeat steps 10–15 to prepare a glass tube. Measure the laser output power at each angle. Fill out Table 13.3.
17. Repeat steps 10–15 to prepare a glass rod and glass tube together. Measure the laser output power at each angle. Fill out Table 13.4.
18. Repeat the above steps for the incandescent and halogen light sources. Table 13.2, Table 13.3, and Table 13.4.
19. Turn on the lights of the laboratory.

13.2.9.4 Light Passing through a Spiral Bar

The following steps illustrate the experimental lab setup:
1. Figure 13.9 shows the apparatus setup using an incandescent light source.
2. Get three types of light sources ready on the table.
3. Turn off the lights of the laboratory.
4. Turn on the incandescent source power. Measure the incandescent input power (P_{in}). Fill out Table 13.5.
5. Prepare a spiral bar by cleaning it using a swab dampened with denatured ethanol. Follow the optical components' cleaning procedure.
6. Mount the spiral bar into the bar holder/positioner.
7. Place the spiral bar and spiral bar holder/positioner in front of the incandescent light source at as close a distance as possible.
8. Place the light sensor at the output end of the spiral bar. Keep the distance between the end of the spiral bar and the light sensor the same for other light sources.
9. Measure the incandescent light output power (P_{out}) from the spiral bar. Fill out Table 13.5.
10. Repeat steps 4–9 using two types of light sources. Measure the light input and output powers. Fill out Table 13.5.
11. Turn on the lights of the laboratory.

FIGURE 13.9 Visible light passing through a spiral bar.

FIGURE 13.10 Visible light passing through a fibre-optic cable bundle.

13.2.9.5 Light Passing through a Fibre-Optic Cable Bundle

The following steps illustrate the experimental lab setup:

1. Figure 13.10 shows the apparatus setup using an incandescent light source.
2. Get three types of light sources ready on the table.
3. Turn off the lights of the laboratory.
4. Turn on the incandescent source power. Measure the incandescent input power (P_{in}). Fill out Table 13.6.
5. Prepare a fibre-optic cable bundle by cleaning its input and output ends using a swab dampened with denaturated ethanol. Follow the optical components' cleaning procedure.
6. Mount the fibre-optic cable bundle into the fibre-optic cable bundle holder/positioner.
7. Place the fibre-optic cable bundle and fibre-optic cable bundle holder/positioner in front of the incandescent light source at a close distance as possible.
8. Place the light sensor at the output end of the fibre-optic cable bundle. Keep the distance between the output end of the fibre-optic cable bundle and the light sensor the same for other light sources.
9. Measure the incandescent light output power (P_{out}) from the fibre-optic cable bundle. Fill out Table 13.6.
10. Repeat steps 4–9 using two types of light sources. Measure the light input and output powers. Fill out Table 13.6.
11. Turn on the lights of the laboratory.

13.2.10 Data Collection

13.2.10.1 Light Passing through Optical Components from the Laser Optics Kit

1. Observe image output of all coloured optical components using laser, incandescent, and halogen light sources. Report your observations in Table 13.1.

13.2.10.2 Laser Light Passing through Optical Components from the Ray Optics Laser Set

1. Draw the rays' paths on a paper for each case of the three combinations using two or more optical components.

TABLE 13.1
Image Comparison

	Colored Optical Components								
Light Source	**Dark Blue**	**Green**	**Light Blue**	**Orange**	**Red**	**Violet**	**Yellow**	**Gold**	**Black**
Laser Light									
Incandescent Light									
Halogen Light									

13.2.10.3 Light Passing through a Glass Rod and Tube

1. Record the input power (P_{in}) without a glass rod or tube in front of the laser light.
2. Record the output power (P_{out}) of the laser light exiting from the glass rod for every 15 degrees.
3. Repeat steps 1–2 for each of the incandescent and halogen light sources.
4. Fill out Table 13.2 for part one of Case (c).
5. Repeat the steps 1–5 for a glass rod. Fill out Table 13.3 for part two of Case (c).
6. Repeat the steps 1–5 for a glass rod and tube together. Fill out Table 13.4 for part three of Case (c).

TABLE 13.2
Light Passing through a Glass Rod and Tube

Laser Light Power Input (unit)	P_{in}						
Incandescent Light Power Input (unit)	P_{in}						
Halogen Light Power Input (unit)	P_{in}						

	Laser Light		Incandescent Light		Halogen Light	
Angle of Glass Microscope Slide with the Normal (degrees)	**Power Output P_{out} (unit)**	**Loss (dB)**	**Power Output P_{out} (unit)**	**Loss (dB)**	**Power Output P_{out} (unit)**	**Loss (dB)**
0						
15						
30						
45						
60						
75						
90						

TABLE 13.3
Light Passing through a Glass Tube

Laser Light Power Input P_{in} (unit)	
Incandescent Light Power Input P_{in} (unit)	
Halogen Light Power Input P_{in} (unit)	

Angle of Glass Microscope Slide with the Normal (degrees)	Laser Light		Incandescent Light		Halogen Light	
	Power Output P_{out} (unit)	Loss (dB)	Power Output P_{out} (unit)	Loss (dB)	Power Output P_{out} (unit)	Loss (dB)
0						
15						
30						
45						
60						
75						
90						

TABLE 13.4
Light Passing through the Glass Rod and Tube Together

Laser Light Power Input P_{in} (unit)	
Incandescent Light Power Input P_{in} (unit)	
Halogen Light Power Input P_{in} (unit)	

Angle of Glass Microscope Slide with the Normal (degrees)	Laser Light		Incandescent Light		Halogen Light	
	Power Output P_{out} (unit)	Loss (dB)	Power Output P_{out} (unit)	Loss (dB)	Power Output P_{out} (unit)	Loss (dB)
0						
15						
30						
45						
60						
75						
90						

13.2.10.4 Light Passing through a Spiral Bar

1. Record the input power (P_{in}) without a spiral bar in front of the incandescent light source.
2. Record the output power (P_{out}) of the incandescent light exiting from the spiral bar.
3. Repeat steps 1–2 for each of the laser and halogen light sources.
4. Fill out Table 13.5.

TABLE 13.5
Light Passing through a Spiral Bar

Laser Light Power Input (unit)		P_{in}			
Incadescent Light Power Input (unit)		P_{in}			
Halogen Light Power Input (unit)		P_{in}			
Laser Light		Incandescent Light		Halogen Light	
Power Output P_{out} (unit)	Loss (dB)	Power Output P_{out} (unit)	Loss (dB)	Power Output P_{out} (unit)	Loss (dB)

13.2.10.5 Light Passing through a Fibre-Optic Cable Bundle

1. Record the input power (P_{in}) without a fibre-optic cable bundle in front of the incandescent light source.

TABLE 13.6
Light Passing through a Fibre-Optic Cable Bundle

Laser Light Power Input (unit)		P_{in}			
Incandescent Light Power Input (unit)		P_{in}			
Halogen Light Power Input (unit)		P_{in}			
Laser Light		Incandescent Light		Halogen Light	
Power Output P_{out} (unit)	Loss (dB)	Power Output P_{out} (unit)	Loss (dB)	Power Output P_{out} (unit)	Loss (dB)

2. Record the output power (P_{out}) of the incandescent light exiting from the fibre-optic cable bundle.
3. Repeat steps 1–2 for each of the laser and halogen light sources.
4. Fill out Table 13.6.

13.2.11 CALCULATIONS AND ANALYSIS

13.2.11.1 Light Passing through Optical Components from the Laser Optics Kit

No calculations and analysis are required in this case.

13.2.11.2 Laser Light Passing through Optical Components from the Ray Optics Laser Set

No calculations and analysis are required in this case.

13.2.11.3 Light Passing through a Glass Rod and Tube

1. Calculate the power loss (dB) of the laser, incandescent, and halogen light sources when they are passing through a glass rod, a tube, or both, using Equation 13.1. Fill out Table 13.2, Table 13.3, and Table 13.4.
2. Compare power loss (dB) of the three types of light sources among the three parts involved in this case.

13.2.11.4 Light Passing through a Spiral Bar

1. Calculate the power loss (dB) of the laser, incandescent, and halogen light sources when they are passing through a spiral bar using Equation 13.1. Fill out Table 13.5.
2. Compare power loss (dB) of the three types of light sources involved in this case.

13.2.11.5 Light Passing through a Fibre-Optic Cable Bundle

1. Calculate the power loss (dB) of the laser, incandescent, and halogen light sources when they are passing through a fibre-optic cable bundle using Equation 13.1. Fill out Table 13.6.
2. Compare power loss (dB) of the three types of light sources involved in this case.

13.2.12 RESULTS AND DISCUSSIONS

13.2.12.1 Light Passing through Optical Components from the Laser Optics Kit

1. Discuss your observations when using the three types of light sources with all the coloured optical components.
2. Discuss the outputs of each coloured optical component when using different light sources.

13.2.12.2 Laser Light Passing through Optical Components from the Ray Optics Laser Set

1. Discuss your observations for each of the three combinations using two or more optical components.
2. Describe the reflections, refractions, and intersections of the rays that occurred through the optical components.

3. Discuss the combinations of the optical components that were used in the laboratory work.

13.2.12.3 Light Passing through a Glass Rod and Tube

1. Report the power loss (dB) of the light sources.
2. Report the incident angle of the light where no power output is recorded.
3. Compare and discuss the power loss in each component.

13.2.12.4 Light Passing through a Spiral Bar

1. Report the power loss (dB) of the light sources.
2. Compare and discuss the power loss for each light source.

13.2.12.5 Light Passing through a Fibre-Optic Cable Bundle

1. Report the power loss (dB) of the light sources.
2. Compare and discuss the power loss for each light source.

13.2.13 Conclusion

Summarize the important observations and findings obtained in this laboratory experiment.

13.2.14 Suggestions for Future Lab Work

List any suggestions for improvements using different experimental equipment, procedures, and techniques for any future laboratory work. These suggestions should be theoretically justified and technically feasible.

13.3 LIST OF REFERENCES

List any references that were used in the report. Use one format in writing the references. Never mix reference formats in a report.

13.4 APPENDIX

List all of the materials and information that are too detailed to be included in the body of the report.

FURTHER READING

Beiser, A., *Physics*, 5th ed., Addison-Wesley, Reading, MA, 1991.

Blaker, W. J., *Optics: The Matrix Theory*, Marcel Dekker, New York, 1972.

Blaker, J. W. and Schaeffer, P., *Optics an Introduction for Technicians and Technologists*, Prentice-Hall, Upper Saddle River, NJ, 2000.

Chen, K. P., In-fiber light powers active fiber optical components, *Photonics Spectra*, April, 78–90, 2005.

Cutnell, J. D. and Johnson, K. W., *Physics*, 5th ed., Wiley, New York, 2001.

Dutton, H. J. R., *Understanding Optical Communications*, IBM/Prentice-Hall, Englewood Cliffs, NJ/Research Triangle Park, NC, 1998.

Edmund Industrial Optics, *Optics and Optical Instruments Catalog*, 2004.

EPLAB, *EPLAB Catalogue 2002*, The Eppley Laboratory, Inc., 12 Sheffield Avenue, P.O. Box 419, Newport, RI 02840, USA, 2002.

Ghatak, A. K., *An Introduction to Modern Optics*, McGraw-Hill, New York, 1972.

Halliday, D., Resnick, R., and Walker, J., *Fundamental of Physics*, 6th ed., Wiley, New York, 1997.

Hecht, J., *Understanding Fiber Optics*, 3rd ed., Prentice-Hall, Upper Saddle River, NJ, 1999.

    11. Hecht, E., *Optics*, 4th ed., Addison-Wesley/Longman, Reading, MA, 2002.

Hewitt, P. G., *Conceptual Physics*, 8th ed., Addison-Wesley, Reading, MA, 1998.

Jenkins, F. W. and White, H. E., *Fundamentals of Optics*, McGraw-Hill, New York, 1957.

Jones, E. and Childers, R., *Contemporary College Physics*, McGraw-Hill Higher Education, Boston, MA, 2001.

Kao, C. K., *Optical Fiber Systems: Technology, Design and Applications*, McGraw-Hill, New York, 1982.

Kolimbiris, H., *Fiber Optics Communications*, Prentice-Hall, Upper Saddle River, NJ, 2004.

Naess, R. O., *Optics for Technology Students*, Prentice-Hall, Upper Saddle River, NJ, 2001.

Newport Corporation, *Projects in Fiber Optics Applications Handbook*, Newport Corporation, 1986.

Newport Corporation, *Optics and Mechanics Section, the* Newport Resources 1999/2000 Catalog, Newport Corporation Irvine, CA, 1999/2000.

Nolan, P. J., *Fundamentals of College Physics*, Wm. C. Brown Publishers, Inc., Dubuque, IA, 1993.

Okamoto, K., *Fundamentals of Optical Waveguides*, Academic Press, San Diego, 2000.

Pedrotti, F. L. and Pedrotti, L. S., *Introduction to Optics*, 2nd ed., Prentice-Hall, Englewood Cliffs, NJ, 1993.

Pritchard, D. C., *Environmental Physics: Lighting*, Longmans, Green, London, 1969.

Salah, B. E. A. and Teich, M. C., *Fundamentals of Photonics*, Wiley, New York, 1991.

Serway, R. A. and Jewett, J. W., *Physics for Scientists and Engineers with Modern Physics*, 6th ed., volume 2, Thomson Books/Cole, USA, 2004.

Shamir, J., *Optical Systems and Processes*, SPIE Optical Engineering Press, Bellingham, WA, 1999.

Sterling, D. J. Jr., *Technician's Guide to Fiber Optics*, 2nd ed., Delmar Publishers, Albany, NY, 1993.

Tippens, P. E., *Physics*, 6th ed., Glencoe/McGraw-Hill, Westerville, OH, 2001.

White, H. E., *Modern College Physics*, 6th ed., Van Nostrand Reinhold Company, New York, 1972.

Woods, N., *Instruction's Manual to Beiser Physics*, 5th ed., Addison-Wesley, Reading, MA, 1991.

Yeh, C., *Handbook of Fiber Optics: Theory and Applications*, Academic Press, San Diego, 1990.

14 Optical Instruments for Viewing Applications

14.1 INTRODUCTION

The previous chapters presented the principles of image formation by optical components, such as mirrors and lenses. This chapter presents the use of these optical components in building common optical instruments. Such optical instruments include the eye, camera, projector, microscope, telescope, and binoculars, which are used for different vision enhancement applications. These optical instruments help people to perform ordinary and sophisticated tasks associated with vision, magnification, and image formation. The ray diagram method is used in image formation in these devices. Image formation by the mirrors, lenses, lenses combinations, and prisms is discussed in detail in the previous chapters. There are many optical devices and instruments using basic optical components that are covered in other chapters.

This chapter includes an experiment involving the operation, image formation, functionality, disassembly, and reassembly of the optical components of an optical viewing instrument.

14.2 OPTICAL INSTRUMENTS

There is a wide variety of optical devices, instruments, and systems that use mirrors and lenses, for example: cameras, microscopes, projectors, and telescopes. There are also optical instruments that use prisms, gratings, and fibre optics, along with one type of lens or mirror. These instruments will be described later in the optics and optical fibre sections of this book. Such optical components are commonly used in manufacturing optical fibre devices, which are used in building communication systems, medical instruments, scanning and imaging processors, optical spectrum analysers, etc.

14.3 THE CAMERA

Cameras are among the most common of optical instruments. The camera shown in Figure 14.1 is a simple optical instrument. Figure 14.2 illustrates a cross-sectional view of a simple camera. The basic elements of a camera are: a light-tight box, a converging lens, an aperture, a shutter, and a light sensitive film. The converging lens produces a real and an inverted image of the object being photographed. The shutter allows the light from the lens to strike the film for a prescribed length of time. The aperture controls the diameter of the cone of light from the lens onto the film. The light-sensitive film records the image, which is formed by the lens.

Focusing of the image is accomplished by varying the distance between lens and film using a mechanical arrangement in most cameras. When the camera is in proper focus, the position of the film coincides with the position of the real image formed by the lens, as shown in Figure 14.2. The resulting photograph then will be as sharp as possible. When using a converging lens, the image distance increases as the object distance decreases. Hence, to focus the camera, the lens is moved closer to the film for a distant object and farther from the film for a nearby object. This is achieved by rotating the threaded ring, which holds the lens. This rotation controls the linear movement of the lens. The shutter, located behind the lens, is a mechanical device that is opened for selected time intervals. With this arrangement, one can photograph moving objects by using short exposure times, or dark scenes by using long exposure times.

FIGURE 14.1 Camera.

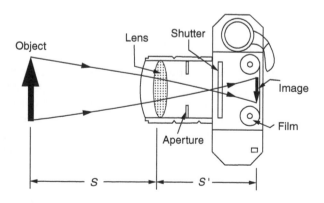

FIGURE 14.2 Cross-sectional view of a simple camera.

The major cause of blurred images is the movement of the camera or the object while the shutter is open. To avoid blurred images, a short time exposure and/or a tripod can be used, even for stationary objects. Most expensive cameras have compound lenses to correct the aberration that normally occurs in lenses. Most cameras have an aperture of adjustable diameter (either behind or in between the lenses) to provide further control of the intensity of the light reaching the film. When an aperture of small diameter is used, only light from the central portion of the lens reaches the film and so the aberration is reduced to a minimum.

The choice of the focal length f for a camera lens depends on the film size and desired angle of view. A lens of long focal length captures a small angle of view and produces a large image of a distant object; such a lens is called a telephoto lens. A lens of short focal length captures a small image and produces a wide angle of view; such a lens is called a wide-angle lens. Again, the focal length of a lens is defined as the distance from the lens to the image, when the object is infinitely far way from the lens. In general, for any distant object, using a lens of longer focal length produces a small image distance. This also increases the height of the image. In other words, the lens produces a magnified image of the object. As was discussed in the chapters on spherical mirrors and lenses the lateral magnification is defined as the ratio of the image height h' to the object height h. The lateral magnification is also equal to the ratio of the image distance S to the object distance S', as given in the following equation:

$$M = \frac{h'}{h} = -\frac{S'}{S} \tag{14.1}$$

When using a lens of short focal length, the lateral magnification is small, and a distant object produces only a small image. When a lens with a long focal length is used, the image of this same object may entirely cover the area of the film.

The intensity of light reaching the film is proportional to the area viewed by the camera lens and the effective area of the lens. The light intensity I will be proportional to the area of the lens. The area of the lens is proportional to the square of the angle of view of the lens so it also is nearly proportional to $1/f^2$. The effective area of the lens is controlled by means of an adjustable lens aperture, or diaphragm, a nearly circular hole with variable diameter D. The effective area is proportional to D^2. The intensity of light reaching the film is proportional to f^2/D^2. Thus, it follows that the light intensity is also proportional to $1/f^2$. The ratio f/D is defined as the f-number of the lens and given by the following equation:

$$f\text{-number} = \frac{\text{Focal Length}}{\text{Aperture Diameter}} = \frac{f}{D} \tag{14.2}$$

For a lens with a variable-diameter aperture, increasing the diameter by a factor of $\sqrt{2}$ changes the f-number by $1/\sqrt{2}$ and increases the intensity at the film by a factor of 2. Adjustable apertures usually have scales labeled with successive numbers related by factors of $\sqrt{2}$, such as: $f/2, f/2.8, f/4, f/5.6, f/8, f/11$, and $f/16$.

A useful consequence of the f-number method of classifying lens openings is that for the same shutter speed, lenses of different focal lengths give proper exposure at the same f-numbers. This applies independent of the focal length for the lens in place. This result is due to two compensating factors. First, the amount of the light that passes through the aperture is proportional to its area and thus to the square of its diameter, D^2. Second, the light per unit area that reaches the film depends inversely on the area of the image.

For the usual situation in which the object distance is large compared with the focal length, the magnification is proportional to the focal length f of the lens, so the area of the image is proportional to f^2. The rate at which a photographic image is formed, or the speed of the lens, is then given by the following equation:

$$\text{Speed of Lens} \propto \frac{1}{(f/D)^2} \propto \frac{1}{(f\text{-number})^2} \qquad (14.3)$$

The f-number is a measure of the light intensity and the speed of the lens. A fast lens has a small f-number. Fast lenses, with an f-number as low as about 1.4, are more expensive because it is more difficult to keep aberrations acceptably small. Cameras lenses often are marked with various f-numbers, such as: $f/2$, $f/2.8$, $f/4$, $f/5.6$, $f/8$, $f/11$, and $f/16$. The various f-numbers are obtained by adjusting the aperture, which effectively changes D. The smallest f-number corresponds to the case where the aperture is wide open and the full lens area is in use.

14.4 THE EYE

Vision is perhaps the most important sense of living beings. The eye of living creatures is one of the most familiar optical instruments. It is also one of the most complex parts of a living creature. The optical behaviour of the eye is similar to the optical principle of a simple camera. The essential external and internal parts of the human eye are shown in Figure 14.3 and Figure 14.4, respectively. The eye is nearly spherical in shape, and its diameter is typically less than 3 cm. There are a number of similarities between a human eyeball and a simple camera. The eye is basically a light-tight box, whose is outer walls are formed by the hard white sclera. There is a two-element lens system, consisting of the outer cornea and the inner crystalline lens or eyelens. The eyelens is acting as a converging lens. The cornea is a curved transparent tissue. The retina is a light sensitive lining, which corresponds to the film in the camera. The lens system forms an inverted image on the retina at the back of the eyeball, as shown in Figure 14.4. The coloured iris corresponds to the diaphragm in a camera. Its pupil, the dark and circular hole through which the light enters, corresponds to the camera aperture. The iris can change the pupil diameter, which controls the amount of light entering the eye. The eyelid is like a lens cover, protecting the cornea from dirt and objects. The eye or tear fluid cleans and moistens the cornea and can be compared to a lens cloth or brush. The volume of the eye is not empty, like the eye of the camera, but instead is filled with two transparent jellylike liquids, the vitreous humour and aqueous humour, which provide nourishment to the eyelens and

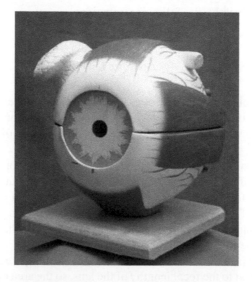

FIGURE 14.3 External parts of the human eye.

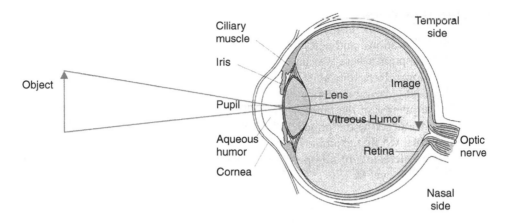

FIGURE 14.4 Cross section of the human eye.

cornea. Additionally, their internal pressure helps to hold the shape of the eyeball, which is necessary for stable vision. The lack of blood vessels in the cornea and eyelens means that, when surgically transplanted, these elements are less likely to be rejected by the immune system, which resides in the blood. Nowadays, corneal transplants are performed routinely.

The eye focuses on an object by varying the shape of the flexible crystalline lens through an amazing process called accommodation. An important component in accommodation is the ciliary muscle, which is attached to the lens. When the eye is focused on distant objects, the ciliary muscle is relaxed. For an object distance of infinity, the focal length of the eye (the distance between the lens and the retina) is about 1.7 cm. The eye focuses on nearby objects by tensing the ciliary muscle. This action effectively decreases the focal length of the eyelens by slightly decreasing the radius of the curvature of the lens, which allows the image to be focused on the retina. This lens adjustment takes place inside the eye automatically.

The retina of the eye is composed of two types of photosensitive cells, called rods and cones numbering about a hundred million. These cells are packed together like the chambers of a honeycomb. Near the centre of the retina, in a small region called the fovea, there are a great number of cones but no rods. This region is responsible for most precise vision and also for colour vision. Away from the fovea, the rods predominate. At the extremes of the retina, corresponding to peripheral vision, there are only a few cones. The more numerous rods have a greater sensitivity to light and can distinguish between low light intensities for twilight (black and white) vision. The cones respond selectively to certain colours of light, some to one colour and others to other colours. Cones are considerably less sensitive to light than are rods. This is why humans cannot see colour in very dim light. The rods and cones of the retina are connected to optic nerve fibres, from which the light stimulates signals to the brain.

In the region where the optic nerve enters the eyeball, there are no rods and cones. As a result, the human eye has a blind spot for which there is no optical response. Binocular (two-eyed) vision compensates for the blind spots. Binocular vision also accounts for some depth perception, which depends on slight differences between the shapes and positions of the images on the retina of the two eyes. This occurs because the eyes are set several centimetres apart and get slightly different views of objects.

The extremes of the range over which eye vision is possible are known as the far point and near point. The far point is the farthest distance that the normal eye can see clearly. The far point is assumed to be infinity. The near point is the position closest to the eye at which objects can be seen clearly. The near point depends on the extent that the eyelens can be deformed by accommodation. The range of accommodation decreases with age as the eyelens loses its elasticity.

260 Light and Optics: Principles and Practices

14.4.1 Defects of Vision

Figure 14.5(a) shows the image formation in a normal eye. As mentioned above, the focusing of the image is controlled by the ciliary muscle, which is attached to the lens. When the eye is focused on distant objects, the ciliary muscle is relaxed. The eye focuses on nearby objects by tensing the ciliary muscle.

Three common defects of vision are myopia (nearsightedness), hyperopia (farsightedness), and astigmatism. In myopia, the eyeball is too long, or the cornea is too curved. Light from a very distant object comes to a focus in front of the retina, as shown in Figure 14.5(b). This occurs even though the ciliary muscle is completely tensed. Accommodation permits nearby objects to be seen clearly but not more distant ones. A diverging lens of the proper focal length can correct this condition, as shown in Figure 14.5(b). In hyperopia, the eyeball is too short or the cornea has insufficient curvature. Light from a very distant object does not come to a focus within the eyeball, even when the ciliary muscle is relaxed. Its power of accommodation permits a hyperopic eye to focus on distance objects, but the range of accommodation is not enough for nearby objects to be seen clearly. In this condition, light from a distant object is focused behind the retina. The correction for hyperopia is a converging lens, as shown in Figure 14.5(c).

Another common eye defect, astigmatism, occurs when the cornea is not spherical but is more curved in one plane then in another. As a result, the focal length of the astigmatism eye is different in one plane than in the perpendicular plane. When light rays that lie in one plane are in focus on the

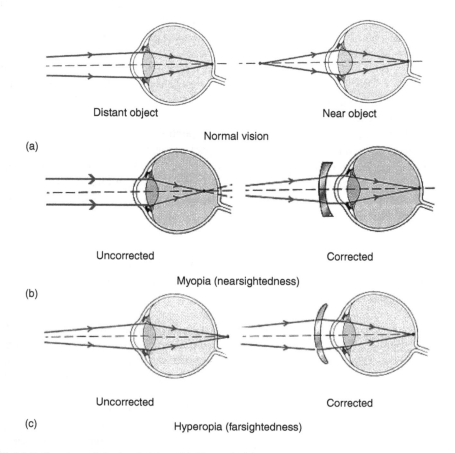

FIGURE 14.5 Common defects of vision. (a) Normal vision, (b) Myopia (nearsightedness), (c) Hyperopia (farsightedness).

retina of an astigmatism eye, those in the perpendicular plane will be in focus either in front or in back of the retina. Astigmatism is corrected by a lens that converges (or diverges) rays in one plane, while not effecting rays in the perpendicular plane. Such a lens is a cylindrical lens, which is curved in one direction but not in the perpendicular direction, as if cut out of a cylinder of glass. More modern techniques, such as laser surgery, are being used to correct common vision defects.

The eye is also subject to several diseases. One disease, which usually occurs in old age, is the formation of cataracts occurring when the lens becomes partially or totally opaque. The common remedy for cataracts is surgery performed on the lens. Another disease, called glaucoma, arises from an abnormal increase in the fluid pressure inside the eyeball. The pressure increase can lead to a swelling of the lens and also to strong myopia. A chronic form of glaucoma can lead to blindness. If the disease is discovered at an early stage, it can be treated with medicine or surgery.

14.4.2 Colour Vision

Colour vision is actually a physiological sensation of the brain in response to the light excitation of the cones reception in the retina. The cones in human eye are sensitive to light with wavelengths between about 400–700 nm (7.5×10^{14}–4.3×10^{14} Hz frequencies). Different wavelengths of light are perceived by the brain as having different colours. The association of colours with particular light wavelengths is subjective. Monochromatic light has a particular wavelength, but perception of its colour may vary from one person to another.

14.5 THE MAGNIFYING GLASS

Many optical devices are used to produce magnified images of objects. The simplest optical magnifying device is the magnifying lens, or simple microscope with an objective lens, as shown in Figure 14.6. The magnifying glass is a single converging lens. The lens held near the eye produces an image whose projected size on the retina is larger than the real size of the object (when compared to unaided eye). Maximum magnification of the object can be found by adjusting the distance between the lens and object. Varying this distance, one can find a clear image. The magnifying glass is considered a passive optical device. The magnifying glass is not only used to enlarge a fine object; it is also used in lens compensation, which is used in a compound microscope. In this arrangement, the magnifying glass is called the eyepiece, and the other lens near to an object is called the objective lens. The compound microscope instrument will be explained in detail in the next section. The primary function of magnifying glasses is to increase the angular size of the image while viewing with a relaxed eye.

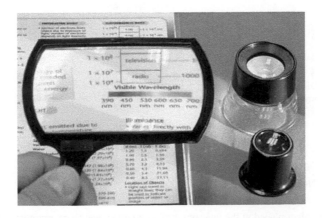

FIGURE 14.6 Simple magnifying lenses.

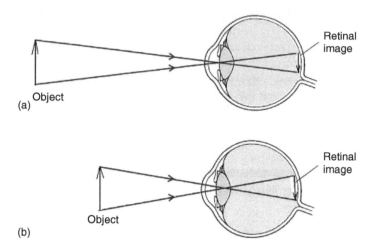

FIGURE 14.7 (a) The object placed a distance from the eye. (b) The object placed closer to the eye.

The apparent size of an object is determined by the size of its image on the retina. To an unaided eye, an object placed a distance from the eye creates an image covering part of the retina, as shown in Figure 14.7(a). If the object comes closer to the eye, its image covers a larger part of the retina, as shown in Figure 14.7(b). The greater the angle viewed, the larger the image appears.

In Figure 14.8(a) the object is located at the near point, where it subtends an angle θ at the eye. In Figure 14.8(b), a converging lens is used to form a virtual image that is larger and farther from

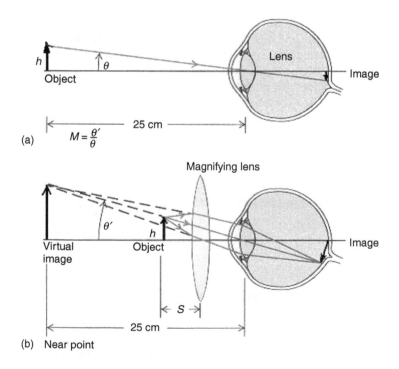

FIGURE 14.8 (a) An object placed at the near point of an unaided eye. (b) A magnifying glass placed close to the eye.

the eye than the object. The image formation by converging lenses is widely explained in the lenses chapter. A lens used to enlarge an object is called a simple magnifier, or sometimes is called a magnifying glass. In Figure 14.8(b) a magnifier in front of the eye forms an image at the near point, with the angle θ' subtended at the magnifier. The magnification power of the magnifier is defined by the ratio of the angle θ' (viewed with the magnifier) to the angle θ (viewed without the magnifier). This ratio is called the angular magnification M_a, as defined by:

$$M_a = \frac{\theta'}{\theta} \tag{14.4}$$

Notice here the difference between the definition of angular magnification M_a and the lateral magnification M. The angular magnification M_a is defined in Equation (14.4). The lateral magnification M is defined as the ratio of the height of an image to the height of the corresponding object. Lenses chapter presents a detailed definition of lateral magnification.

To find the value of the angular magnification, consider angles small enough that each angle measured in radians is equal to its sine and its tangent values. Figure 14.8(a) shows an object of height h placed at the standard near the point of 25 cm from the eye, where it subtends at an angle of θ. In Figure 14.8(b) the object is moved close to the eye.

If the angle θ is small, it can be defined as:

$$\tan \theta \cong \theta = \frac{h}{25} \tag{14.5}$$

where both the near point and the object height h are expressed in centimetres.

Similarly, Figure 14.8(b) shows the angle θ' subtended by the virtual image is approximately (ignoring the small distance between the magnifier lens and the eye) as:

$$\tan \theta' \cong \theta' = \frac{h}{S} \tag{14.6}$$

To calculate the object distance S, using the thin lens equation with focal length f and virtual image at a distance of -25 cm, obtains S by:

$$\frac{1}{S} + \frac{1}{-25} = \frac{1}{f} \tag{14.7}$$

$$S = \frac{25f}{(25 + f)} \tag{14.8}$$

Substituting Equation (14.8) for S in Equation (14.6), from Equation (14.4) gives:

$$M_a = \frac{\theta'}{\theta} = \frac{25}{25f/(25 + f)} \tag{14.9}$$

which simplifies for the case of an image at the near point to:

$$M_a = 1 + \frac{25}{f} \tag{14.10}$$

If the object is held at or just inside the focal point of the lens, the image forms very far away, essentially at infinity, rather than at the near point. This far point corresponds to the most comfortable viewing distance because the eye is relaxed. In this case, $\theta' = h/f$, and Equation (14.6) can be rewritten so the angular magnification M_a for image at infinity is given by:

$$M_a = \frac{25}{f} \tag{14.11}$$

From Equation (14.11), a maximum angular magnification power can be obtained by using lenses with short focal lengths. In fact, the lens aberrations limit the practical range of a single magnifying glass to about 3X and 4X, or a sharp image magnification of three or four times the size of the object when used normally. The manufacturers of the magnifying glasses commonly use Equation (14.11) to specify the magnification power of a magnifying glass.

14.6 THE COMPOUND MICROSCOPE

In many magnification applications, a system using two or more lenses is better than a single lens. For example, an ordinary magnifying glass cannot produce sharp images enlarged more than 3X and 4X. However, two converging lenses combined together as a microscope can produce sharp images magnified hundreds of times. Such an optical instrument, called a compound microscope, is shown in Figure 14.9.

Figure 14.10 shows image formation in a compound microscope explained by the ray method. The objective lens is a converging lens of short focal length that forms a real and enlarged image of the object. An object to be viewed is placed at O just beyond the focal length of the objective lens. The objective lens forms a real and enlarged image at I_1. This image is enlarged further by the eyepiece lens, which acts as a simple magnifier to form an enlarged virtual image at I_2. In this lens combination, the first image at I_1 produced by the objective acts as an object of the eyepiece, and the second image at I_2 has been magnified twice. As shown in Figure 14.10, in a properly designed instrument the first image at I_1 lies just inside the focal length of the eyepiece. The position of the second image at I_2 may be anywhere between the near and far points of the eye.

The total angular magnification M_{total} of a lens combination in a microscope is the product of magnifications produced by two lenses. The first magnification factor is the lateral magnification M_1 of the objective; M_1 determines the linear size of the real image I_1. The magnification factor M_2 is the angular magnification of the eyepiece. M_2 relates the angular size of the virtual image I_2 (seen through the eyepiece) to the angular size that the real image I_1 would have (if viewed without the eyepiece). The lateral magnification factor of the objective is given by:

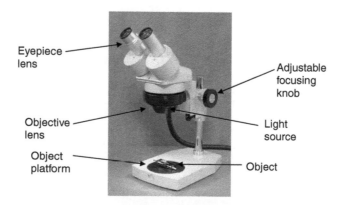

FIGURE 14.9 Compound microscope.

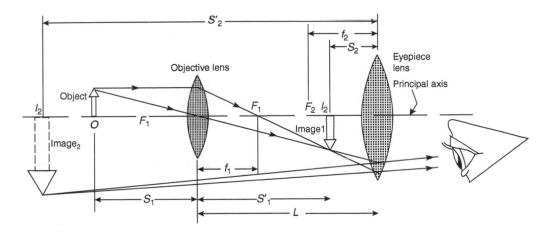

FIGURE 14.10 Image formation by a compound microscope.

$$M_1 = -\frac{S_1'}{S_1} \qquad (14.12)$$

where S_1' and S_1 are the object and image distances, respectively, for the objective lens.

Under normal operation, the object location is adjusted to be very close to the focal point f_1 of the objective lens. In this case, the image distance S_1' is very great in comparison to the focal length f_1 of the objective lens. Thus S_1 is approximately equal to f_1, and the lateral magnification M_1 of the objective lens can be written as:

$$M_1 = -\frac{S_1'}{f_1} \qquad (14.13)$$

Under normal operation, the eyepiece is adjusted so that the real image is just inside the focal point F_2 of the eyepiece lens. The eyepiece angular magnification M_2, can be written as:

$$M_2 = \frac{25}{f_2} \qquad (14.14)$$

where f_2 is the focal length of the eyepiece.

Again, the total angular magnification M_{total} of a lens combination in a microscope is the product of magnifications produced by two lenses.

$$M_{\text{total}} = M_1 \times M_2 = -\frac{25S_1'}{f_1 f_2} \qquad (14.15)$$

where S_1', f_1, and f_2 are measured in centimetres. The second image is inverted with respect to the object.

Microscope manufacturers usually specify the values of the magnification powers M_1 and M_2 rather than the focal lengths of the objective and eyepiece lenses, respectively. These magnification values assume that the object and first image I_1 is near to the focal point of the objective and eyepiece, respectively. This is the normal operation of the microscope. Interchangeable eyepieces with different magnification power from about 5X to over 100X are available. Most microscopes often are equipped with rotating turrets, which usually contain three or more objectives for different focal lengths, so that the same object can be viewed at different magnification powers. These objectives and eyepieces can be used in various combinations to provide common magnifications from 50X to about 900X. The maximum magnification power available for compound microscopes is about 2000X. All microscopes are equipped with a light source to illuminate the object with the required light intensity to view focused images. There are advanced compound microscopes that produce higher magnifications that are required by researchers and investigators.

14.7 ADVANCED MICROSCOPES

Many types of advanced microscopes are now available in the market. Most of them are operated by a laser source and controlled by software, which analyses and displays the data captured by a set of optical components. These microscopes are very expensive and sophisticated but also easy to use. They are used for viewing microscopic materials and particles, such as human cells, DNA, micro-organisms, and biological and medical test samples. Some of the highly sophisticated advanced microscopes are used in research and development activities.

Figure 14.11 shows the main parts and attachments of a confocal scanning laser microscope (CSLM). Figure 14.12 illustrates the basic components and operation of the CSLM.

The beam of approximately 1 mm in diameter emerges from the HeNe laser source mounted under the optical plate and then is deflected by a fixed mirror upwards towards a mirror, mounted on a kinematic mount. The beam is deflected horizontally from left-to-right towards a 15X beam expander. A beam 5 mm in diameter emerges at the output of the beam expander. The 15X beam expander consists of two positive lenses of 10 mm and 150 mm focal lengths, respectively. The beam then is defected by a mirror on a kinematic mount towards the front of the instrument. A broadband (range 400–700 nm) beamsplitter (50% reflection and 50% transmission) deflects the beam to the left towards an XY set of galvanometric mirrors. The beam is deflected by the galvo scanners to the front towards a unitary (1X) beam expander that consists of two identical 100 mm focal length positive lenses. The unitary telescope is designed to include a mirror at 45 degrees on a fixed mount between the two lenses. The beam passes through the first lens and is deflected by the mirror vertically downwards through the second lens of the unitary beam expander into the objective lens. The unitary telescope's function is to transfer a stationary beam at the galvos, as it is

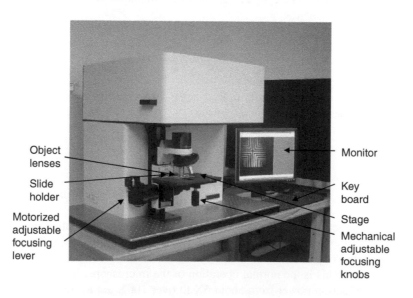

FIGURE 14.11 Confocal microscope.

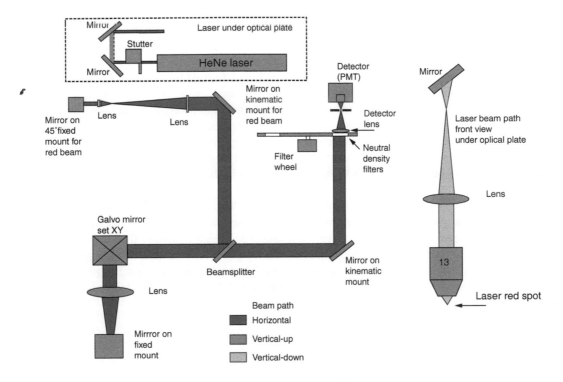

FIGURE 14.12 Basic components and operation of a confocal microscope.

scanned along the x and y directions, into a stationary beam at the entrance pupil of the objective lens. The objective lens focuses the laser beam onto a sample mounted on a motorized stage.

Light from the sample is reflected back, collected by the objective lens, passes through the unitary beam expander, and de-scanned by the XY galvo mirror set towards the beam splitter. The reflected light beam that passes through the beamsplitter is deflected by a mirror on a kinematic mount towards a neutral density (ND) filter mounted on a motorized filter wheel. The reflected light that passes through the ND filter is focused by the detector lens (100 mm focal length) onto a pinhole. Light that passes through the pinhole is incident on a detector type called a photomultiplier tube (PMT) and is measured. The PMT is a current generating device.

A current-to-voltage preamplifier converts the current to voltage and amplifies it. The voltage is transferred from the microscope to the computer by a coaxial cable where it is converted into a 16-bit digital number, by an analogue-to-Digital converter board that resides at one of the PCI slots of the host computer, and is displayed on the screen.

14.8 THE TELESCOPE

The optical instrument known as the telescope applies the principles of lenses and mirrors. The telescope is designed to aid in viewing distant objects, such as the planets and stars in the solar system. Basically, there are two types of telescopes: the refracting telescope, which uses a combination of lenses to form an image, and the reflecting telescope, which uses a curved mirror and a lens to form an image. The refracting telescope depends on the converging and reflecting of light by lens combinations. The reflecting telescope depends on the converging and reflecting of light, by lens and mirror combinations.

14.8.1 THE REFRACTING TELESCOPE

A simple refracting telescope, sometimes called astronomical telescope, is shown in Figure 14.13. The refracting telescope consists of a long tube with an objective lens toward the object and an adjustable eyepiece lens toward the viewer. The final image formed by this telescope is inverted, but the inverted image poses no problem in astronomical work. Image formation in the telescope is similar to a compound microscope. In both instruments, the image formed by an objective lens is viewed through an eyepiece lens. As in a microscope, two lenses are involved, with an eyepiece to enlarge the image produced by the objective. The telescope objective has a long focal length, whereas that of a microscope is very short. Therefore, the telescope is used to view large objects at long distances in outer space, such as stars and galaxies in the universe.

Figure 14.14 shows image formation, explained by the ray method, in a refracting telescope. The objective is a converging lens of long focal length. The object being viewed is far away compared with the focal length of the converging lens. Incoming rays from a distant object are nearly parallel. These rays form a real, inverted image at I_1, which is smaller than the object. The image appears near the focal point of the objective. The eyepiece lens enlarges this image further to form an enlarged virtual image at I_2. In a properly designed telescope, the first image I_1 must be at the first focal point of the eyepiece, then the final image at I_2 formed by the eyepiece is at infinity.

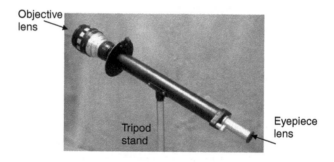

FIGURE 14.13 Simple refracting telescope.

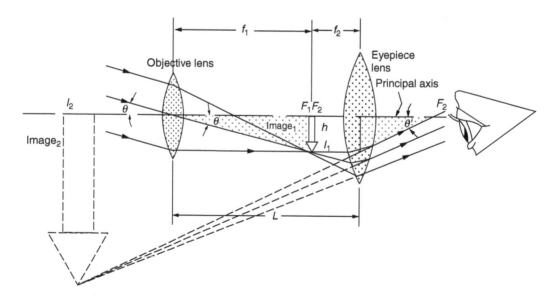

FIGURE 14.14 Image formation by a refracting telescope.

The eyepiece can be adjusted to move the image closer to the viewer's eyes. The distance between the objective lens and eyepiece lens, which is the length of the telescope L, is the sum of the focal length of objective and eyepiece lenses, f_1 and f_2.

The angular magnification M of a telescope is defined as the ratio of the angle θ' subtended by the object (when viewed through the telescope) to the angle θ subtended (when the object is viewed with the unaided eye). From Figure 14.14, the tangent of the angles θ and θ' can be written as:

$$\tan \theta \cong \theta = \frac{-h}{f_1} \tag{14.16}$$

$$\tan \theta' \cong \theta' = \frac{h}{f_2} \tag{14.17}$$

and the angular magnification M_a of a telescope is defined as the ratio of the focal length of the objective lens to the focal length of the eyepiece lens. The negative sign shows that the final image is inverted.

$$M_a = \frac{\theta'}{\theta} \cong -\frac{f_1}{f_2} \tag{14.18}$$

14.8.2 Terrestrial Telescopes

A telescope in which the final image is formed is erect is called a terrestrial telescope. Terrestrial telescopes are more convenient than astronomical telescopes, for viewing an upright image. This is useful for viewing objects on Earth. There are two types of terrestrial telescopes. Figure 14.15 shows one type of terrestrial telescope, which is called a Galilean telescope after the Italian astronomer and physicist Galileo Galilei (1564–1642), who built one in 1609.

In the Galilean telescope, the converging eyepiece is replaced by a diverging lens, so that an erect image is formed, as shown in Figure 14.16. Incoming parallel rays from a distant object will form a real inverted image from the objective lens. A diverging eyepiece lens is adjusted to be just within the focal length of the objective lens, so that the rays striking the eyepiece emerge parallel to each other. In this case an erect, virtual image is formed. The angular magnification M_a is again defined by Equation (14.18) as the ratio of the focal lengths. Note in this case, the focal length of the diverging lens is negative $(-f_2)$. Therefore, the angular magnification of a Galilean telescope is positive, which indicates the final image is erected and enlarged.

Galilean telescopes have disadvantages, such as very narrow fields of view and limited magnification, compared to other telescope types. A better type of terrestrial telescope is shown in Figure 14.17. In this type, a third lens, called the intermediate erecting lens, is used between the objective and the eyepiece to produce an erect final image, as illustrated in Figure 14.17. If the first image formed by the objective at a distance is twice the focal length of the intermediate erecting lens (at $2f_e$), then the erecting lens inverts the image without magnification. The angular magnification M_a of this telescope type is still defined by Equation (14.18) as the ratio of the angles.

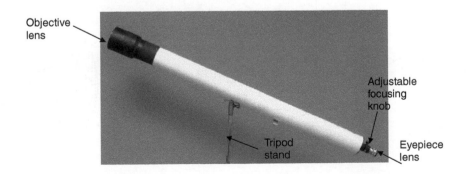

FIGURE 14.15 Galilean telescope.

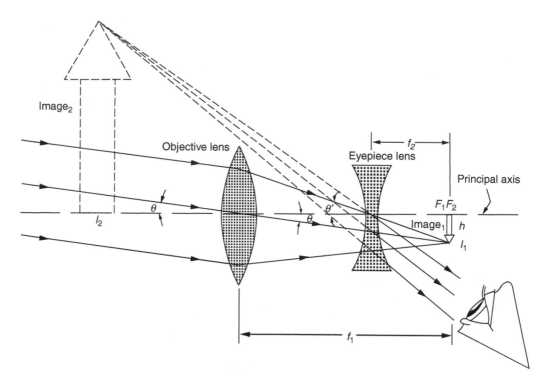

FIGURE 14.16 Image formation by a Galilean telescope.

The disadvantage of this type of telescopes that the telescope tube is longer than a telescope with similar magnification. A better design shortens the instrument length by using a pair of prisms that both invert the image so that the image is erect. This shorter arrangement is used in the design of binoculars. The principles of binoculars will be explained in a later section.

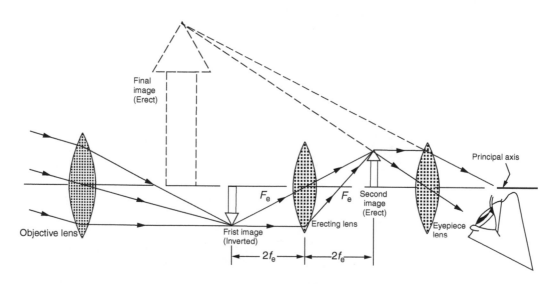

FIGURE 14.17 In a terrestrial telescope an intermediate lens is used to form an erect image.

14.8.3 THE REFLECTING TELESCOPE

In astronomical applications, such as observing nearby astronomical objects (such as the sun, moon, and planets), the magnification achieved by most refracting telescopes is sufficient. However, stars are so far away that they always appear as small points of light. Large research telescopes with very high magnification are used to study very distant objects. The higher the magnification of a telescope, the greater in diameter the objective lens must be, in order to gather in enough light for the image to be visible. Large lenses for refraction telescopes are difficult and expensive to manufacture. Another difficulty with large lenses is that their large weight requires a big telescope structure. This causes an additional source of lens aberration, dispersion, and misalignment. These problems can be solved by replacing the objective lenses with reflecting, concave parabolic mirrors. Such a mirror produces a real image of a distant object and is subject to less aberration.

Figure 14.18 illustrates a design for a typical reflecting astronomical telescope. This type of telescope is called Newtonian focus, because it was Newton who developed and constructed it in 1670. Incoming light rays pass down the barrel of the telescope and are reflected by a concave parabolic mirror at the base. These rays converge toward a small secondary flat mirror that reflects the light toward an opening in the side of the tube. Then, the rays pass into an eyepiece for viewing or more often for photographing.

Figure 14.19 shows another arrangement, which is called a Cassegrain focus reflecting telescope. This telescope uses a small mirror to form the image below the main parabolic mirror at the base. The mirror reflects the light toward an opening in the base. Then, the rays pass into an eyepiece for viewing or more often for photographing.

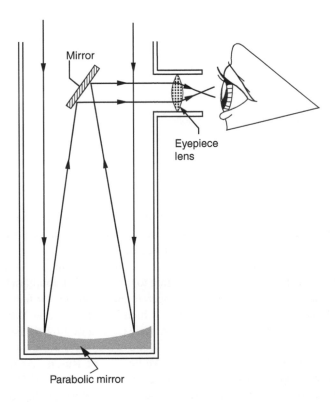

FIGURE 14.18 Newtonian focus reflecting telescope.

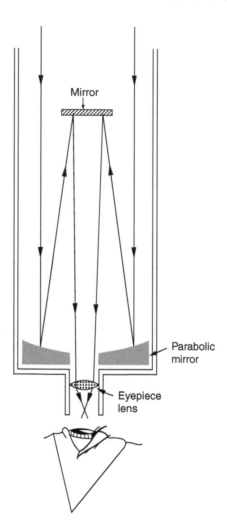

FIGURE 14.19 Cassegrain focus reflecting telescope.

14.8.4 FAMOUS TELESCOPES

The Hubble Space Telescope is in orbit outside of the Earth's atmosphere to enable observations without the distortion caused by refraction. In this way the telescope can be diffraction limited, and used for coverage in the ultraviolet (UV) and infrared ranges.

The Very Large Telescope (VLT) is currently the record holder in size; it has four telescopes, each 8 m in diameter. The four telescopes, belonging to the European Southern Observatory (ESO) and located in the Atacama desert in Chile, can operate independently or together.

The Largest refracting telescope in the world, which is located at the Yerkes Observatory in Williams Bay, Wisconsin in the United State of America, has a of 1.02-m diameter.

In contrast, the largest single-mirror optical telescope in the world is the 6-m diameter reflecting telescope on Mount Pastukhov in the Caucasus Mountains of Russia. At the same site is the RATAN 600 radio telescope, consisting of radio reflectors in a circle of 600-m diameter. Both instruments are operated by the Special Astrophysical Observatory of the Russian Academy of Sciences in St Petersburg.

Next in size is a 5-m reflector at Mount Palomar in California that enlarged the horizons of astronomy since 1948. The 5-m Hale telescope at Mt. Palomar is a conventional research telescope that was the largest for many years. It has a single borosilicate mirror that was famously difficult to construct. The equatorial mounting is also unique, permitting the telescope to image near the north celestial pole.

The 2.5-m Mt. Wilson telescope was used by Edwin Hubble to discover galaxies, and the redshift. It is now part of a synthetic aperture array with several other Mt. Wilson telescopes, and is still useful for advanced research.

The 60-cm telescope currently utilized by the Telescope In Education (TIE) programme has a distinguished history. Designed and built by Caltech in the early 1960s. This telescope was used to investigate the moon and prove that the lunar surface was solid rather than covered by a thick layer of dust. A thick layer of fine dust would have been a hazard to astronauts landing on the moon during the Apollo missions. This telescope was designed as a 60-cm (f/16) Cassegrain telescope. There were two secondary mirrors made: One was larger and used for visual and photographic observations. The second mirror was smaller because it was housed in a special fixture that could move the mirror for accurate infrared observations. The first secondary mirror was destroyed in an accident during an observing run. The second mirror, though slightly undersized, continues to function well on the instrument. This 60-cm telescope has been thoroughly refurbished and reconfigured to f/3.5 with a Newtonian focus, in order to operate with digital cameras.

The latest generation of astronomical telescopes do not rely on single large mirrors; practical difficulties limit the mirror diameter to at most 8 m. Instead, a number of individual mirrors are linked to produce a single image. In one approach the Keck telescope in Hawaii in the United State of America uses hexagonal segments to give a collecting surface 10 m across. In another approach separate circular mirrors are used. A proposed new American telescope will have four 7.5-m mirrors mounted in one structure for the equivalent of a 15-m mirror. A third scheme, which is being developed in Europe, uses four individual 8-m telescopes whose images are added together electronically.

14.8.5 Research Telescopes

In recent years, some technologies have been adapted to overcome the negative effect of atmosphere on ground-based telescopes. These technologies have achieved successful results. Some research telescopes have several instruments to choose from for different applications, such as imagers of different spectral responses, spectrographs useful in different regions of the spectrum, and polarimeters that detect light polarization.

Most large research telescopes can operate as either a Cassegrainian (longer focal length, and a narrower field with higher magnification) or Newtonian (brighter field). In a new era of telescope making, a synthetic aperture composed of six segments synthesizes a mirror of 4.5 m in diameter. Another company has developed a telescope with a synthetic-aperture 10 m in diameter.

The current generation of ground-based telescopes being constructed has a primary mirror of between 6 and 8 m in diameter. In this generation of telescopes, the mirror is usually very thin and is kept in an optimal shape by an array of actuators. This technology has driven new designs for future telescopes with diameters of 30, 50 and even 100 m.

Initially the detector used in telescopes was the human eye. Later, the sensitized photographic plate took its place. Thus, the spectrograph was introduced, allowing the gathering of spectral information. After the photographic plate, successive generations of electronic detectors, such as CCDs, have been perfected, each with more sensitivity and resolution.

The phenomenon of optical diffraction sets a limit to the resolution and image quality that a telescope can achieve, which is the effective area of the Airy disc, which limits how close to place two such discs. This absolute limit is called Sparrow's resolution limit. This limit depends on the wavelength of the studied light (so that the limit for red light comes much earlier than the limit for blue light) and on the diameter of the telescope mirror. This means that a telescope with a certain mirror diameter can resolve up to a certain limit at a certain wavelength, so if more resolution is needed at that very wavelength, a wider mirror must be built.

There are many plans for even larger telescopes; one of them is the Overwhelmingly Large Telescope (OWL), which is intended to have a single aperture of 100 m in diameter.

14.9 THE BINOCULARS

A pair of common binoculars shown in Figure 14.20 is shorter in length than any type of telescope. To shorten the optical instrument length, the common binoculars design uses a pair of prisms that both invert the image so that the image is erect to the viewer.

FIGURE 14.20 Pair of binoculars.

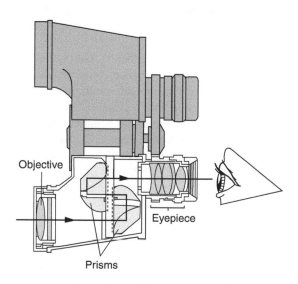

FIGURE 14.21 Optical components of binoculars.

The arrangement of the prisms and lenses in binoculars is shown in Figure 14.21. Common binoculars have three major parts. An objective lens produces an inverted image and focuses the image onto a prism. A set of prisms turns the image upright, and an eyepiece magnifies the image.

There are two types of prisms that are used in making binoculars: Porro prisms and roof prisms. There are advantages to each prism type. The Porro prism design is simpler and more light efficient, and its images show better contrast. Roof prism binoculars appear simpler than Porro prism binoculars. But inside, they have a more complex light path and require much greater optical precision in manufacturing. As a result, roof prism binoculars cost more to make. Though modern eyepieces and objective lenses are each comprised of multiple elements, their basic functions remain unchanged.

14.10 THE SLIDE PROJECTOR

A slide projector is an opto-mechanical device, as shown in Figure 14.22, used to view photographic slides, as shown in Figure 14.22. It has four main elements: a fan-cooled electric light bulb or other light source, a focusing lens to direct the light to cover the area of the slide, a holder for the slide, and magnifying lenses, as shown in Figure 14.23. Light passes through the transparent slide and magnifying lenses; the resulting image is enlarged and projected onto a perpendicular flat screen so the audience can view its reflection. Alternatively, the image may be projected onto a translucent rear projection screen using continuous automatic display for close viewing.

Common in the 1950s and 1960s households as an alternate to television or movie entertainment, family members and friends would gather, darken the living room and show slides of recent holidays or vacations. In-home photographic slides and slide projectors have largely been replaced by low cost paper prints, digital cameras, DVD media, video display monitors, and digital projectors.

FIGURE 14.22 Slide projector.

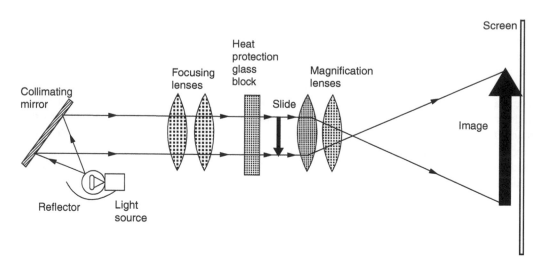

FIGURE 14.23 Image formation by a slide projector.

14.11 THE OVERHEAD PROJECTOR

The overhead projector is another common type of viewing device, as shown in Figure 14.24. The overhead projector is a display system used to display images to an audience. It typically consists of a large box containing a very bright lamp and a fan to cool it, on top of which is a large Fresnel lens that collimates the light. Above the box, typically on a long arm, is a magnifyer convex lens and a

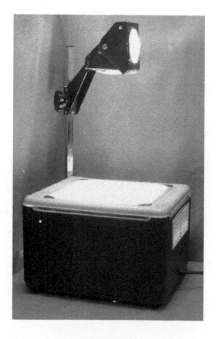

FIGURE 14.24 Overhead projector.

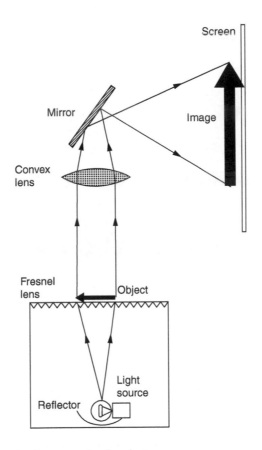

FIGURE 14.25 Image formation by an overhead projector.

mirror that redirects the light forward onto the screen. Figure 14.25 shows an image formation by an overhead projector.

Overhead projectors were a common fixture in most classrooms and business conference rooms, but today are being replaced by larger computer monitors and dedicated computer projection systems.

14.12 THE LIQUID CRYSTAL DISPLAY (LCD) PROJECTOR

A liquid crystal display (LCD) projector is a device for giving presentations generated on a computer. They are the modern equivalent to the slide projector and overhead projector. LCD projectors place a small LCD panel, almost always colour, in front of the bright lamp used in all overhead systems. The imagery on the LCD is provided by an attached computer. Currently, LCD projectors tend to be smaller and much more portable than older systems.

Figure 14.26 shows a computer display crystal screen. The LCD units can be used as a display panel in cellular phones, calculators, pocket television screens, computer screens, and diagnostic and measuring devices. LCDs give excellent readability from any angular position to within a few degrees of the horizontal plane, even in the most intense, direct sunlight.

FIGURE 14.26 Liquid crystal display.

14.13 THE LIGHT BOX

A light box is made of hardwood with a translucent acrylic diffuser panel and double-thick glass for durability, as shown in Figure 14.27. A special light diffuser illuminates the working surface with uniform colour-shifted light for bright, sharp images. The light box features a metal side rail for a T-square. Ventilation channels on the side keep the unit cool. This box is used for making engineering drawings.

Another design for a light box is used for slide viewing, as shown in Figure 14.28. Fluorescent illumination is used for tracing, opaquing, stripping, transparency and slide viewing, and slide sorting. The surface viewing area stays cool. The light box features durable, lightweight construction, and a frosted acrylic top.

A standard x-ray illuminator is another type of light box. They are used for x-ray viewing by medical staff in clinics and hospitals, as shown in Figure 14.29. All standard x-ray illuminators are constructed from cold rolled steel and finished inside and out with a non-yellowing baked white enamel finish. The x-ray illuminators provide the best viewing conditions for diagnostic purposes. The viewing area of the light box is one-piece, white, translucent, shatterproof plastic.

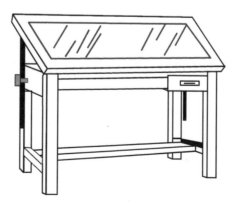

FIGURE 14.27 Light box with a support frame.

FIGURE 14.28 Light box for slide viewing.

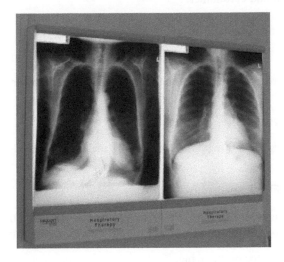

FIGURE 14.29 Standard x-ray illuminator.

14.14 EXPERIMENTAL WORK

The theory behind this experiment is based on the laws of light. Optical components, such as mirrors, lenses, and prisms, work on the basis of image formation by reflection and refraction. These optical components are used in building different optical viewing instruments.

In this experiment, the student will examine, operate, dismantle, and reassemble a variety of optical viewing instruments. These instruments are composed of different types of optical components that are used for image formation. First, the student will operate and observe image formation by optical instruments. Second, the student will examine the internal components and determine the principle operation of the instrument.

The student will operate, observe image formation, disassemble, examine the functionality of the optical components, and reassemble an optical viewing instrument. The following cases of optical viewing instruments are considered for this experiment:

a. a slide projector;
b. a film projector.

   c. an overhead projector.

   d. a camera.

   e. a compound microscope.

   f. a telescope.

14.14.1 TECHNIQUE AND APPARATUS

Appendix A presents the details of the devices, components, tools, and parts.

1. Slide projector, film projector, overhead projector, camera, compound microscope, and telescope.
2. Slides, transparent film, overhead transparency, and a small object.
3. Screen.
4. Toolbox (screwdriver kits, etc.).
5. Black/white card and cardholder.
6. Ruler.

14.14.2 PROCEDURE

Follow the laboratory procedures and instructions given by the professor and/or instructor.

14.14.3 SAFETY PROCEDURE

Follow all safety procedures and regulations regarding the use of optical viewing instrument.

14.14.4 APPARATUS SET-UP

14.14.4.1 A Slide Projector: Operate, Observe Image Formation, Disassemble to Examine the Functionality of the Optical Components, and Reassemble

1. Place a slide projector on the table.
2. Have the toolbox ready for use.
3. Inspect the slide projector condition and optical parts.
4. Load the slide tray with a few slides.
5. Connect the slide projector to the power supply.
6. Turn on the slide projector.
7. Turn off the lights of the lab.
8. Try to project and focus the image of the slide on the screen.
9. Turn on the lights of the lab.
10. Turn off the slide projector and disconnect the power cable from the wall outlet.
11. Take a screwdriver and unscrew the screws on the side/back cover of the slide projector, as shown in Figure 14.30.
12. Examine the optical components that are used in building the slide projector.
13. Illustrate the locations of the light source, mirror, lenses, slide, and image in a diagram using the ray tracing method.
14. Put back the side/back cover and screw the screws back in place.
15. Test the slide projector again.
16. Put the slide projector and the parts back into the slide projector case.

For Cases (b), (c), (d), (e), and (f), repeat the procedure described in Case (a), using the appropriate optical viewing instrument.

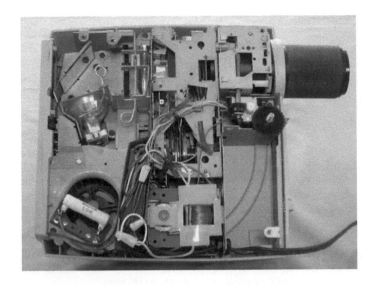

FIGURE 14.30 Disassembled slide projector.

14.14.5 DATA COLLECTION

No data collection is required.

14.14.6 CALCULATIONS AND ANALYSIS

No calculations and analysis are required for any case.

14.14.7 RESULTS AND DISCUSSIONS

14.14.7.1 A Slide Projector: Operate, Observe Image Formation, Disassemble to Examine the Functionality of the Optical Components, and Reassemble

1. Illustrate the locations of the light source, lens, object, and image in a diagram using the ray tracing method.
2. Discuss the characteristics of the image.

For Cases (b), (c), (d), (e), and (f), repeat the steps for the results and discussions in Case (a).

14.14.8 CONCLUSION

Summarize the important observations and findings obtained in this lab experiment.

14.14.9 SUGGESTIONS FOR FUTURE LAB WORK

List any suggestions for improvements using different experimental equipment, procedures, and techniques for any future lab work. These suggestions should be theoretically justified and technically feasible.

14.15 LIST OF REFERENCES

List any references that were used in the report. Use one format in writing the references. Never mix reference formats in a report.

14.16 APPENDICES

List all of the materials and information that are too detailed to be included in the body of the report.

FURTHER READING

Beiser, A., *Physics*, 5th ed., Addison-Wesley Publishing Company, Inc., USA, 1991.

Blaker, W. J., *Optics: The Matrix Theory*, Marcel Dekker, New York, 1972.

Blaker, J. W. and Schaeffer, P., *Optics: An Introduction for Technicians and Technologists*, Prentice Hall, Englewood Cliffs, NJ, 2000.

Cutnell, J. D. and K, W., *Physics Physics*, 5th ed., John Wiley and Sons, Inc., New york, 2001.

Damaskinos, S., *Confocal Scanning Laser Microscope*, MACROscope®—the widefield confocal™, 550 Parkside Dr., Unit A12, Waterloo, Ontario, Canada, N2L 5V4, 2005.

Edmund Industrial Optics, *Optics and Optical Instruments Catalog, 2004*, Edmund Industrial Optics, Barrington, NJ, 2004.

Ghatak, A. K., *An Introduction to Modern Optics*, McGraw-Hill Book Company, New York, 1972.

Halliday, D., Resnick, R., and Walker, J., *Fundamental of Physics*, 6th ed., John Wiley and Sons, Inc., New York, 1997.

Hecht, E., *Optics*, 4th ed., Addison-Wesley Longman, Inc., Reading, MA, 2002.

Hewitt, P. G., *Conceptual Physics*, 8th ed., Addison-Wesley, Inc., Reading, MA, 1998.

Jenkins, F. W., H., E., and White, *Fundamentals of Optics*, McGraw Hill, New York, 1957.

Jones, E. and Childers, R., *Contemporary College Physics*, 3rd ed., McGraw-Hill Higher Education, USA, 2001.

Naess, R. O., *Optics for Technology Students*, Prentice Hall, Englewood Cliffs, NJ, 2001.

Newport Corporation, Optics and Mechanics Section, *the Newport Resources 1999/2000 Catalog*, Newport Corporation, Irvine, CA, USA, 1999/2000.

Nolan, P. J., *Fundamentals of College Physics*, Wm. C. Brown Publishers, Inc., Dubuque, IA, USA, 1993.

Pedrotti, F. L., L., S., and Pedrotti, *Introduction to Optics*, 2nd ed., Prentice Hall, Englewood Cliffs, NJ, 1993.

Pritchard, D. C., *Environmental Physics: Lighting*, Longmans, Green and Co. Ltd., London, 1969.

Serway, R. A. and Jewett, J. W., *Physics for Scientists and Engineers with Modern Physics*, 6th ed., volume 2, Thomson Books/Cole, USA, 2004.

Shamir, J., *Optical Systems and Processes*, SPIE-The International Society for Optical Engineering, SPIE Press, USA, 1999.

Sterling, D. J. Jr., *Technician's Guide to Fiber Optics*, 2nd ed., Delmar Publishers Inc., Albany, New York, 1993.

Tippens, P. E., *Physics*, 6th ed., Glencoe McGraw-Hill, Westerville, OH, USA, 2001.

White, H. E., *Modern College Physics*, 6th ed., Van Nostrand Reinhold Company, New York, USA, 1972.

Woods, N., *Instruction's Manual to Beiser Physics*, 5th ed., Addison-Wesley Publishing Company, Reading, MA, 1991.

15 Polarization of Light

15.1 INTRODUCTION

Polarization is another interesting wave property of transverse light waves. Polarization is the principle applied to polarized sunglasses used in protecting the eyes from the sun's rays under a clear sky. Light waves are propagated in an electromagnetic wave as vibrating electric and magnetic fields, which are perpendicular to the direction of the wave propagation. The human eye cannot distinguish between polarized and unpolarized light. Therefore, an analyser is needed to detect polarized light.

Polarization states are linear, circular, or elliptical according to the paths traced by electric field vectors in a propagating wave. Unpolarized light, such as light from an incandescent fluorescent tube, is a combination of all polarization states. Randomly polarized light, in reference to laser output, is composed of two orthogonally linearly polarized beams.

The polarization phenomenon is used in many optical applications, such as testing plastic or glass under mechanical stress. Polarizer materials, such as beamsplitters and prisms, are used as components in polarization beamsplitter devices that are used in fibre communication systems. In this chapter, along with the theoretical presentation, four experimental cases demonstrate the principles of the polarization of light.

15.2 POLARIZATION OF LIGHT

Light can be considered to have a dual nature. In some cases light acts like a wave, and in others light acts like a particle. Classical electromagnetic wave theory provides an adequate explanation of light propagation and of the effects of interference. On the other hand, the photoelectric effect, involving the interaction of light with matter, is best explained by considering light as a particle.

An electromagnetic wave consists of an electric field and magnetic field oriented perpendicular to one another. Both fields have a direction and strength (or amplitude). The propagation axis of

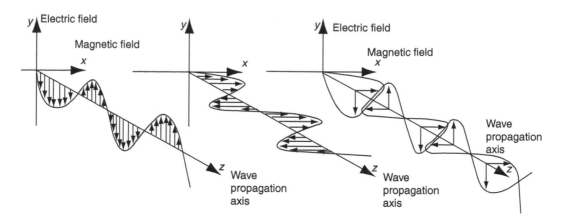

FIGURE 15.1 Electric and magnetic fields in a light wave.

the wave is perpendicular to the plane containing the electric and magnetic fields. As shown in Figure 15.1, if the electric field is oriented along the y-axis, and the magnetic field along the x-axis, then the wave propagation is along the z-axis. This means that the vibrations in the electric field are parallel to one another at all points in the wave, so that the electric field forms a plane called the plane of vibration. Similarly, all points in the magnetic field component of the wave lie in a plane that is perpendicular to the electric field plane.

The electric and magnetic fields vary sinusoidally with time. The curved line represents field strength or amplitude. A field starting at a maximum in one direction decays to zero and then builds up in the other direction until it reaches a maximum again. The electric and magnetic fields oscillate in phase; this means that the electric and magnetic fields reach their peaks (and their troughs) at exactly the same time. The rate of oscillation represents the frequency of the wave. The distance traveled during one period of oscillation represents the wavelength.

15.3 FORMS OF POLARIZATION OF LIGHT

Light from an ordinary light source consists of many waves emitted by the atoms or molecules of the light source. Each atom produces a wave with its own orientation of the electric and magnetic vibration; the orientation corresponds to the direction of the atomic vibration. Figure 15.2 shows a light source with electric and magnetic field vectors. The direction of propagation is coming out of the page in this figure. When the field vectors of a light source are randomly oriented, the light is called unpolarized, as shown in Figure 15.2(a). If there is some partial preferential orientation of the field vectors, the light is called partially polarized, as shown in Figure 15.2(b). If the field vectors are laid on the plane of polarization, the light is called linearly polarized, plane polarized, or sometimes called simply polarized. The plane of the linearly polarized light could be vertical or horizontal, as shown in Figure 15.2(c).

There are three forms of polarization in light: linear polarization, elliptical polarization, and circular polarization, as shown in Figure 15.3. In linearly polarized light, the field vectors vibrate in one axis. In the case of elliptical polarization, both electric and magnetic field vectors do not have the same amplitudes; while with circular polarization, the two fields do have equal amplitudes.

The analyser absorbs or transmits the polarized light, depending on the orientation of the analyser. When the analyser has the same orientation as the polarizer, light is transmitted. When the polarizer is rotated 90°, the analyser absorbs the polarized light, and then no light is transmitted.

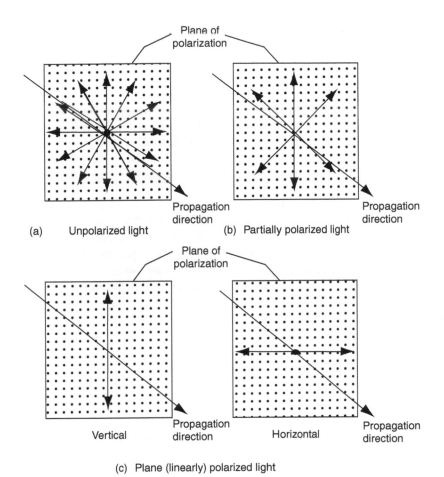

FIGURE 15.2 An unpolarized and a polarized light of transverse waves.

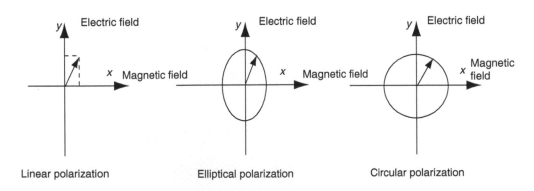

FIGURE 15.3 Forms of polarization of light.

15.4 OCCURRENCE OF POLARIZATION

Ordinary light (unpolarized) can become polarized in several ways. Polarization may result from the reflection of light from various surfaces, from refraction and double refraction of light through a crystal, from selective absorption of light in a crystal, and scattering by particles of the medium. Polarization by these means will be discussed in detail in the following subsections.

15.4.1 POLARIZATION BY REFLECTION

Consider unpolarized light incident on a smooth optical surface, such as glass. The light is partially reflected and partially transmitted according to the laws of light, as shown in Figure 15.4. The reflected light is either completely polarized, partially polarized, or unpolarized, depending on the angle of incidence. If the angle of incidence is either 0 or 90° to the surface, then the reflected light is unpolarized. If the angle of incidence is between 0 and 90°, then the reflected light is partially polarized.

For one particular angle of incidence, the reflected light is completely polarized while the refracted light is partially polarized, as shown in Figure 15.5. This angle of incidence is called polarizing angle (θ_p) or the Brewster's angle after Sir David Brewster (1781–1868), a Scottish physicist. He discovered that when the angle of incidence is equal to this polarizing angle, the reflected light and the refracted light are perpendicular to each other. In other words, Brewster's angle is that incident angle where the angle between the reflected light and the refracted light is 90°. The angle of refraction (θ_{refr}) is the complement of (θ_p), so $\theta_{refr} = 90° - \theta_p$.

Applying the law of refraction (Snell's Law):

$$n_1 \sin \theta_p = n_2 \sin \theta_{refr} \tag{15.1}$$

Substituting for θ_{refr} gives:

$$n_1 \sin \theta_p = n_2 \sin(90° - \theta_p) = n_2 \cos \theta_p \tag{15.2}$$

Dividing the Equation 15.2 by $\cos \theta$ gives:

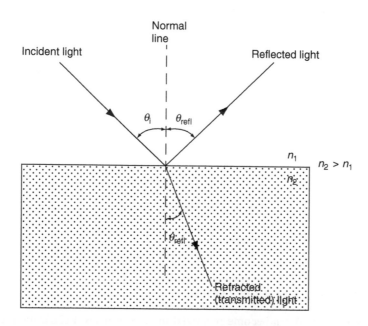

FIGURE 15.4 Light incident, reflected, and refracted by an optical material.

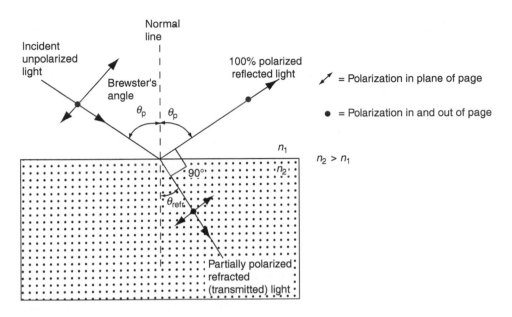

FIGURE 15.5 Polarization at Brewster's angle.

$$\frac{\sin \theta_p}{\cos \theta_p} = \frac{n_2}{n_1} = \tan \theta_p \tag{15.3}$$

For light incidence in air $(n_1) \approx 1.00$, then:

$$\tan \theta_p = n \tag{15.4}$$

Then, Brewster's angle is:

$$\theta_p = \tan^{-1} n \tag{15.5}$$

This relation is known as Brewster's law. For example, Brewster's angle for crown glass $(n = 1.52)$ has the value $\theta_p = \tan^{-1}(1.52) = 56.7°$. Brewster's angle is a function of the index of refraction (n) and varies to some degree with the wavelength of the incident light. This is because (n) varies with wavelength for a given optical material, as explained in the laws of light chapter.

Polarization by reflection is a common phenomenon. Sunlight that is reflected from water, glass, smooth asphalt, or snow surfaces is partially polarized. When the sunlight is reflected from a horizontal surface, the electric field vector of the reflected light has a strong horizontal component. Sunglass manufacturers make the polarizing axis of the lens material vertical, to absorb the strong horizontal component of the reflected light. The sunglasses can reduce the overall intensity of the light transmitted through the lenses to a certain value depending on the quality of the lens material. Other types of polarizing filters are widely used in different industrial applications.

15.4.2 Polarization by Double Refraction

An Iceland spar (calcite) crystal placed on a page shows two refracted images of the print, as shown in Figure 15.6. This double refraction of light by calcite was discovered by Erasmus Bartholinus, a Scottish Physician, in 1669, and later studied by Huygens and Newton.

(a) Calcite crystal   (b) Double refraction phenomenon

FIGURE 15.6 Calcite crystal. (a) Calcite crystal; (b) double refraction phenomenon.

Most crystalline substances, such as calcite, quartz, mica, sugar, topaz, aragonite, and ice, are now known to exhibit double refraction. The double refraction depends on the purity of the crystal. In other words, it depends on the index of refraction of the crystal and the wavelength of the light. Calcite (Calcium Carbonate $CaCo_3$) and quartz (Silicon Dioxide SiO_2) are the most important crystals used in manufacturing optical devices and instruments. Calcite is a natural, birefringent material and always has the shape shown in Figure 15.6. The calcite must be cut, ground, and polished to exact angles with respect to its optical axis. Each face of the calcite crystal is a parallelogram whose angles are 78 and 102°. The two opposite surfaces of the calcite crystal are parallel to each other.

When a beam of unpolarized light passes at an angle to the calcite crystal's optical axis, the beam is doubly refracted, dividing into two light beams as it enters the surface of the crystal. Both refracted light beams are found to be plane polarized and their planes of polarization perpendicular to each other, as shown in Figure 15.7. One light beam, called the ordinary light beam, is linearly polarized with its vibrations in one plane. The ordinary beam passes straight through without deviation. This beam is the horizontal component of the light. The second light beam, called the extraordinary light beam, is linearly polarized with its vibrations in a plane

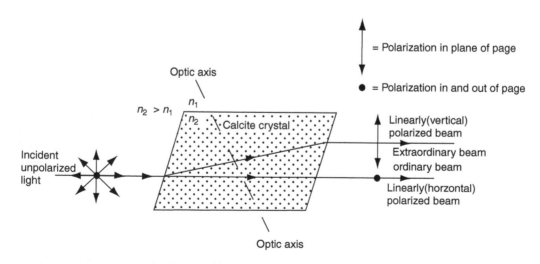

FIGURE 15.7 Polarization by double refraction in a calcite crystal.

perpendicular to the first plane. This beam is the vertical component of the light. The ordinary beam follows the law of refraction, Snell's law, and the extraordinary beam does not. Because the two opposite surfaces of the calcite crystal are always parallel to each other, the two refracted beams always emerge parallel to the incident light beam and are parallel to each other. One important property of the calcite crystal is that there is only one direction through the crystal in which there is no double refraction. This particular direction, called the optic axis, is indicted by dashed lines shown in Figure 15.7.

15.4.3 POLARIZATION BY SCATTERING

When light is incident on atmospheric particles, such as the molecules of air (nitrogen and oxygen) and dust, some of the light can be absorbed and reradiated. This process, called scattering of light is shown in Figure 15.8. Atmospheric scattering causes the sky's light to be polarized. In order for the scattering to occur, the sizes of the air molecules present in the atmosphere must be smaller than the wavelengths of visible light. The molecules scatter light inversely proportional to the fourth power of the wavelength ($1/\lambda^4$). This wavelength relation discovered by Lord Rayleigh (1842–1919), a British physicist, is due to Rayleigh scattering. Blue light has a shorter wavelength than red light. Therefore, blue light is scattered toward the ground more than red light. This is the reason why the sky appears blue. At sunset, the sun's rays pass through a maximum length of atmosphere. Much of the blue has been removed by scattering. The light that reaches the surface of the Earth is thus lacking of blue, which is why sunsets appear reddish.

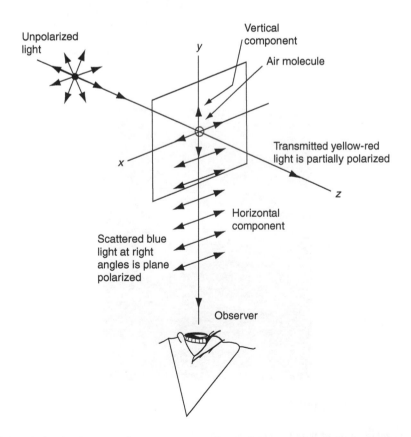

FIGURE 15.8 Polarization by scattering.

It is possible to produce polarized light by filtering ordinary light through a polarizer, which is a material that transmits only those waves that vibrate in a single plane. Polarizers will be explained in the next sections. Polarizers can be used as analysers or filters. When an ordinary beam of light strikes a polarizer, part of the light is absorbed and part is transmitted, as shown in Figure 15.9(a). The light emerging from this polarizer is completely polarized along the direction of propagation. If a second polarizer acting as analyser is placed behind the first and aligned in the same orientation, light is still transmitted. However, there is less transmission because of the light absorbed in the two polarizers. If the second polarizer is rotated by 90° about the axis of the light beam, then no light passes through, as shown in Figure 15.9(b). When the emerging polarized light from the first polarizer falls on the second polarizer, all the light is absorbed. This is because the second polarizer is turned so that its polarization axis is 90° away from that the first polarizer.

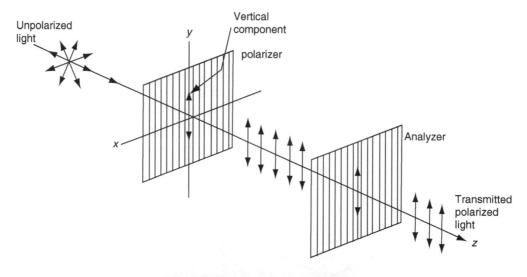

(a)

FIGURE 15.9. Polarization of light by two polarizers.

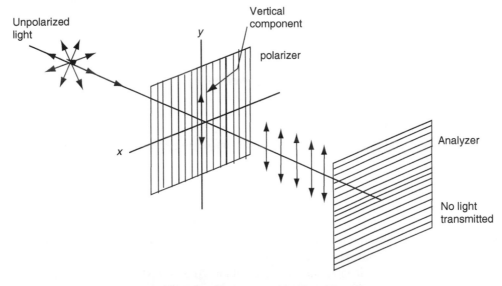

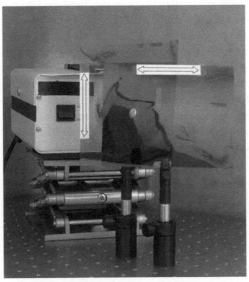

(b)

FIGURE 15.9   (*continued*)

15.4.4 Polarization by Selective Absorption

When ordinary light enters a crystal of tourmaline, double refraction takes place in much the same way as with a calcite crystal, as explained above, but with one difference. The tourmaline crystal entirely absorbs the ordinary beam while the extraordinary beam passes through. This phenomenon is called selective absorption because the crystal absorbs light waves vibrating in one plane and absorbs light waves vibrating in the other plane. The common type of tourmaline crystal is the Nicol prism. When two such crystals are lined up parallel, with one behind the other, the plane polarized light from the first crystal passes through the second with a very low light loss. If either crystal is rotated to be perpendicular to the other, then the light is completely absorbed and none of the light passes through. The result is the same as explained above in the polarizer case.

15.5 POLARIZING MATERIALS

The most common technique for obtaining polarized light is to use a polarizing material. These materials transmit waves when the electric field components vibrate in a plane parallel to a specific direction. The material absorbs the waves whose electric field components vibrate in the other direction. Any substance that transmits light with the electric field component vibrating in only one direction and absorbs the other component during transmission is referred to as a dichroic substance.

Another dichroic crystal is quinine sulfide periodide, also called herapathite, after W. Herapath, an English physician who discovered its polarizing properties in 1852. In 1928, Edwin Land, an American scientist, discovered such a dichroic material, which is now called Polaroid. In 1938, Land invented H-sheet, which is now the most widely used linear polarizer. The H-sheet polarizes light through selective absorption in linearly oriented molecules. This material is fabricated in thin sheets of long-chain hydrocarbons such as polyvinyl alcohol. The sheets are stretched during manufacturing, so that the long-chain molecules align. After the sheet has been dipped into a solution containing iodine, the molecules become conductive. The conduction takes place along the hydrocarbon chains because the valence electrons of the molecules can move easily only along the chains (recall that valence electrons are free electrons that can readily move through the conductor). As a result, the molecules readily absorb light in which the electric field component is parallel to the chain. Likewise, the polarizing effect also causes the molecules to transmit light when the electric field component is perpendicular to the chain. It is common to refer to the direction perpendicular to the molecular chains, as the transmission axis. In an ideal polarizer, all light with the electrical field parallel to the transmission axis is transmitted, and all light with the electrical field perpendicular to the transmission axis is absorbed. Figure 15.10 shows a polarizing sheet, which is used in experiments to identify the polarization axis of a light beam.

FIGURE 15.10 Polarizing sheet and sheet holder.

15.6 POLARIZING OPTICAL COMPONENTS

There are many types of polarizing optical components that are used in splitting light into two beams. These optical components are prisms, rectangular beamsplitters, polarizers, etc. They work in different wavelength ranges. These components are explained in detail in prisms and beamsplitters in chapters. Figure 15.11 shows polarized light passing through a polarizing cube beamsplitter.

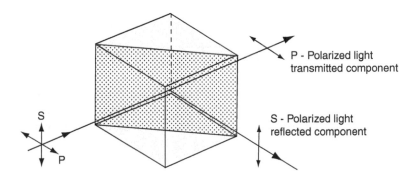

FIGURE 15.11 Polarizing cube beamsplitter.

15.7 THE LAW OF MALUS

Figure 15.12 shows a polarized light beam of intensity I_o passing through a polarizer whose axis is at an angle θ relative to the polarized light axis. The transmitted light intensity varies as the square of the transmitted light amplitude. The intensity of the transmitted polarized light varies as:

$$I = I_o \cos^2\theta \tag{15.6}$$

where

I_o is the intensity of the polarized light incident on the polarizer in (W/m^2) and
θ is the angle between two transmission axes of the polarized light and the polarizer in (degrees).

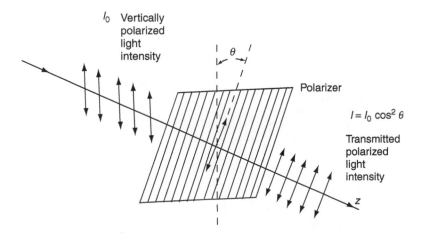

FIGURE 15.12 Transmission of polarized light through a polarizer—Malus's law.

This formula is known as the law of Malus, after the French Engineer Etienne-Louis Malus (1775–1812). Applying this formula, the transmitted intensity is unchanged if $\theta = 0$, and is zero if $\theta = 90°$. The orientation of the polarizer can be adjusted to give a beam of transmitted light of variable intensity and polarization.

Consider vertically polarized light with an intensity (I_o) of 500 W/m^2 passing through a polarizer oriented at an angle of 45° to the vertical. The transmitted light intensity, (I), from the polarizer calculated by Malus's law, Equation (15.6), gives:

$$I = I_o \cos^2\theta = 500 \times \cos^2(45) = 250 \text{ W/m}^2$$

Figure 15.13 shows an unpolarized light beam with intensity, (I_o), incident on a polarizer. The light that passes through the polarizer is vertically polarized. In this case, there is no single angle θ; instead, to obtain the transmitted intensity, the $\cos^2\theta$ over all angles must be averaged. The average of $\cos^2\theta$ is one-half. Thus, the transmitted light intensity varies as the square of the transmitted light amplitude. The transmitted intensity of a polarized light is half of the intensity of an unpolarized light, as given:

$$I = \frac{I_o}{2} \tag{15.7}$$

Figure 15.14 shows an unpolarized light beam incident on the first polarizer, where the transmission axis is vertically indicated. The light that passes through the polarizer is vertically polarized. A second polarizer is lined up after the first polarizer, called the analyser. The analyser axis is at an angle θ to the polarizer axis. The intensity of the transmitted polarized light varies as Malus's law.

Consider unpolarized light with an intensity (I_o) of 600 W/m^2 passing through a polarizer and analyser. The axis of the polarizer is vertical. The analyser axis is at an angle 45° to the polarizer axis. The transmitted intensity (I_1) through the polarizer calculated using Equation (15.7) gives:

$$I_1 = \frac{I_o}{2} = \frac{600}{2} = 300 \text{ W/m}^2$$

The transmitted intensity through the analyser calculated by Malus's law, Equation (15.6), gives:

$$I_2 = I_1 \cos^2\theta = 300 \times \cos^2(45) = 150 \text{ W/m}^2$$

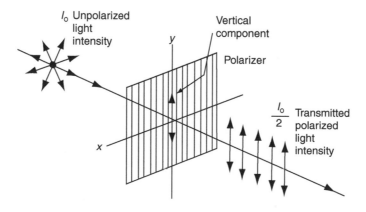

FIGURE 15.13 Transmission of unpolarized light through a polarizer.

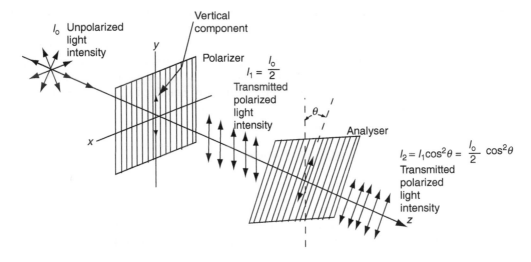

FIGURE 15.14 Transmission of unpolarized light through a polarizer and analyser.

15.8 OPTICAL ACTIVITY

Some optical materials, such as quartz, sugar, tartaric acid, and turpentine, are known as optically active because they rotate the plane of polarized light. The French physicist Dominique F.J. Arago first observed this phenomenon in 1811. The rotation of the polarized axis is due to the molecular structure of the optically active material. Figure 15.15 shows a linearly polarized light undergoing a continuous rotation as it propagates along the optic axis of a quartz material. The rotation may be clockwise or counterclockwise, depending on the molecular orientation. A water solution of cane sugar (sucrose) rotates the plane of polarization to clockwise. For a given path length through the solution, the angle of rotation is proportional to the concentration of the solution. Scientists have found numerous applications for the optically active materials in analytical procedures. For example, instruments for measuring the angle of rotation under standard conditions are known as polarimeters. These instruments are used specially for measuring the concentration of sugar solution and are known as saccharimeters. A standard method for determining the concentration of sugar solutions is to measure the rotation angle produced by a fixed length of the solution.

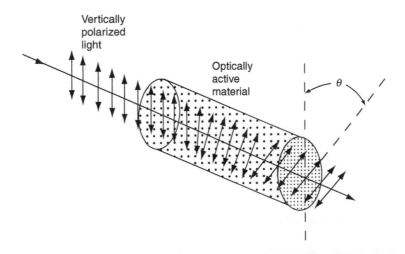

FIGURE 15.15 An optically active material rotates the direction of the polarization.

A Faraday rotator is an optical component that rotates the polarization of light due to the Faraday effect. The Faraday rotator works because one polarization of the input light is in ferromagnetic resonance with the material, which causes its phase velocity to be higher than the other. Specifically, given two rays of circularly polarized light, one with left-hand and the other with right-hand polarization, the phase velocity of the one with the polarization in the same sense as the magnetization is greater. In other words, the plane of linearly polarized light is rotated, when a magnetic field is applied parallel to the propagation direction.

The angle of rotation can be calculated by:

$$\theta = \nu B d \tag{15.8}$$

where

    θ is the angle of rotation in (radians),
    ν is the Verdet constant for the material in (arc minutes $\text{cm}^{-1}\,\text{Gauss}^{-1}$),
    B is the magnetic flux density in the direction of the propagation in (teslas), and
    d is the length of the path in (metres).

This equation is proportionally constant (in radians per tesla per metre) varies with wavelength of the light and temperature. The angle θ also varies with various materials.

Faraday rotators (for fixed and tunable wavelenghts) are used in optical isolators to prevent undesired back reflection of light in reverse direction from disrupting or damaging an optical system.

15.9 PHOTOELASTICITY

Some optical materials become birefringent when they are subjected to mechanical stress. Materials, such as glass and plastic, become optically active when placed under any kind of

(a) Specimen under low stress.   (b) Specimen under high stress.

FIGURE 15.16 Photoelastic stress analysis for a photoelastic specimen. (a) Specimen under low stress. (b) Specimen under high stress.

stress. Photoclasticity is the basis of a technique for studying the stress analysis in both transparent and opaque materials, which was discovered by Sir David Brewster in 1816. If polarized light passes through an unstressed piece of plastic and then through an analyser with an axis perpendicular to that of the polarizer, none of the polarized light is transmitted. However, if the plastic is subjected to a stress, the regions of greatest stress produce the largest angles of rotation of polarized light. Therefore, a series of light and dark bands can be observed, as shown in Figure 15.16. Manufacturers of automobiles, bridges, machine parts, and building materials use models made of photoeleastic materials to study the effects of stresses on various structures, because the strains in the structures show clearly when analysed by polarized light. Very complicated stress distributions can be studied by these optical techniques.

15.10 LIQUID CRYSTAL DISPLAY

As briefly explained in the optical instruments for viewing chapter, the optical liquid crystal display (LCD) is used for display units. The LCD units can be used as a display panel in cellular phones, calculators, pocket television screens, computer screens, diagnostic and measuring devices, and so forth, as shown in Figure 15.17.

In 1888, the Austrian botanist named Friedrich Reinitzer (1857–1927) observed that cholesterol benzoate has a two-melting-point phonomena, one at which the crystal changed into a cloudy liquid and another where the crystal became transparent at high temperature. This crystal is know now as a liquid crystal (LC). Liquid crystals have long cigar-shaped molecules that can move about like ordinary liquids. There are three types of liquid crystal distinguished by the way in which its molecules align. The common liquid crystal is the nematic, where the molecules tend to be more or less parallel, even though their positions are fairly random.

To prepare a parallel nematic cell, one face of each to two pieces of flat glass is coated with a transparent electrically conducting metallic film. These two windows will also serve as the electrodes where a controlled voltage can be applied. These windows are etched carefully by indium tin oxide producing parallel microgrooves. This allows the LC molecules in contact with the windows to be oriented parallel to the glass and to each other. When a thin space about a few micrometers between two each prepared glass windows is filled with nematic LC, the molecules

(a) LCD for numberic display.

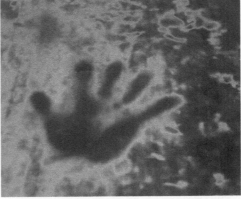

(b) LCD for colour display.

FIGURE 15.17 Liquid crystal display. (a) LCD for numberic display. (b) LCD for color display.

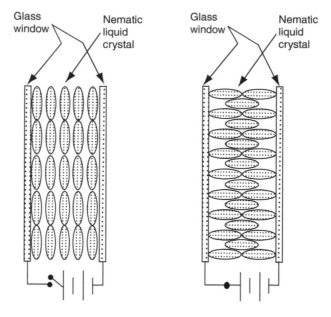

(a) A Nematic LC between two   (b) With applied voltage, the molecules
    transparent electrodes.   rotate into alignment with the field.

FIGURE 15.18 Liquid crystal operation. (a) A Nematic LC between two tranparent electrodes (b) With applied voltage, the molecules rotate into alignment with the field.

in contact with the mircogrooves anchor themselves parallel to the ridges. The LC molecules then essentially drag each other into alignment. Soon the entire liquid is similarly oriented, as shown in Figure 15.18(a). The direction in which the molecules of a liquid crystal are aligned is called the director.

When a voltage is applied across the LC, an electric field is created perpendicular to the glass windows. Electric diploes are either present or induced, and the LC molecules experience torques that cause them to rotate into alignment with the field. As the the voltage increases more and more molecules turn toward the direction of the field, as shown in Figure 15.18(b).

Figure 15.19 illustrates the operation of an LCD. The LCD consists of a liquid crystal sandwiched between crossed polarizers sheets (first vertical axis and second horizontal axis) and backed by a mirror. Figure 15.19(a) shows that unpolarized light enters the first polarizer (vertical axis), immediately linearly vertical polarized light exits, is rotated 90°, passes the second polarizer, is incident on the mirror and reflected, and again rotated 90° by liquid crystal, and leaves the LCD. After the return trip through the liquid crystal, the polarization direction of the light is the same of the first polarizer. Thus, the light is transmitted and leaves the display unit. Hence, the display appears lighted when illumimated.

When a voltage is applied to the liquid crystal (or heat is applied in the case of coloured LC), the liquid crystal reorients itself and loses its ability to rotate the plane of polarization, as shown in Figure 15.19(b). Vertical polarized light enters and leaves the liquid cell unchanged, only to be completely absorbed by the second polarizer. Thus, the entrance window is now black and no light emerges.

Since very little energy is required to give the voltage necessary to turn a liquid crystal cell on, a LCD is very energy efficient. In addition, the LCD uses light already present in the environment. A LCD does not need to produce its own light, as do similar displays.

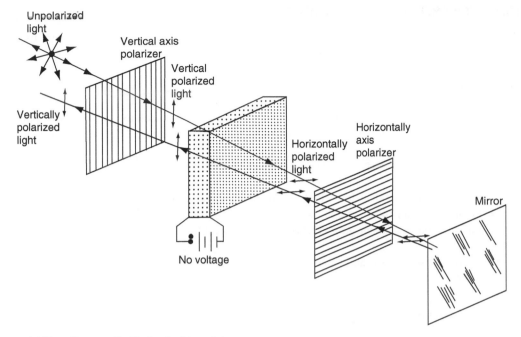

(a) No voltage applied to the liquid crystal

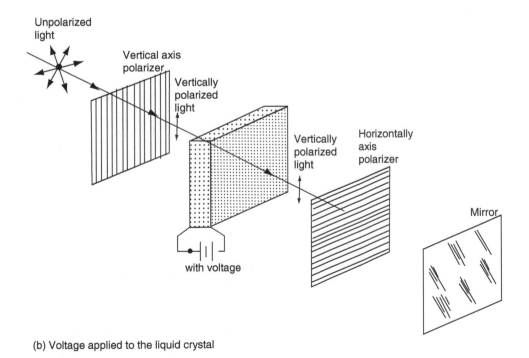

(b) Voltage applied to the liquid crystal

FIGURE 15.19 Liquid crystal display operation. (a) No voltage applied to the liquid crystal. (b) Voltage applied to the liquid crystal.

15.11 POLARIZATION MAINTAINING FIBRES

Design, construction, and applications of the polarization maintaining (PM) fibres are very important in fibre optic cable technology and communication systems.

15.12 POLARIZATION LOSS

Polarization losses play an important role in many devices, such as isolators, modulators, polarization beamsplitters, amplifiers, and polarization maintaining fibres. The polarization losses include:

- Polarization mode dispersion (PMD) in optical fibres,
- Polarization dependent loss (PDL) in passive optical components,
- Polarization dependent modulation (PDM) in electro-optic modulators, and
- Polarization dependent gain (PDG) in optical amplifiers.

Some of these losses and measurements are generally explained in detail in fibre optic testing.

15.13 EXPERIMENTAL WORK

In this experiment the student will perform the following cases:

- a. Light passing through a polarizing cube beamsplitter. Refer to the materials presented in prisms and beamsplitters chapters.
- b. Glan Thompson polarizing beamsplitter. Refer to the materials presented in prisms and beamsplitters chapters.
- c. Light passing through a calcite material. The students will test light passing through a birefringence of calcite material, as shown in Figure 15.20. They will also measure the power of the laser beams exiting from the calcite.
- d. The law of Malus.

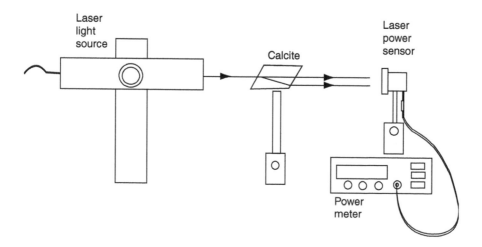

FIGURE 15.20 Light passes through a calcite.

In this case, the student will verify the law of Malus. This law states that when linearly polarized light goes through a polarizer, the transmitted light intensity is dependent on the axis angle of the polarizer with respect to the direction of the polarized light axis, as shown in Figure 15.12. This relation is given by Equation (15.6).

15.13.1 TECHNIQUE AND APPARATUS

Appendix A presents the details of the devices, components, tools, and parts.

1. 2×2 ft. optical breadboard.
2. HeNe laser light source and power supply.
3. Laser light sensor.
4. Laser light power meter.
5. Laser mount assembly.
6. Hardware assembly (clamps, posts, screw kits, screwdriver kits, sundry positioners, etc.).
7. Polarizing sheet and sheet holder, as shown in Figure 15.10.
8. Polarizing cube beamsplitter and holder/positioner assembly, as shown in Figures 15.21 and 15.24.
9. Glan Thompson polarizing beamsplitter and holder/positioner assembly, as shown in Figure 15.22.
10. Two disc polarizers and holder/positioner assemblies, as shown in Figure 15.23.
11. Calcite material and calcite holder/positioner assembly, as shown in Figure 15.26.
12. Black/white card and cardholder.
13. Protractor.
14. Ruler.

FIGURE 15.21 Polarizing cube beamsplitter.

FIGURE 15.22 Glan Thompson polarizing beamsplitter.

FIGURE 15.23 Disc polarizer and holder.

15.13.2 Procedure

Follow the laboratory procedures and instructions given by the professor and/or instructor.

15.13.3 Safety Procedure

Follow all safety procedures and regulations regarding the use of optical components and instruments, light source devices, and optical cleaning chemicals.

15.13.4 APPARATUS SET-UP

15.13.4.1 Light Passing through a Polarizing Cube Beamsplitter

1. Figure 15.24 shows the apparatus set-up.
2. Bolt the laser short rod to the breadboard.
3. Bolt the laser mount to the clamp using bolts from the screw kit.
4. Put the clamp on the short rod.
5. Place the HeNe laser into the laser mount and tighten the screw.
6. Turn on the laser device. Follow the operation and safety procedures of the laser device in use. Allow sufficient time to stabilize the laser beam polarization.
7. Check the laser alignment with the line of bolt holes and adjust when necessary.
8. Mount a cube beamsplitter holder/positioner to the breadboard.
9. Mount the polarizing cube beamsplitter on the beamsplitter holder/positioner, as shown in Figure 15.24.
10. Carefully align the laser beam so that the laser beam impacts normally to the centre of the centre of the cube beamsplitter face.
11. Try to capture the laser beam exiting the cube beamsplitter on the black/white card, as shown in Figure 15.24.
12. Turn off the lights of the lab before taking measurements.
13. Measure the power of the laser beam at the input face and output faces of the cube beamsplitter. Fill out Table 15.1 for Case (a).
14. Place the polarizing sheet to intersect the laser beam at the input face and output faces of the prism.
15. Observe the laser beam orientation using the polarizing sheet at the input face and output faces of the prism. Fill out Table 15.1 for Case (a).

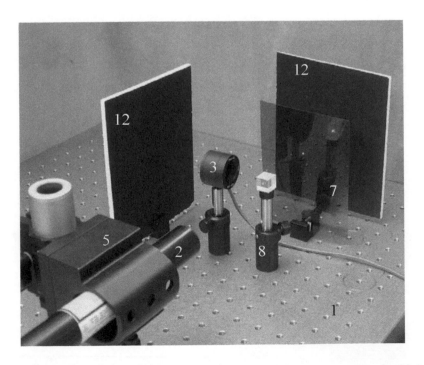

FIGURE 15.24 Polarizing cube beamsplitter apparatus set-up.

16. Rotate the polarizing sheet 90° and observe the laser beam orientation exiting from the polarizing sheet at the input face and output faces of the prism.
17. Turn on the lights of the lab.

15.13.4.2 Glan Thompson Polarizing Beamsplitter

Figure 15.25 shows the apparatus set-up for Case (b). Repeat the procedure from step 1 to 8 in Case (a) using a Glan Thompson polarizing beamsplitter; add the following steps:

1. Mount a Glan Thompson polarizing beamsplitter holder/positioner to the breadboard.
2. Mount a Glan Thompson polarizing beamsplitter on the beamsplitter holder/positioner, as shown in Figure 15.25.
3. Carefully align the laser beam so that the laser beam impacts normally to the centre of the face of the Glan Thompson polarizing beamsplitter.
4. Try to capture the laser beam exiting from the face of the Glan Thompson polarizing beamsplitter on the black/white card. Two laser spots can be caught on the black/white card of the laser beam components exiting from the output faces of the Glan Thompson polarizing beamsplitter.
5. Turn off the lights of the lab before taking measurements.
6. Measure the laser power at the input face and output faces of the Glan Thompson polarizing beamsplitter. Fill out Table 15.1 for Case (b).
7. Place the polarizing sheet to intersect the laser beam at the input face and output faces of the prism and observe the laser beam on the card.
8. Observe the laser beam orientation using the polarizing sheet at the input face and output faces of the beamsplitter. Fill out Table 15.1 for Case (b).
9. Rotate the polarizing sheet 90° and observe the laser beams on the card.
10. Turn on the lights of the lab.

FIGURE 15.25 Glan Thompson polarizing beamsplitter apparatus set-up.

FIGURE 15.26 Calcite material apparatus set-up.

15.13.4.3 Light Passing through a Calcite Material

Figure 15.26 shows the apparatus set-up for Case (c). Repeat the procedure from step 1 to 8 in Case (a) using a calcite material; add the following steps:

1. Mount a calcite holder/positioner to the breadboard.
2. Mount the calcite on the calcite holder/positioner, as shown in Figure 15.26.
3. Carefully align the laser beam so that the laser beam impacts normally to the centre of the calcite face.
4. Try to capture the laser beams exiting the calcite on the black/white card, as shown in Figure 15.27.
5. Turn off the lights of the lab before taking measurements.

FIGURE 15.27 Laser beams exiting from the calcite material.

6. Measure the power of the laser beam at the input face and the two laser beams at the output face of the calcite.
7. Mount the polarizing sheet in a sheet holder.
8. Place the polarizing sheet to intersect the laser beam at the input face and output face of the calcite.
9. Observe the laser beam orientation using the polarizing sheet at the input face and output face of the calcite. Fill out Table 15.1 for Case (c).
10. Rotate the polarizing sheet 90° and observe the laser beam orientation exiting from the polarizing sheet at the input face and output face of the calcite.
11. Turn on the lights of the lab.

15.13.4.4 The Law of Malus

Figure 15.28 shows the apparatus set-up for Case (d). Repeat the procedure from step 1 to 8 in Case (a), using unpolarized laser light source or any unpolarized light source. Add the following steps:

1. Mount two polarizer holders/positioners to the breadboard at a distance of 15 cm apart. The two polarizers are lined up parallel to each other and perpendicular to the laser beam.
2. Mount one polarizer on each polarizer holder/positioner.
3. Mount unpolarized laser light source to the breadboard.
4. Carefully align the laser beam so that the laser beam impacts normally to the centre of the first and second polarizer face.
5. Turn off the lights of the lab before taking measurements.
6. Turn on the laser source power.
7. Measure the laser beam power from the laser source. Fill out Table 15.2.
8. Place the laser sensor after the first polarizer. Rotate the polarizer slowly and watching the laser power meter reading of the laser light.

FIGURE 15.28 The law of Malus apparatus set-up.

9. Rotate the polarizer until the power meter reading is at the maximum. Fix the position of the polarizer. Measure the power of the laser light beam. Fill out Table 15.2.
10. Measure the power of the laser beam exiting form the second polarizer. Fill out Table 15.2.
11. Rotate the second polarizer in $10°$ increments for a total of $180°$, using the rotating knob of the polarizer holder/positioner. Measure the power of the laser beam at each angle. Fill out Table 15.2.
12. Turn on the lights of the lab.

15.13.5 Data Collection

15.13.5.1 Light Passing through a Polarizing Cube Beamsplitter

1. Measure the laser beam power at the input face and output faces of the cube beamsplitter.
2. Observe the laser beam orientation using the polarizing sheet at the input face and output faces of the cube beamsplitter.
3. Fill out Table 15.1 with the collected data for Case (a).

TABLE 15.1
Polarizing Beamsplitters and Calcite

Case	Beamsplitter	Laser Power Input (unit)	P Component Power (unit)	P Component Orientation Axis	S Component Power (unit)	S Component Orientation Axis	Splitting Ratio
(a)	Polarizing Cube Beamsplitter						
(b)	Glan Thompson Polarizing						
(c)	Calcite Material						

15.13.5.2 Glan Thompson Polarizing Beamsplitter

1. Measure the laser beam component power at the input face and output faces of the Glan Thompson polarizing beamsplitter.
2. Observe the laser beam orientation, using the polarizing sheet at the input face and output faces of the Glan Thompson polarizing beamsplitter.
3. Fill out Table 15.1 with the collected data for Case (b).

15.13.5.3 Light Passing through a Calcite Material

1. Measure the laser beam component power at the input face and the two laser beams at the output face of the calcite.
2. Observe the laser beam orientation, using the polarizing sheet at the input face and output face of the calcite.
3. Fill out Table 15.1 with the collected data for Case (c).

TABLE 15.2
Application of the Law of Malus

Laser Light Beam Power I_0 (unit)	Measured Laser Light Beam Power I_1 (unit)	Calculated Laser Light Beam Power I_1 (unit)	Rotation Angle θ (degrees)	Measured Laser Light Beam Power I_2 (unit)	$\cos^2\theta$	Calculated Laser Light Beam Power $I_2 = I_1\cos^2\theta$ (unit)
			0			
			10			
			20			
			30			
			40			
			50			
			60			
			70			
			80			
			90			
			100			
			110			
			120			
			130			
			140			
			150			
			160			
			170			
			180			

15.13.5.4 The Law of Malus

1. Measure the power of the laser beam before and after exiting from the first and second polarizer.
2. Fill out Table 15.2 with the collected data for this case.

15.13.6 CALCULATIONS AND ANALYSIS

15.13.6.1 Light Passing through a Polarizing Cube Beamsplitter

1. Compare the laser beam power at the input face and output faces of the cube beamsplitter.
2. Calculate the splitting ratio of the cube beamsplitter. Fill out Table 15.1 with the collected data for Case (a).

15.13.6.2 Glan Thompson Polarizing Beamsplitter Cube

1. Compare the laser beam power at the input face and output faces of the Glan Thompson polarizing beamsplitter.
2. Calculate the splitting ratio of the Glan Thompson polarizing beamsplitter. Fill out Table 15.1 with the collected data for Case (b).

15.13.6.3 Light Passing through a Calcite Material

1. Compare the laser beam power at the input face and the two laser beams at the output face of the calcite.
2. Calculate the splitting ratio of the calcite material. Fill out Table 15.1 with the collected data for Case (c).

15.13.6.4 The Law of Malus

1. Calculate the power of the light beam exiting from the first polarizer. Fill out Table 15.2.
2. Calculate $\cos^2\theta$ and the transmitted power I_2 from the second polarizer using Malus's law, Equation (15.6). Fill out Table 15.2.
3. Plot I_2 the measured transmitted power of the laser beam against angle θ.
4. Plot I_2 the calculated transmitted power of the laser beam against angle θ.

15.13.7 RESULTS AND DISCUSSIONS

15.13.7.1 Light Passing through a Polarizing Cube Beamsplitter

1. Report the laser beam power measurements at the input face and output faces of the cube beamsplitter and the splitting ratio.
2. Verify the results with the technical specifications provided by the manufacturer.
3. Determine the polarization orientation of the laser beam at the input face and output faces of the cube beamsplitter.

15.13.7.2 Glan Thompson Polarizing Beamsplitter

1. Report the laser beam power measurements at the input face and output faces of the Glan Thompson polarizing beamsplitter and the splitting ratio.
2. Verify the results with the technical specifications provided by the manufacturer.
3. Determine the polarization orientation of the laser beam at the input face and output faces of the Glan Thompson polarizing beamsplitter.

15.13.7.3 Light Passing through a Calcite Material

Report the laser beam power measurements at the input face and output faces of the calcite material and the splitting ratio.

Verify the results with the technical specifications provided by the manufacturer.

Determine the polarization orientation of the laser beam at the input face and output faces of calcite material.

15.13.7.4 The Law of Malus

Report the power of the light beam exiting from the first polarizer using Equation (15.7) with the measured data.

Present the plotted curves with the measured and calculated data. Comment on the shape of the curves.

15.13.8 CONCLUSION

Summarize the important observations and findings obtained in this lab experiment.

15.13.9 SUGGESTIONS FOR FUTURE LAB WORK

List any suggestions for improvements using different experimental equipment, procedures, and techniques for any future lab work. These suggestions should be theoretically justified and technically feasible.

15.14 LIST OF REFERENCES

List any references that were used in the report. Use one format in writing the references. Never mix reference formats in a report.

15.15 APPENDICES

List all of the materials and information that are too detailed to be included in the body of the report.

FURTHER READING

Astle, T.B., Gilbert, A. R., Ahmed, A., and Fox, S., Optical components—The planar revolution? Merrill Lynch & Co., The Global Telecom Equipment—Wireline, In-depth Report, U.S.A., May 17, 2000.

Alkeskjold, T. T., Laegsggaard, J., Bjarklev, A., Hermann, D., Anawati, A., Broeng, J., Li, J., and Wu, S. T., All optical modulation in dye-doped nematic liquid crystal photonic bandgap fibers, *Optics Express*. vol. 12(24), pp. 5857–5871, November 20, 2004.

Blaker, W. J., *Optics: The Matrix Theory*, Marcel Dekker, New York, 1972.

Brower, D. L., Ding, W. X., and Deng, B. H., *Fizeau Interferometer for Measurement of Plasma Electron Current*, Electrical Engineering Department, University of California, Los Angeles, U.S.A., vol. 75(10), pp. 3399–3401, October 2004.

Callister, W. D. Jr., *Materials Science and Engineering—An Introduction*, 3rd ed., John Wiley & Sons, New York, U.S.A., 1994.

Chee, J. K. and Liu, J. M., Polarization-dependent parametric and raman processes in a birefringent optical fiber, *IEEE J. Quantum Electron.*, 26, 541–549, 1990.

Chen, J. H., Su, D. C., and Su, J. C., Holographic spatial walk-off polarizer and its application to a 4-port polarization independent optical circulator, *Opt. Express*, 11 (17), 2001–2006, 2003.

Chen, J. H., Su, D. C., and Su, J. C., Multi-port polarization-independent optical circulators by using a pair of holographic spatial- and polarization- modules, *Opt. Express*, Vol. 12 (4), 601–608, 2004.

Drisoll, W. G. and Vaughan, W., *Handbook of Optics*, McGraw Hill Book Company, New York, 1978.

EPLAB, *EPLAB Catalogue 2002*, The Eppley Laboratory, Newport, RI, 2002.

Fedder, G. K. and Howe, R. T., Multimode digital control of a suspended polysilicon microstructure, *IEEE J. MEMS*, 5 (4), 283–297, 1996.

Francon, M., *Optical Interferometry*, Academic Press, New York pp. 97–99, 1966.

Hariharan, P., Modified Mach-Zehnder interferometer, *Appl. Opt.*, 8 (9), 1925–1926, 1969.

Hecht, J., *Understanding Fiber Optics*, 3rd ed., Prentice Hall, Inc., Englewood Cliffs, NJ, U.S.A., 1999.

Hecht, E., *Optics*, 4th ed., Addison-Wesley Longman, Inc., San Diego, CA, U.S.A., 2002.

Hood, D. C. and Finkelstein, M. A., In *Sensitivity to Light Handbook of Perception and Human Performance Vol. 1 of Sensory Processes and Perception*, Boff, K. R., Kaufman, L., and Thomas, J. P., Eds., Wiley, Toronto, 1986.

Lambda Research Optics, Inc., *Lambda Catalog 2004*, Lambda Research Optics, Inc., Costa Mesa, California, 2004.

Lerner, R. G. and Trigg, G. L., *Encyclopedia of Physics*, 2nd ed., VCH Publishers, Inc., New York, 1991.

Melles Griot, The practical application of light, Melles Griot, Inc., *Melles Griot Catalog*, Rochester, NY, 2001.

Newport Corporation, *Projects in Fiber Optics Applications Handbook*, Newport Corporation, Newport, RI, U.S.A., 1986.

Newport Corporation. Optics and mechanics section, the Newport Resources 1999/2000 Catalog, Newport Corporation, Irvine, CA, 1999/2000.

Newport Corporation, Photonics Section, the Newport Resources 2004 Catalog, Newport Corporation, Irvine, CA, U.S.A., 2004.

Ocean Optics, Inc., *Product Catalog 2003*, Ocean Optics, Inc., Florida, U.S.A., 2003.

Palais, J. C., *Fiber Optic Communications*, 4th ed., Prentice Hall, Inc., Englewood Cliffs, NJ, 1998.

Salah, B. E. A. and Teich, M. C., *Fundamentals of Photonics*, Wiley, New York, 1991.

Shamir, J., *Optical Systems and Processes*, SPIE Optical Engineering Press, Washington, DC, 1999.

Shen, L. P., Huang, W. P., Chen, G. X. et al., Design and optimization of photonic crystal fibers for broad-band dispersion compensation, *IEEE Photon. Technol. Lett.*, 15, 540–542, 2003.

Smith, W. J., *Modern Optical Engineering*, McGraw-Hill Book Company, New York, 1966.

Von Weller, E. L., Photonics crystal fibers advances in fiber optics, *Appl. Opt.*, March 1, 2005.

Woods, N., *Instruction's Manual to Beiser Physics*, 5th ed., Addison-Wesley Publishing Company, San Diego, CA, 1991.

Yariv, A., *Optical Electronics*, Wiley, New York, 1997.

Zhang, L. and Yang, C., Polarization celective coupling in the holey fiber with Aasymmetric dual cores. LEOS2003, Tucson, U.S.A., paper ThP1, Oct 26–30, 2003.

Zhang, L. and Yang, C., Polarization splitter based on photonic crystal fibers, *Opt. Express*, 11 (9), 1015–1020, 2003.

Zhang, L. and Yang, C., Polarization splitting with dual-core holey fibers., CLEO/PACIFIC RIM 2003, Taipei Taiwan, paper W4C-(14)-3, 2003.

16 Optical Materials

16.1 INTRODUCTION

Optical materials are used in the construction of devices that guide light in some way. These materials refract light. This can be observed when light passes through a prism or a lens. The angle through which light is refracted is a function of the wavelength of the light, so that when more than one frequency is present the light can be dispersed into different components. Most of us have observed the rainbow produced when white light is passed through a prism.

Optical materials can also be used to attenuate, or decrease the intensity of light. Under the right conditions, some can also amplify, increasing the intensity of light. Optical materials are used to transmit light, not only visible, but also light in the infrared and ultraviolet ranges. Historically, glass was the first optical material used in the fabrication of imaging systems, such as telescopes and microscopes. As modern applications have evolved, other materials better suited to extended wavelength ranges and other design considerations have been developed.

16.2 CLASSES OF MATERIALS

The following sections introduce different classes of materials used in the fabrication of optical devices and instruments.

16.2.1 GLASS

Glass is characterized by an amorphous molecular structure, where the chemical bonds holding the material together do not have a regular lattice, such as is found in crystals. Most glass used in optical applications is based on silica (i.e., silicon dioxide) with other materials mixed in to control various optical properties. This creates a new glass type with different index of refraction and sometimes a different colour. Pure silica has a melting point closer to 2000°C. Often soda or potash is added to lower the melting point to about 1000°C. When soda is added to glass, the mixture becomes slightly soluble in water. Lime (calcium oxide) needs to be added to the mix to return the material to an insoluble form.

Most glass types will transmit light well through the visible range, into the near infrared. In order to build optical systems with good transmission in the ultraviolet, it is necessary to employ more exotic materials, such as quartz, fused silica, and sapphire, as shown in Figure 16.1. Some of

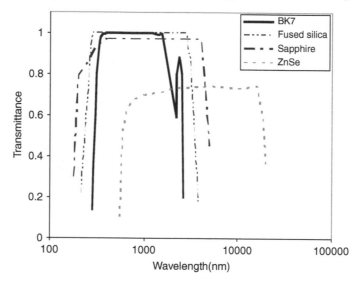

FIGURE 16.1 Transmission of various optical materials.

FIGURE 16.2 Glass lens formed by grinding and polishing with diamond tools.

the motivation for developing materials with improved ultraviolet transmission comes from the photolithography industry, which needs to work at shorter and shorter wavelengths.

At longer wavelengths, it is sometimes necessary to use other materials, which have transmission through to the far infrared. These materials include zinc selenide (ZnSe), zinc sulphide (ZnS), and germanium (Ge).

Usually, optical components are fabricated by grinding and polishing these optical materials. This process involves rotating the optic on a spindle and bringing a grinding or polishing tool in contact with it, as shown in Figure 16.2. The tool also rotates, and an abrasive is delivered in the form of a slurry. By reducing the grit size of the abrasive at various stages, the optical surface of the part can be made smoother.

16.2.2 DOPANTS

As discussed in the Section 16.2.1, dopants can be added to glass to change its optical, mechanical and thermal properties. Dopants can also be added to semiconductor materials to change the conductivity by selectively increasing the number of free electrons or holes in the material. For example, when silicon is doped with arsenic, there is an electron left over after the atom forms bonds with the silicon crystal. This electron can be easily promoted into the valence band by thermal energy, thus becoming a free carrier. In this case, the arsenic is a donor, and the material is called n-type.

If silicon is doped with boron, boron leaves one unpaired electron when it forms a bond in the silicon lattice. This unpaired electron can accept an electron from the valence band, which leaves behind a hole. A hole can move around in the lattice under the influence of an applied electric field. Here boron is an acceptor, and the material is termed p-type. Some dopants are shown in Figure 16.3.

FIGURE 16.3 Dopant materials.

A designer can manipulate the regions where different kinds of doping occurs, to design devices, such as light emitting diodes, diode lasers, as well as photo-detectors.

16.2.3 CO-DOPANTS

In some instances, both electrons donors and acceptors are doped into the optical material. Co-dopants can be added to glass to change its optical, mechanical and thermal properties.

Co-dopants can also be added to semiconductor materials to change the conductivity by selectively increasing the number of free electrons or holes in the material. Some of the free electrons will become trapped in the holes, and the free carriers will tend to cancel each other out. The net effect being that the dopant with the higher concentration will determine whether the material is n-type or p-type. In this case, the dopant with the lower concentration is called the co-dopant.

16.2.4 POLYMERS IN PHOTONICS

Optical polymers are transparent plastics, which can be used for lenses, optical storage media, such as compact discs, and for lightguides, metallized reflectors, optical fibres, and couplers. Materials commonly used include polycarbonates (PC), acrylics (polymethyl methacrylate (PMMA)), polystyrene (PS), allyl diglycol carbonate (ADC), and cyclic olefin copolymer (COC). These materials are lighter and cheaper to fabricate than glass, but have some additional challenges to overcome for demanding applications. A summary of some of the properties of these materials is given in Table 16.1.

TABLE 16.1
Properties of Some of the Optical Polymers

Property	PC	PMMA	PS	ADC	COC
Transmittance (%) in the 400–700 nm range of a 3 mm thick sample	85–91	92	87–90	89–91	92
Index of refraction	1.586	1.491	1.589	1.5	1.533
Abbe number	30.0–30.3	57.4	31	58	58
Density g/cm^3	1.2	1.18	1.06	1.31	1.01

These materials are usually injection-molded or cast, but can be diamond turned as well. Their operating temperature range is much smaller than that of glass, and extra care must be taken in applications where large optical power is transmitted.

16.3 APPLICATIONS

The following sections introduce the types of lenses used in building optic and optical fibre devices and instruments.

16.3.1 REFLECTORS

Reflections typically occur at the interface between two dissimilar materials, a common example is air and a coated or uncoated mirror. In the applications of reflectors, the bulk optical properties of the reflector material are unimportant. The properties considered will be the environmental properties of the material and its manufacturability.

Typically the surface accuracy of an optic is measured in terms of the wavelength of light. The HeNe laser's wavelength at 632 nm is most commonly used. In some applications, the optic is required to perform over a wide range of temperatures. In order to achieve the required performance of the optic, it is not only necessary to manufacture the optic within its required surface accuracy, but also to use a material that maintains that surface accuracy over the specified temperature range. Some materials undergo thermal expansion when heated, and in the case of reflectors, may undergo

FIGURE 16.4 Glass is often delivered as rough cut pieces or blanks, which must be ground or pressed to the approximate shape before polishing is started.

surface distortion, which may compromise the performance of the optic. Clearly, this is more of an issue for a reflecting mirror in a space telescope than for a flashlight reflector.

Figure 16.4 shows that glass is often delivered as roughly cut pieces or blanks, which must be ground or pressed to the approximate shape before polishing.

A variety of techniques can be used to fabricate a reflector. An inexpensive method is by molding. This process is suitable for plastics and glass. Glass may also be formed by pressing it in a metal mold at a temperature where it has softened.

Molding and pressing are not capable of producing high surface accuracy. Molded plastic optics will typically shrink by up to about 1%, after being released from the mold. Manufacturers can take this into account by using a mold with larger dimensions than required for the finished optic. Even so, surface irregularities of a few waves are typical. This means that the difference between the desired surface shape and the actual surface shape of the optic can be a few times the HeNe laser wavelength.

To fabricate optics with higher surface accuracy, other processes are used. Conventional grinding and polishing of glass will yield surfaces well below one wave in accuracy, and in some instances better than 1/50 wave.

Single point diamond turning is another process used to manufacture accurate surfaces. In this process, the optic substrate is rotated on a spindle, while a computer controlled diamond cutting tool is moved to cut the desired surface. Some metals, such as nickel, are suitable for diamond turning.

16.3.2 LENSES

The refractive indices of all materials change with wavelength. This can lead to optical distortions, such as chromatic aberration, occurring when the light source in the system is not monochromatic. An optical designer can reduce these distortions by using a combination of different glass types. The lens combination balances these refractive deviations in such a way that light of different

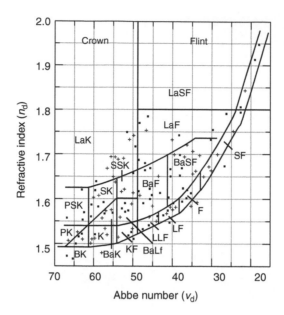

FIGURE 16.5 Index of refraction versus the Abbe number for different glasses.

wavelengths will focus at the correct position in the image plane. The glass map available to optical designers, shown in Figure 16.5, is a plot of the index of refraction versus the Abbe number for different glasses. The Abbe number is a measure of the chromatic dispersion of the glass: larger values correspond to lower dispersion.

16.3.3 Fibre Optics

Fibre optics used in communication systems suffer from chromatic aberration effects. In an optical fibre transmitting digital information in the form of light pulses, chromatic aberration manifests itself as dispersion. It is impossible to form light pulses from purely monochromatic light. In fact, there is an inverse relationship between how short a pulse can be made, and how much bandwidth or spread in frequencies is required from light source. In order to increase the data rate travelling through a fibre optic system, it is desirable to operate with shorter pulses, which in turn require a larger range of frequencies in a pulse. As these pulses travel through the fibre, the longer wavelength components tend to travel slightly faster, since they encounter a slightly lower index of refraction. This causes the pulse to be stretched in time, and the amount it stretches increases with fibre length. At some point in a long fibre, the tail end of one pulse will overlap the leading edge of the next pulse, which limits the ability of the system to transmit information accurately.

In order to build systems that are able to transmit shorter pulses over long distances, fibre manufacturers attempt to use materials with low dispersion. In order to achieve low dispersion, dopants are added to the glass, to reduce dispersion in the wavelength range in which the fibre will be used.

Another important requirement for fibres used in long haul communication applications is low absorption. Most of the intrinsic absorption that occurs in fibres results from hydroxyl impurities in the glass core. Fibre manufacturers work hard to control their manufacturing processes, in order to minimize these impurities.

Light is guided in fibres by engineering an index of refraction difference between the core and the cladding of the fibre. In general, the core of the fibre needs to have a higher index of refraction in order to achieve this effect. Germanium, titanium or phosphorus can be added to glass, to increase the index of refraction. Boron or fluorine may be added to decrease the index of refraction.

Optical fibres can also be doped for other purposes. The fibre may be doped to provide a mechanism for the fibre to act as an optical amplifier. A fibre doped with erbium is used to act as an amplifier in the important communication wavelength range around 1550 nm.

16.3.4 MECHANICAL COMPONENTS

The materials used to make up the mechanical components of an optical system should also receive careful consideration. Strength and stiffness have ramifications in applications where the optical assembly may be subject to vibration and mechanical shock.

Some applications may have a weight limit; sometimes significant weight savings can be made by utilizing, for example, aluminum housings instead of steel. In order to reduce weight further, sometimes designers must use more exotic materials, such as titanium and carbon fibre composites.

Some systems must be designed to perform over large temperature ranges. This can be a particular challenge for infrared systems, as the index of refraction of many materials can be very temperature dependent. To compensate for the resulting change in focal lengths of components in these systems, one method uses materials with predictable thermal expansion properties. With careful mechanical design, it is possible to compensate for temperature dependent optical variations, by introducing temperature dependent mechanical changes. The optical and mechanical variables cancel each other out.

There are materials available to move optical components with the application of a bias voltage. One class of these materials is called piezoelectric. These materials are crystals, which acquire a charge when they are mechanically compressed. Conversely, a piezoelectric undergoes a dimensional change when a voltage is applied across it. Piezoelectric can be used to act as transducers for adjusting the position of optics. They have also been used to build telescopes with improved performance when viewing through distortions caused by atmospheric turbulence. Piezoelectric transducers (PZTs) are installed on the backs of mirrors used in these telescopes in such a way that by adjusting voltages the optical surface of the mirror can be distorted. This distortion can be used to compensate for distortion in the atmosphere, such that the two distortions cancel each other out, and a clearer view of celestial objects can be made. This field of adaptive optics has revolutionized the use of earth-based telescopes for astronomy.

16.4 EXPERIMENTAL WORK

In this experiment, the voltage coefficient of a piezoelectric transducer will be measured. Hysteresis will also be observed.

16.4.1 TECHNIQUE AND APPARATUS

Appendix A presents the details of the devices, components, tools, and parts.

1. 2×2 ft. optical breadboard.
2. HeNe laser source and power supply.
3. Laser mount assembly.
4. Hardware assembly (clamps, posts, screw kits, screwdriver kits, sundry positioners, etc.).
5. A Michelson interferometer will be required. The end mirror on one of the arms should be mounted on a spring-loaded single axis translation stage, as shown in Figure 16.6.
6. Beamsplitter and Beamsplitter holder/positioner assembly.
7. Diverging lens and lens holder/positioner assembly, as shown in Figure 16.6.
8. Piezoelectric stack and translation stage, as shown in Figure 16.7.

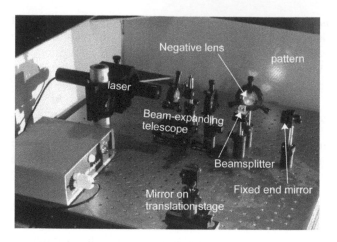

FIGURE 16.6 Interferometer set-up.

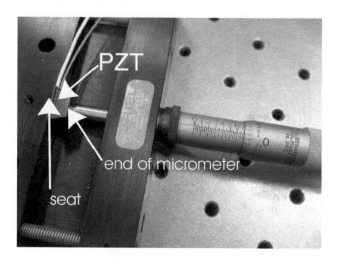

FIGURE 16.7 Installation of PZT into a translation stage.

    9. DC voltage supply, with sufficient range to adjust the PZT stack through several microns of travel, as shown in Figure 16.7.
   10. Voltmeter.
   11. Black/white card and cardholder.
   12. Protractor.
   13. Viewing screen.
   14. Ruler.

16.4.2 PROCEDURE

Follow the laboratory procedures and instructions given by the professor and/or instructor.

16.4.3 SAFETY PROCEDURE

Follow all safety procedures and regulations regarding the use of optical instruments and measurements, and light source devices.

16.4.4 Apparatus Set-up

16.4.4.1 Set Up of Interferometer

1. Figure 16.6 shows the experimental set-up for the interferometer.
2. Connect the PZT stack to the DC voltage supply.
3. Install the PZT stack in the spring-loaded single axis translation stage, as shown in Figure 16.7. The PZT should be placed in between the end of the spindle of the micrometre and the seat of the translation stage. Make sure that the wires do not interfere with the motion of the stage.
4. Connect the voltmeter to the DC voltage supply, to monitor the voltage applied to the PZT.
5. Install the single axis stage in the interferometer such that one of the arms could be lengthened by adjusting it.
6. Align the interferometer to the monochromatic source. Adjust the tilt of the end mirrors to get a few fringes at the output.
7. Install the diverging lens such that the output projected onto the viewing screen has a fringe spacing of a couple of centimetres.
8. Place the ruler perpendicular to the fringes on the screen.
9. Record the fringe spacing as measured on the viewing screen.
10. Setup a table for recording fringe displacement as a function of power supply voltage. Plan to take about 20 measurements through the voltage range of the DC voltage supply. Do not exceed the maximum voltage that the PZT can handle.
11. Mark the starting position of one of the fringes on the viewing screen. Turn on the power supply and ramp the voltage to the first point in the table. Do not reverse the direction of voltage adjustment. Record the distance the fringe has shifted from its original position and record the voltage.
12. Repeat for all of the voltages, up to the maximum range of the DC supply or the PZT, which ever is less.
13. Reverse the direction of voltage adjustment. Measure displacements at the same intervals, all the way down to zero voltage.
14. Reverse the positive and negative connections to the PZT at the power supply.
15. Repeat the measurements of fringe displacement through the range of the power supply in both the upward and downward direction. Remember that these voltages should be recorded with the opposite sign compared to the earlier measurements.

16.4.5 Data Collection

1. Record the distance the fringe has shifted from its original position and record the voltage. Repeat for all voltages.
2. Convert the displacement column from screen displacement to PZT displacement. Remember that the fringe spacing corresponds to one wavelength of extra travel in the arm of the interferometer (and that the light travels double pass through that arm).

16.4.6 Calculations and Analysis

1. Plot the data. Is there any observed hysteresis in the data?
2. Calculate the slope of the curve and determine the voltage coefficient.

16.4.7 Results and Discussions

1. Present and discuss the plot of the data and the observation of a hysteresis in the data.
2. Present the slope of the curve and the voltage coefficient.

16.4.8 CONCLUSION

Summarize the important observations and findings obtained in this lab experiment.

16.4.9 SUGGESTIONS FOR FUTURE LAB WORK

List any suggestions for improvements using different experimental equipment, procedures, and techniques for any future lab work. These suggestions should be theoretically justified and technically feasible.

16.5 LIST OF REFERENCES

List any references that were used in the report. Use one format in writing the references. Never mix reference formats in a report.

16.6 APPENDIX

List all of the materials and information that are too detailed to be included in the body of the report.

FURTHER READING

Agrawal, G. P., *Nonlinear Fiber Optics Optics and Photonics*, 2nd ed., Academic Press, New York, 1995.

Becker, P. M., et al., *Erbium-Doped Fiber Amplifiers: Fundamentals and Technology*, Elsevier, Amsterdam, 1999.

Charschan, S., *Lasers in Industry*, Van Nostrand, New York, 1972.

Davis, C. C., *Lasers and Electro-Optics, Fundamental and Engineering*, Cambridge University Press, New York/Melbourne, 1996.

Digonnet, M. J. F., *Rare-Earth Fiber Lasers and Amplifiers*, Marcel Dekker, New York, 2001.

Edmund Industrial Optics, *Optics and Optical Instruments Catalog*, New Jersey, USA, 2004.

Eggleton, B. J., et al., Microstructured optical fiber devices, *Opt. Express*, December 17, 698–713, 2001.

Fedder, G. K., Iyer, S., and Mukherjee, T., Automated optimal synthesis of microresonators, In *Technical Digestion of the IEEE International Conference on Solid-State Sensors and Actuators (Transducers 97)*, Vol. 2, Chicago, IL, pp. 1109–1112, June 16–19, 1997.

Ghatak, A. K., *An Introduction to Modern Optics*, McGraw-Hill Book Company, New York, 1972.

Griffel, G., et al., Low-threshold InGaAsP ring lasers fabricated using bi-level dry etching, *IEEE Photon. Technol. Lett.*, 12, 146–148, 2000.

Hecht, E., *Optics*, 4th ed., Addison-Wesley Longman, Inc., Reading, MA, 2002.

Heidrich, H., et al., Passive mode converter with a periodically tilted InP/GaInAsP rib waveguide, *IEEE Photon. Technol. Lett.*, 4, 34–36, 1992.

Hurvich, L. M. and Jameson, D., A psychophysical study of white I adoption as variant, *J. Opt. Soc. Am.*, 41, 521–527, 1951.

Hurvich, L. M. and Jameson, D., A psychophysical study of white III adoption as variant, *J. Opt. Soc. Am.*, 41, 701–709, 1951.

Jeong, Y., et al., Ytterbium-doped double-clad large-core fibers for pulsed and CW lasers and amplifiers, *SPIE*, 140–150, 2004.

Kao, C. K., *Optical Fiber Systems: Technology, Design, and Applications*, McGraw-Hill, New York, 1982.

Koga, M., Compact quartzless optical quasi-circulator, *Electron. Lett.*, 30, 1438–1440, 1994.

Lerner, R. G. and Trigg, G. L., *Encyclopedia of Physics*, 2nd ed., VCH Publishers, Inc., New York, 1991.

Malacara, D., *Geometrical and Instrumental Optics*, Academic Press Co, Boston, MA, 1988.

McComb, G., *The Laser Cookbook- 88 Practical Projects*, Tab Book, Division of McGraw-Hill, Inc., New York, 1988.

Newport Corporation, *Optics and Mechanics*, Newport, 1999/2000 Catalog.

Salah, B. E. A. and Teich, M. C., *Fundamentals of Photonics*, Wiley, New York, 1991.

SCIENCETECH, *Designers and Manufacturers of Scientific Instruments Catalog*, London, Ontario, Canada, 2003.

Serway, R. S., *Serway: Physics for Scientists and Engineers*, 3rd ed., Saunders Golden Sunburst Series, London, 1990.

Shashidhar, N., *Lensing Technology*, Corning Incorporated, Fiber Product News, pp. 14–15, 2004.

Shen, L. P., Huang, W. P., Chen, G. X., and Jian, S. S., Design and optimization of photonic crystal fibers for broad-band dispersion compensation, *IEEE Photon. Technol. Lett.*, 15, 540–542, 2003.

Smith, W. J., *Modern Optical Engineering*, McGraw-Hill Book Co., New York, 1966.

Thompson, G. H. B., *Physics of Semiconductor Laser Device*, Wiley, Chichester, 1980.

Vail, E., et al., GaAs micromachined widely tunable Fabry–Perot filters, *Electron. Lett.*, 31 (3), 228–229, 1995.

Weisskopf, V. F., How Light Interacts with Matter, Scientific American, pp. 60–71, 1968.

Yariv, A., *Optical Electronics*, Wiley, New York, 1997.

Yariv, A., Universal relations for coupling of optical power between microresonators and dielectric waveguides, *Electron. Lett.*, 36, 321–322, 2000.

Yeh, C., *Handbook of Fiber Optics: Theory and Applications*, Academic Press, San Diego, CA, 1990.

Yeh, C., *Applied Photonics*, Academic Press, New York, 1994.

Zirngibl, M., Joyner, C.H., and Glance, B., Digitally tunable channel dropping filter/equalizer based on waveguide grating router and optical amplifier integration, *IEEE Photon. Technol. Lett.*, 6, 513–515, 1994.

17 Photonics Laboratory Safety

17.1 INTRODUCTION

Our lives are filled with hazards created by electrical power supply, lasers, chemicals, and a diversity of equipment used in laboratories and classrooms. While studying science and engineering and performing experiments, students will learn to identify hazards and to protect themselves. Students will also learn to take care of their health and safety while working in laboratories. A safer and healthier learning and working environment should be created so that students have the opportunity to live safely and more healthily.

The following list of safety reminders is a brief compilation of generally accepted practices and should be adopted or modified to suit the unique aspects of each working environment, school policy, and local and/or set of Provincial and Federal codes. The intent of this chapter is to stimulate thinking about important safety considerations for students in laboratories.

17.2 ELECTRICAL SAFETY

The importance of electrical safety cannot be overstated. Electrical accidents can result in property damage, personal injury, and sometimes death. Ensuring electrical safety in laboratories and classrooms is important for students and staff. Students can learn to have a healthy respect for electricity and to spot potential electrical hazards anywhere. Respecting electricity does not mean that one should fear it; rather, one should just use it properly and wear personal protective equipment.

17.2.1 FUSES/CIRCUIT BREAKERS

The most common protection against property damage from circuit overloads (too much current) and overheating is the use of fuses and circuit breakers. All electrical circuits in laboratories are required to be protected by these means. When too much current flows in a circuit, the circuit

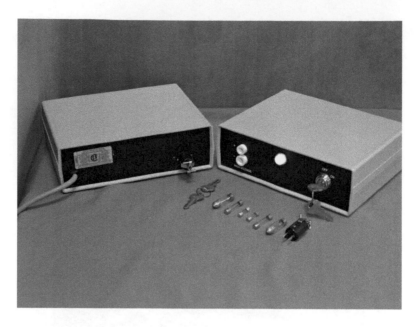

FIGURE 17.1 Types of fuses.

becomes hot and could melt the wire insulation, emit caustic fumes, and start a fire. An overload may also burn out and damage devices and instruments. Electronic equipment commonly has fuses to protect the components from overloads. A fuse is essentially a short strip of metal with a low melting point. When the current in a fused circuit exceeds the fuse rating, for example, 3, 5, 15, or 20 amps, the heat melts or vaporizes the fuse strip. The fuse blows, and the circuit is opened. Figure 17.1 shows types of fuses used in most laboratory electric instruments.

Fuses and circuit breakers should be the correct current rating for the circuit. If the correct rating is unknown, a certified electrician can identify and label it. A fuse should always be replaced with another of the same rating. Determine the reason why a fuse blew or a circuit breaker tripped, before replacing the fuse or resetting the breaker. Figure 17.1 also shows a few types of fuses and a power supply. The fuse will need to be plugged in on the back of the power supply. Plug in the electrical cord; turn on the key switch on the front panel; and turn the power supply on, as shown in Figure 17.1. After finishing with the power supply, remember to turn off the key and unplug the fuse.

A common problem is that the insulation may become worn on, for example, an extension cord, device wire, or instrument cord. If bare wires touch each other, or if a high-voltage or hot wire touches ground, this is called a short circuit, since the path of the circuit is effectively shortened. A low-resistance path to ground is created, causing a large current, which blows the protecting fuse.

Circuit breakers are more commonly used today instead of fuses in large equipment and houses, as shown in Figure 17.2. If the current in a circuit exceeds a certain value, the breaker is activated, and a magnetic relay (switch) breaks or opens the circuit. The circuit breaker switch can be reset or closed manually.

In either case, whether a circuit is opened when a fuse blows or when a circuit breaker trips, steps should be taken to remedy the cause. Remember, fuses and circuit breakers are safety devices. When fuses blow and open a circuit, they are indicating that the circuit is overloaded or shorted. Or they may be indicating the presence of another problem. In any case, a certified technician must investigate the source of the problem.

FIGURE 17.2 A circuit breaker panel.

17.2.2 Switches ON/OFF

Figure 17.3 shows samples of ON/OFF switches, which are used in computers, lighting systems, and instruments.

17.2.3 Plugs

Switches, fuses, and circuit breakers are always placed on the hot (high-voltage) side of the line, to interrupt power flow to the circuit element. Fuses and circuit breakers may not, however, always protect from electrical shock. To prevent shock, a grounding wire is used. The circuit is then completed (shorted) to ground, and the fuse in the circuit is blown. This is why many electrical

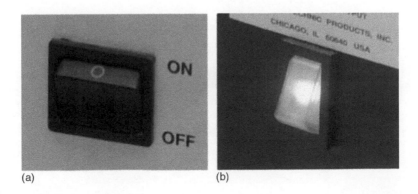

(a)   (b)

FIGURE 17.3 Examples of switches.

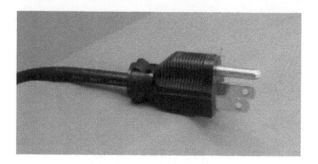

FIGURE 17.4 A three-prong plug.

tools and appliances have three-prong plugs, as shown in Figure 17.4. In the wall receptacle, this connection runs to ground.

When trying to plug in a two-prong plug that will not fit, do not use force. Instead, turn the plug over and try again. Figure 17.5 shows a two-prong plug. One of the prongs is bigger than the other, making the plug polarized. Polarizing in the electrical sense refers to a method of identification by which proper connections can be made. The original purpose of these types of plugs was to act as a safety feature. The small slit in the receptacle is the hot side, and the large slit is the neutral or ground side, if properly connected. The housing of an appliance could then be connected to the ground side all the time via a three-prong plug. A receptacle or appliance not wired (polarized) properly can be dangerous. The polarization is ensured with a dedicated third grounding wire as in a three-prong plug system, which is the accepted safety system. The original two-prong polarized plug system remains as a general backup safety system, provided it is wired properly.

Ensure the plug type fits the receptacle. Never remove the ground pin (the third prong) to make a three-prong plug fit into a two-conductor outlet; doing so could lead to an electrical shock. Never force a plug into an outlet if it does not fit. Plugs should fit securely into outlets. Avoid overloading electrical outlets with too many devices.

17.2.4 WALL OUTLETS

Figure 17.6 shows a wall outlet, which is used to connect computer and extension cords. Avoid using wall outlets with loose fitting plugs. They can overheat and lead to fire. Ask a certified technician to replace any missing or broken wall plates.

FIGURE 17.5 Two-prong plugs.

FIGURE 17.6 A wall outlet.

17.2.5 CORDS

Ensure the cords are in good condition. Check cords for cut, broken, or cracked insulation. Protect flexible cords and cables from physical damage. Ensure they are not placed in traffic areas. Cords should never be nailed or stapled to the wall, table, baseboard or to another object. Do not place cords under a device or computer; do not rest them under any object. Cords can create tripping hazards and may be damaged if walked upon. Allow slack in flexible cords to prevent tension on electrical terminals.

Check that extension power bars are not overloaded, as demonstrated in Figure 17.7. Figure 17.7(a) shows an overloaded extension power bar, while Figure 17.7(b) shows a bar not overloaded. Additionally, extension power bars should only be used on a temporary basis; they are not intended for use as permanent wiring. Ensure that the extension power bars have safety closures.

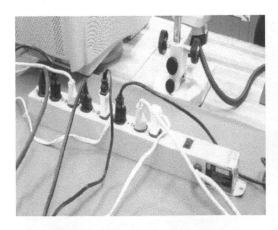
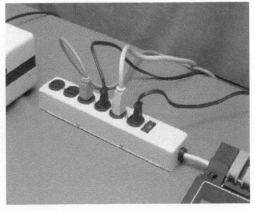

(a)   Extension power bar is overloaded.   (b)   Extension power bar is not overloaded.

FIGURE 17.7 An extension power bar.

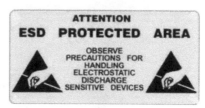

FIGURE 17.8 (See colour insert following page 200.) Electrostatic discharge warning symbols and signs.

17.2.6 GROUND FAULT CIRCUIT INTERRUPTERS

Ground fault circuit interrupters (GFCIs) can help prevent electrocution. They should be used in any area where water and electricity may come into contact, especially near a sink or basin. Water and electricity do not mix; they create an electrical shock. When a GFCI senses current leakage in

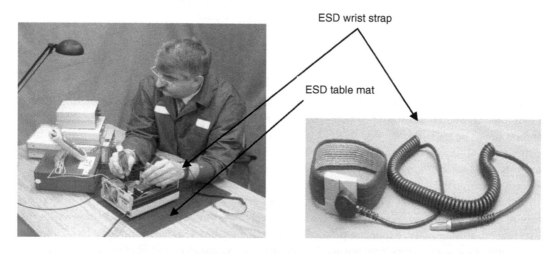

FIGURE 17.9 Working with ESD wrist strap and table mat.

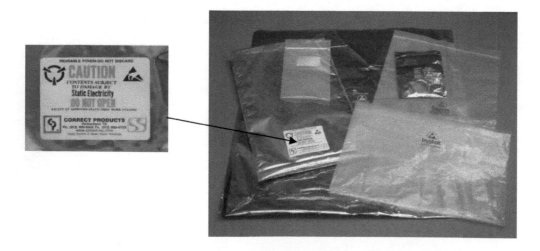

FIGURE 17.10 (See colour insert following page 200.) ESD bags.

an electrical circuit, it assumes that a ground fault has occurred. It then interrupts power quickly enough to help prevent serious injury due to electrical shock. GFCIs should be regularly tested according to the manufacturer's instructions to ensure they are working properly. Some benches are connected to true ground to be electrostatic discharge (ESD) compliant. This compliance is very important for devices and equipment that are very sensitive to ESD. Figure 17.8 shows ESD warning symbols and signs.

Figure 17.9 shows an ESD wrist strap and table mat used in handling an ESD sensitive device. The straps and mats should be connected to the true ground before handling a sensitive device. The strength of the charge on a human body is enough to destroy an ESD sensitive device. Each person should discharge his or her electrostatic charge before entering an environment sensitive to ESD. The discharge devices are usually located at the entrance of sensitive areas. An ESD heel strap is also available to wear when handling devices and walking in an environment sensitive to ESD.

Figure 17.10 shows ESD bags used to package devices sensitive to ESD. Available in various sizes, the bags have printed labels.

17.3 LIGHT SOURCES

The wattage rating should be checked for all bulbs in light fixtures, table lamps, and other light sources, to make sure they are the correct rating for the fixture. Bulbs must be replaced with another of the same wattage rating; bulbs' wattage rating must not be higher than recommended. If the correct wattage is unknown, check with the manufacturer of the fixture. Ensure that the bulbs are screwed in securely; loose bulbs may overheat. Different gas light sources (e.g., hydrogen, mercury, neon), as shown in Figure 17.11, are used in laboratories for light-loss measurements and for spectrometers and optical applications. These lamps operate at much higher temperatures than those of standard incandescent light bulbs. Never place a lamp where it could come in contact

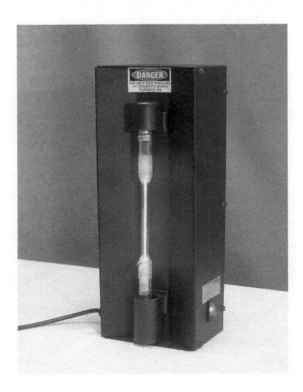

FIGURE 17.11 A mercury light source.

with any combustible materials or the skin. Be sure to turn the lamp off before leaving the laboratory for an extended period of time. Note that laser light sources have special provisions and, therefore, special precautions must be taken to operate them.

17.4 DEVICES AND EQUIPMENT

If a device or piece of equipment repeatedly blows a fuse or trips a circuit breaker, or if it has given you a shock, report the incident immediately to your supervisor/instructor. Unplug the device and remove it to have it repaired or replaced.

17.5 AUDIO–VISUAL AND COMPUTER PERIPHERALS

Audio–visual and computer equipment must be checked and kept in good working condition. Ask the technician to load the printer with paper and replace the toner. Report the faulty equipment to the technician for repair.

17.6 HANDLING OF FIBRE OPTIC CABLES

Fibre optic cables are made from a glass strand, covered with a polymer jacket. They are very thin and rigid, with sharp ends. Handle fibre optic cable with care during inspection, cleaning, and preparation of the fibre optic cable ends. Fibre optic cables should be cleaned using the cleanser recommended by the manufacturer. Follow the recommended procedure for each fibre optic cable type during cleaning, handling, assembling, packaging, and storage. When cleaving a fibre optic cable, the loose scrap material is hard to see and can be very dangerous. Dispose of loose scrap immediately in a properly designated container. Do not touch the end of a stripped fibre optic cable or a loose (scrap) piece of fibre. Fibre easily penetrates skin, and a fibre shard could break off. Do not rub your eyes when handling fibre optic cables; this would be extremely painful and requires immediate medical attention. Follow all safety procedures and regulations, and always wear the

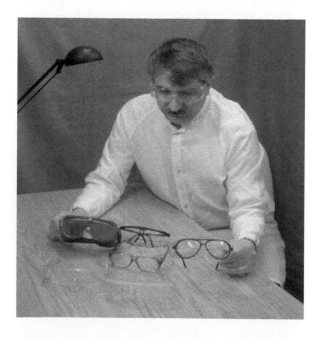

FIGURE 17.12 Safety goggles.

required personal protective safety equipment. Use safety goggles with side shields and wear protective rubber gloves or finger cots, when handling fibre optic cables. Figure 17.12 shows different types of safety goggles. Always treat fibre optic cables as a potential hazard. Never look directly at the fibre optic cable ends during fibre optic assembly and testing.

17.7 EPOXY ADHESIVES AND SEALANTS

Epoxy adhesives and sealants are essential components in the manufacturing of optical devices. There are different types and colours, depending upon the application. Epoxy adhesives come in several forms. One-part, two-part, and ultraviolet (UV) systems are the most common. A graded index (GRIN) lens can be glued to a beamsplitter with an epoxy. Sealant materials are used in the packaging of optical devices.

When using adhesives and sealant materials, be aware of their specifications. Specifications, applications, and handling procedures of these materials are found on the Material Safety Data Sheets (MSDS), which are available from the manufacturer or distributor. They may also be downloaded from a number of web sites. The adhesives and sealants are also very hazardous during storage, handling, and application. Prolonged or repeated exposure may cause eye or skin irritation. If contact does occur, wash the contact area immediately and seek medical help. Use safety goggles with side shields and wear protective rubber gloves or finger cots when handling adhesives and sealants. Follow all safety procedures and regulations, read the MSDS carefully, and wear the required personal protective safety equipment.

17.8 CLEANING OPTICAL COMPONENTS

Optical surfaces have to be clean and free of dust and other particles, which can range in size from tenths to hundreds of microns in diameter. Their comparative size means that they can cover a part of the optical surfaces, and thus degrade the reflection or transmission quality of the data transmission in telecommunication systems. There are many standard procedures for cleaning optical surfaces. Before starting any cleaning procedure, locate the following standard equipment:

1. Cleaning material (Denatured Ethanol)
2. Cotton swabs
3. Tissue
4. Safety goggles
5. Finger cots or rubber gloves
6. Compressed air
7. Disposal container
8. Microscope with a magnification range of about 50X
9. Infrared sensor card
10. Additional cleaning equipment:
    Ultrasonic bath
    Warm water and liquid soap
    Premoistened cleaning wipes
    Polymer film

Some optical components (e.g., lenses, mirrors, prisms, beamsplitters) have special coatings, such as antireflection coatings, that are sensitive to solvents, grease, liquid, and mechanical abrasion. Take extra care and choose appropriate cleaning liquid and swabs when cleaning optical components with these coatings. The following is the preferred cleaning procedure for optical components:

1. Wear rubber gloves and safety goggles.
2. Hold a lens or a mirror by the rim and a prism by the corners. Clean the optical component using a new dry or dampened swab with the recommended solvent. Rub the surfaces of the lens, using small circular movements, or one-directional movement on plane prism surfaces.
3. Blow away any remaining lint with compressed air. This step depends on the optical component size and surface conditions. Check the air quality from the compressor before using to clean optical components.

Some optical devices consisting of several optical components may not always be sealed completely. Therefore, use the recommended procedure to clean optical component surfaces without leaving any residue that could reduce the optical performance.

When cleaning any optical interface, disable all sources of power, such as the end of the ferrule on a fibre connector. Under no circumstances should you look into the end of an optical device in operation. Light from a laser device may not be visible, but it can seriously damage the human eye.

17.9 OPTIC/OPTICAL FIBRE DEVICES AND SYSTEMS

There has been a significant increase in the use of optic/optical fibre devices and systems. As optic/optical fibre devices become more common, it is important to understand the associated hazards. Optical devices typically use a laser as a light source. Not all lasers are created equal. They are classified based on their output wavelength and power. Since they operate over a wide range of wavelengths and power outputs, the hazards arising from their use vary substantially.

Lasers are classified into four classes. Laser sources conformant to Class 1 and Class 2 do not cause serous damage, but the use of eye protection should be taken into consideration. Class 3 and Class 4 lasers are powerful and can cause serious damage. Therefore, it is important to determine the class type of any optical equipment before working with it, assess the associated hazard, and comply with the safety requirements.

It is always a good practice to handle optical devices and measuring instruments with care. Normally, these devices and instruments are very expensive and sensitive, and they may present a potential hazard if not used properly. Follow the recommended procedures for each device or instrument to ensure proper handling during assembly, testing, packaging, and storage.

17.10 CLEANING CHEMICALS

Before the application of an epoxy or sealant, all surfaces should be treated using the recommended cleaning material. When using cleaning materials, be aware of appropriate precautions. Read all the information regarding cleaning materials in the MSDS. All types of cleaning materials are potentially hazardous; they may be flammable (even at low temperatures) and may pose other exposure risks. Use safety goggles with side shields and wear appropriate protective rubber gloves or finger cots. Follow all safety procedures and regulations. Use a ventilation hood when working with cleaning chemicals and epoxy adhesives, sealants, or any material producing fumes.

17.11 WARNING LABELS

There are various types of warning labels used in buildings, transportation, services, and industry to warn users about the level of danger ahead. Warning labels sometimes are called safety signs or safety messages. Safety signs clearly communicate by choosing the proper design and wording to suit safety needs. Standard signs, such as traffic warning signs and construction work labels, are available for general warnings.

Safety signs are divided into three general categories: danger, warning, and caution. They are also available in different sizes and colours, and with different graphics. Sometimes, a standard header can be used to create a new sign to suit a specific need. It is very important to use warning labels in laboratories to alert students to any source of danger. These dangers may come from devices, instruments, chemicals, lasers, sounds, vibrations, and biological hazards. Students should be introduced, in advance, to each source of danger in laboratories and be shown the required personal protective safety equipment. Everybody must remember to consider safety first.

17.12 LASER SAFETY

A laser beam is a parallel, narrow, coherent, and powerful light source. It is increasingly powerful when concentrated by a lens. It is a hazard to human eyes and skin even at very low power.

All lasers are classified based on their potential power. These classifications are from the American National Standards Institute (ANSI Standard Z136.1-1993) entitled American National Standard for Safe Use of Lasers, and Z136.3 (1996), American National Standard for Safe Use of Lasers in Health Care Facilities, the Canada Labor Code, and Occupational and Safety and Health Legislation (L-2-SOR/86-304).

Needing to be adhered to when using laser devices, these standards and codes are universally recognized as definitive documents for establishing an institution, such us a school, factory, or hospital. Their basic classification system has been adopted by every major national and international standards board, including the Center for Devices and Radiological Health (CDRH) in the U.S. Federal Laser Product Performance Standard, which governs the manufacture of lasers in the United States.

Lasers are typed into four clases, with some subclasses: Class 1, Class 2, Class 2a, Class 3a, Class 3b, and Class 4. Higher numbers reflect an increased potential to harm users. Figure 17.13 shows laser warning labels, which are required to identify hazard from laser light sources.

The following criteria are used to classify the hazard level of lasers:

1. Wavelength: If the laser is designed to emit multiple wavelengths, the classification is based on the most hazardous wavelength.
2. Continuous Wave: For continuous wave (CW) or repetitively pulsed lasers, the average power output (Watts) and limiting exposure time inherent in the design are considered.
3. Pulse: For pulsed lasers, the total energy per pulse (Joule), pulse duration, pulse repetition frequency, and emergent-beam radiant exposure are considered.

Details of the laser classifications are listed below:

Class 1 lasers are laser devices with very low output power (between 0.04 and 0.40 mW), and they operate in the lower part of the visible range (450 nm $< \lambda <$ 500 nm). These lasers are generally considered to be safe when viewed indirectly. Some examples of Class 1 laser devices include CD players, scanners, laser pointers, and small measurement equipment. Figure 17.14(a) shows the human eye, while Figure 17.14(b) shows an eye cross-section. Laser light in the visible range entering the human eye is focused on the retina and causes damage. The most likely effect of intercepting a laser beam with the eye is a thermal burn, which destroys the retinal tissue. Never view any Class 1 laser beam directly.

Class 2 lasers are devices with low output power ($<$1 mW of visible CW), and operate in the visible range (400 nm $< \lambda <$ 700 nm). This class of laser could cause eye damage, if the beam is directly viewed for a very short period of time (more than 0.25 s). Some examples of Class 2 lasers include classroom demonstration laser sources and laser-source devices for testing and telecommunications. Never view any Class 2 laser beam directly.

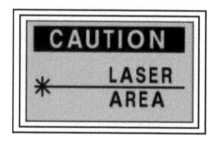

FIGURE 17.13 (**See colour insert following page 200.**) Laser warning labels.

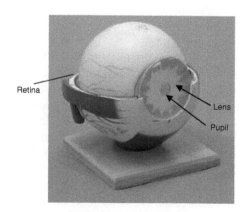

(a)   Eye ball

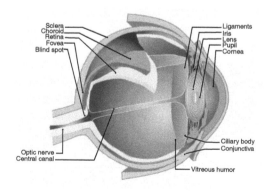

(b)   Eye cross-section

FIGURE 17.14 The human eye.

Class 2a lasers are low-output power devices, which are considered to be visible-light lasers. This class of laser causes injury only when viewed directly for more than 0.25 second. This class must be designed so that intentional viewing of the laser beam is not anticipated. A supermarket bar-code scanner is a typical example of a Class 2a laser device. Never view any Class 2a laser beam directly.

Class 3 lasers are divided into two subgroups (Class 3a and Class 3b lasers).

Class 3a lasers are intermediate power devices; they are allowed to exceed the output power limit of Class 2 lasers by no more than a factor of five, or have visible light power less than 5 mW. They are considered CW lasers. Often they will have an expanded beam diameter so that no more than 1 mW can enter a fully dilated pupil, which is 7 mm in diameter. Some examples of Class 3a laser devices are laser scanners, laser printers, and laser-source devices for testing and telecommunications. Direct viewing of a laser in this class could be hazardous to the eyes. Never view any Class 3a laser beam directly. Although the beam wavelength may not be visible to the human eye, it can cause damage to the eye and skin. Laser safety goggles for appropriate wavelength are required when working with this class.

Class 3b lasers are intermediate power devices; they output between 5 and 500 mW of CW, or else pulsed 10 J/cm^2 power. They are considered to be CW lasers. Scattered energy (diffuse reflection) is not considered hazardous in most situations, unless the laser source is operating near its upper power limit and the diffuse target is viewed at close range. Some examples of Class 3b lasers are laser-source devices for testing. Never view any Class 3b laser beam directly or indirectly (by viewing any reflection from the surrounding surfaces). The laser beam wavelength may not be visible to the human eye, but it causes damage to the eye and skin immediately, with no time to react. Laser safety goggles for the appropriate wavelength are required when working with this class of laser.

Class 4 lasers are high-power devices; they output more than 500 mW of CW, or else pulsed 10 J/cm^2 power. They are considered to be very high-power lasers. Some applications of Class 4 laser devices include the following: surgery, drilling, cutting, welding, and micromachining. For the use of Class 4 lasers, all types of reflections (whether direct, specular, or diffuse) are extremely hazardous to the eyes and skin. Class 4 laser devices can also be a fire hazard. Much greater control is required to ensure the safe operation of this type of laser device. Never view any Class 4 laser beam directly or indirectly (any reflection by surrounding surfaces). Be cautious of this type of laser. The laser beam wavelength may not be visible to the human eye, but it can immediately cause damage to the eye and skin, with no time to react. Laser safety goggles for the appropriate wavelength are required when working with this class of laser.

Always follow all safety procedures and regulations, and wear the required the appropriate personal protective safety equipment when using lasers. Never look directly or indirectly at a laser beam. Each institute should create appropriate safety procedure to guide students and staff toward the creation of a safe working environment. Each laser laboratory has to be controlled by a designated instructor/professor certified in laser safety. All laser safety requirements should be implemented in a laser laboratory. It is recommended to have an introduction course and workshop in laser safety for each laser classification.

Knowing the classification of a particular device and comparing the information in Table 17.1 will usually eliminate the need to measure laser radiation or perform complex analyses of hazard potential.

17.13 LASER SAFETY TIPS

1. Do not enter the Nominal Hazard Zone (NHZ). This zone is established according to the procedures described in ANSI Z136.1-1993. Enter this area accompanied by a designated instructor/professor certified in laser safety. Do not put any body part or clothing in the way of a laser beam.

TABLE 17.1
Institutional Programme Requirements

Class	Power	Class Control Measures	Medical Surveillance	Safety & Training Programme
1	No more than 0.04–0.40 mW	Not applicable	Not applicable	Not required
2	Less than 1 mW of visible, continuous wave light	Applicable	Not applicable	Recommended
2a	Less than 1 mW of visible, continuous wave light	Applicable	Not applicable	Recommended
3a	From 1 to 5 mW of continuous wave light	Applicable	Not applicable	Required
3b	From 5 to 500 mW of continuous wave light	Applicable	Applicable	Required
4	More than 500 mW of continuous wave light	Applicable	Applicable	Required

2. Notice and comply with the signs and labels (shown in Figure 17.13) posted on laboratory door, devices, and equipment.
3. Wear the recommended eyewear and other protective equipment. Use laser safety goggles when you are in a laser laboratory, or in the vicinity of one, as shown in Figure 17.15.
4. Comply with the laser safety controls in the facility.
5. Attend laser safety training and workshops.
6. Update laser safety training and workshops, as needed.
7. While assembling and operating laser devices, it is important to remember that laser beams can cause severe eye damage. Keep your head well above the horizontal plane of the laser beams at all times. Use white index cards to locate beamspots along the various optical paths.
8. When moving optical components, mirrors, or metal tools through the laser beams, the beam may reflect laser light momentarily at your lab partner or you. If there is a possibility of an accidental reflection during a particular step in an operation, then temporarily block or attenuate the laser beam until all optical components are in their proper place.

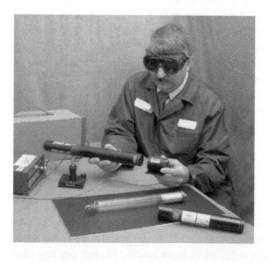

FIGURE 17.15 Wear laser safety goggles.

It is a good policy to be aware of any stray laser beam reflections, and to warn anybody of any danger. If you are unsure of how to proceed safely with a given step in the operation of the laser device, ask the professor/instructor for assistance.

17.14 INDOOR AIR QUALITY

Concerns with indoor air quality (IAQ) have increased since energy conservation measures were instituted in office buildings during the 1970s. These measures minimized the infiltration of outside air and contributed to the buildup of indoor air contaminants. IAQ generally refers to the quality of the air in a work environment. Other terms related to IAQ include indoor environmental quality (IEQ) and sick building syndrome. Complaints about IAQ range from simple complaints, such as the air smelling odd, to more complex situations, where the air quality causes illness and lost work time. It may not be easy to identify a single reason for IAQ complaints because of the number and variety of possible sources, causes, and varying individual sensitivities.

IAQ problems can be caused by ventilation system deficiencies, overcrowding, the presence of tobacco smoke, microbiological contamination, outside air pollutants, and off-gassing from materials in the building and mechanical equipment. Related problems may also include comfort problems caused by improper temperature and relative humidity conditions, poor lighting, and unacceptable noise levels, as well as adverse ergonomic conditions, and study-related psycho-social stressors. Typical symptoms may include headaches, unusual fatigue, itching or burning eyes, skin irritation, nasal congestion, dry or irritated throats, and nausea.

Ventilation is one of the most common engineering controls used to control emissions, exposures, and chemical hazards in the workplace. Other workplace environmental factors, including temperature, humidity, and odours, are also controlled with nonindustrial ventilation systems commonly known as heating, ventilating, and air-conditioning (HVAC) systems.

Management should have created guidelines for:

1. IAQ
2. Building air quality (BAQ)
3. Investigations, recommendations on sampling instrumentation and methods
4. Guidelines for management to prevent or alleviate
5. IAQ problems and take acute health effects of major indoor air contaminants.

Management should have an overview of:

1. Sources of indoor air pollution, and health problems
2. Ventilation, control, ventilation standards and building codes, and ventilation system problems
3. Solutions for air cleaners and resolving problems.

17.15 OTHER CONSIDERATIONS

These considerations apply to all students, staff, and management,

1. Laboratory injuries and illnesses are usually preventable by simply following safety precautions in school throughout the year.
2. Never overload circuits, power bars, or connectors.
3. Lead innovative and cooperative efforts to improve laboratory safety and health and the quality of student life.

FIGURE 17.16 A certified technical staff member.

4. Do not use or work with any device or equipment until it has been checked by qualified and authorized personnel in charge of the laboratory operation, as shown in Figure 17.16.
5. Everyone must wear personal protective equipment (e.g., safety goggles, protective gloves, ground connection, insulated tools) when working with electrical or laser equipment and chemicals.
6. Immediately report any damaged electrical or laser devices and equipment to the professor/instructor for immediate corrective action.
7. Staff should promote safety awareness among students.
8. Staff should teach safe work practices, at the beginning of each new laboratory session.
9. The NHZ should be established for each laser system.
10. Management should create and maintain a safe and healthy work, and study environment.
11. Management, staff, and students should understand the human and economic impact of poor safety and health in laboratories and classrooms.
12. Management should create a safety checklist, and maintenance and auditing programmes for each laboratory.
13. Eye protection should be worn at all times.
14. Eating, drinking, and smoking are not allowed in laboratories.
15. Unauthorized personnel should not be present in the laboratory or area, whether lasers are operating or not.
16. Laboratory coats must be worn when handling cleaning, corrosive, toxic, or flammable materials. Gloves should be worn when necessary, especially when handling corrosive and highly toxic materials.
17. Never work alone in a laboratory or workshop.
18. If a colleague is doing something dangerous, point the action out immediately and inform the supervisor.
19. Know where safety equipment (e.g., eyewash, shower, extinguishers, emergency exits, first aid kit) is located and how to use it.
20. Know where the MSDS and Workplace Hazardous Materials Information System (WHMIS) are located and how to use them.
21. Know where the emergency phones and alarms are located and how to use them.
22. Know how to clean up chemical spills using the appropriate agents.

23. Preplanned experiments and a properly organized work area can eliminate a lot of potential safety problems. Clean-up and decontamination must be routine parts of the experimental procedure for all students.
24. Wash your hands after handling chemicals and before leaving the laboratory.
25. Ensure the laboratory safety programme complements science.

FURTHER READING

Agrawal, G. P. and Dutta, N. K., *Semiconductor Lasers*, 2nd ed., Van Nostrand, New York, 1993.

Black, Eric, *An Introduction to Pound-Drever-Hall Laser Frequency Stabilization LIGO*, California Institute of Technology & Massachusetts Institute of Technology, California, Massachusetts, MA, U.S.A., 2000.

Canadian Health and Safety Legislation, Ecolog Canadian Health and Safety Legislation, Federal, Provincial, and Territorial Acts, Regulations, Guidelines, Codes, Objectives, Workers' Compensation, and WHMIS Legislation, 2000.

Charschan, S., *Lasers in Industry*, Van Nostrand, New York, 1972.

Cornsweet, T. N., *Visual Perception*, Academic Press, New York, 1970.

Davis, Christopher C., *Lasers and Electro-Optics, Fundamental and Engineering*, Cambridge University Press, New York, 1996.

Duarte, F. J. and Piper, J. A., Narrow-linewidth, high prf copper laser-pumped dye laser oscillators, *Appl. Opt.*, 23, 1391–1394, 1984.

Duarte, F. J., *Tunable Lasers Handbook*, Elsevier, Amsterdam, 1999.

Hood, D. C. and Finkelstein, M. A., Sensitivity to light handbook of perception and human performance, In *Sensory Processes and Perception*, Boff, K. R., Kaufman, L., and Thomas, J. P., Eds., Vol. 1, Wiley, Toronto, 1986.

Kuhn, K., *Laser Engineering*, Prentice Hall, Englewood Cliffs, NJ, 1998.

McComb, Gordon, *The laser Cookbook—88 Practical Projects*, McGraw-Hill, New York, 1988.

Nanni, C. A. and Alster, T. S., Laser-assisted hair removal: side effects of Q-switched Nd:YAG, long-pulsed ruby, and alexandrite lasers, *J. Am. Acad. Dermatology*, 2 (1), 165–171, 1999.

Nichols, Daniel R., *Physics for Technology with Applications in Industrial Control Electronics*, Prentice Hall, Englewood Cliffs, NJ, 2002.

Salah, B. E. A. and Teich, M. C., *Fundamentals of Photonics*, Wiley, New York, 1991.

SETON, Signs, labels, tags, and workplace safety, Catalog, 2006.

Tao, W. K. and Janis, R. R., *Mechanical and Electrical Systems in Buildings*, Prentice Hall, Englewood Cliffs, NJ, 2001.

Thompson, G. H. B., *Physics of Semiconductor Laser Device*, Wiley, Chichester, 1980.

Topping, A., Linge, C., Gault, D., Grobbelaar, A., and Sanders, R., A review of the ruby laser with reference to hair depilation, *Ann. Plast. Surg.*, 44, 668–674, 2000.

Venkat, Venkataramanan, *Introduction to Laser Safety*, Photonics Research, Ontario, Canada, 2002.

Yeh, Chai, *Handbook of Fiber Optics: Theory and Applications*, Academic Press, San Diego, 1990.

Glossary

Aberration   Distortion in an image produced by a lens or mirror caused by limitations inherent to some degree in all optical systems.

Absorption of Radiation   The loss of light energy as it passes through a material. Loss is converted to other energy forms, which is usually heat (rise in temperature). The absorption process is dependent on the wavelength of the light and on the absorbing material.

Active Medium   Collection of atoms or molecules which can be stimulated to a population inversion, and emit electromagnetic radiation in a stimulated emission.

Amplification   The process in which the electromagnetic radiation inside the active medium within the laser optical cavity increases by the process of stimulated emission.

Amplitude   The maximum value of a wave, measured from its equilibrium.

Angle of Incident (θ_i)   The angle formed by an incident ray and the normal line to the optical surface at the point of incident.

Angle of Reflection (θ_{refl})   The angle formed by a reflected ray and the normal line to the optical surface at the point of reflection.

Angle of Refraction (θ_{refr})   The angle formed by a refracted ray and the normal line to the optical surface at the point of penetration. The ray is refracted (bent) while passing from one transparent medium to another having different refractive indices.

Angstrom   A unit of measurement, equalling 10^{-10} m or 10^{-8} cm, usually used to express short wavelengths.

Aperture   An adjustable opening in an instrument (like a camera) that controls the amount of light that can enter.

Attenuation   The decrease in magnitude of the power of a signal in transmission media between points, normally measured in decibels (dB). Similarly, attenuation in a fibre optic cable is a measure of how much of the signal injected into an optical fibre cable actually reaches the other end, usually expressed in decibels per kilometre (dB/km).

Backreflection   Reflection of light in the direction opposite to that in which the light was originally propagating.

Beam   A bundle of light rays that are diverging, converging or parallel.

Beamsplitter   An optical device that divides incident light into two components (magnetic and electric).

bel (B)   Unit of intensity of sound, named after Alexander Graham Bell. The threshold of hearing is 0 B (10^{-12} W/m^2). The intensity is often measured in decibels (dB), which is one-tenth of a bel.

Birefringence   The fundamental principle by which polarization-maintaining fibre works.

Birefringent   A birefringent material has distinct indices of refraction. The separation of light beam, as it passes through a calcite crystal object, into two diverging beams, commonly known as ordinary and extraordinary beams.

Candle or Candela (cd)   A unit of luminous intensity.

Chromatic Dispersion   A pulse-broadening and, therefore, bandwidth-limiting phenomenon which occurs because different wavelengths of light travel at different velocities.

Coating   One or more of thin layers of optical material applied to an optical surface to reduce reflection, create a mirror surface, absorb light or protect the surface.

Coherence   A property of electromagnetic waves which are in phase in both time and space. Coherent light has monochromaticity and low beam divergence, and can be concentrated to high power densities. Coherence is needed for interference processes like holography.

Collimate   To cause light rays to become parallel.

Concave Mirror   Mirror that curves inward like a "cave".

Continuous Wave (CW)   It is the output of a laser, which is operated in a continuous rather than pulsed mode.

Control Area   It is an area in which the occupancy and activity of those present is subject to control and supervision for the purpose of protection from hazards like radiation, chemical, electrical, etc.

Conversion Efficiency   In an erbium-doped fibre amplifier (EDFA), it is the ratio between the amplified signal output power and the power input from the pump laser.

Convex Lens   Curved outward. A lens with a surface shaped like the exterior surface of a sphere.

Convex Mirror   Mirror that curves outward. The virtual images formed are smaller and closer to the mirror than the object.

Cornea   It is the transparent outer coat of the human eye, covering the iris and the crystalline lens. The cornea is the main refraction element of the eye.

Cube Beamsplitter   Cube beamsplitters consist of matched pairs of right angle prisms cemented together along their hypotenuses.

dB   Abbreviation for decibel. See bel.

Decibel (dB)   The standard unit used to express loss. Decibel is defined as 10 times the base-10 logarithm of the ratio of the output signal to the input signal power.

Detector   A light-sensitive device that produces electrical signals when illuminated.

Diffraction Grating   A grooved optical element that has been deformed to reflect or transmit light of many colours. It acts like a prism to produce a spectrum.

Diffuse Reflection   It is the change is the spatial distribution of a beam of radiation when the beam is reflected in many directions by a rough surface or by a medium.

Diffusion   It is the flow of particles of a given species from high to low concentration regions by virtue of their random motions.

Dispersion   The separation of a light beams into its various wavelength components. All transparent materials have different indices of refraction for light of different wavelengths.

Distortion   An aberration in an optical element, which causes straight lines in the object, which are off the axis, to appear as curved lines in the image.

Electron Volt (eV) Unit of energy   The amount of energy that the electron acquires while accelerating through a potential difference of 1 V. $1 \text{ eV} = 1.6 \times 10^{-19}$ J.

Extinction Ration   In a polarization-maintaining fibre, it is the ratio between the wanted and unwanted polarization states, expressed in decibels (dB). It is highly dependent upon operating environment.

Focal Length (f)   The distance between the second principal plane or equivalent refracting plane of a lens and the lens focal point when the lens is imaging an object at infinity. In a positive lens, the focal length is measured on the side of the lens opposite to the object. In a negative lens, the focal length is measured on the same side as the object.

Focal Point   The point on the optical axis of a lens where light rays from a distant object point will converge after being refracted by the lens.

Focus   The plane at which light rays from object points form a sharp image after being refracted by a lens.

Fresnel Reflection   The reflection which occurs between parallel optical surfaces or at the interface where two materials have different refractive indices.

Fresnel Reflection Loss   Loss of signal power due to Fresnel reflection.

GRIN Lenses   An acronym for gradient index lenses. They have a cylindrical shape with one end polished at an angle of 2, 6, 8 or 12° and the other end polished at an angle of 2 or 90°.

Hertz (Hz)   The unit used to measure frequency. 1 Hz equals one wave or cycle per second.

Homogeneous   A term used to describe any medium that is uniform in composition throughout.

Image   A likeness of an object formed by an optical element or system.

Image Distance   The distance between the equivalent refracting plane or second principal plane and the focal point measured on the optical axis.

Index of Refraction (n)   The ratio of the speed of light in a vacuum to the speed of light in a material.

Index-Matching Gel   A gel or fluid with a refractive index which is matched to the refractive index of two fibre optic cores. It fills in the air gap between the fibre cable ends and reduces the Fresnel reflection which occurs in the gap.

Intensity   The light energy per unit area.

Laser   An acronym for "light amplification by the stimulated emission of radiation". Lasers produce the coherent source of light for fibre optic telecommunication systems.

Laser Source   An instrument which produces monochromatic, coherent, collimated light.

1. Lens   One or more optical elements having flat or curved surfaces. If used to converge light rays, it is a positive lens; if used to diverge light rays, it is a negative lens. Usually made of optical glass, but may be moulded from transparent plastic. Lenses are sometimes made from a natural or synthetic crystalline substance to transmit very short wavelengths (UV) or very long wavelengths (IR).

Light   The form of electromagnetic radiation with a wavelength ranging from ~ 400 to ~ 700 nm. It generally travels in straight-line and exhibits the characteristics of both a wave and a particle.

Light Ray   The path of a single beam of light. In graphical ray tracing, a straight line represents the path along which the light travels.

Loss (dB)   Attenuation of the power of a signal when it travels through an optical component. Normally measured in decibels (dB).

Lumen (_lm_)   A SI unit of luminous flux. One lumen is the luminous flux emitted per unit solid angle by a light source having an intensity of one candela (cd).

Luminous Intensity (I)   Luminous intensity measures the brightness of a light source. The unit of measure is the candle or candela (cd).

Lux (_lx_)   A unit of luminance equal to one lumen per square metre.

Microbending   Tiny bends in a fibre optic cable, which allow light to leak out the core and introduce loss.

Micrometre (μm)   One-millionth of a metre.

Microscope   An optical instrument used to inspect small objects, which provide power magnification.

Mirror   An optical element with a smooth, highly polished surface (plane or curve) for reflecting light. The reflecting surface is produced by a thin coating of gold, silver or aluminium.

Monochromatic Light   Light is at one specific wavelength. The light out of a laser device is the monochromatic light.

Nanometre (nm)   One-billionth of a metre. The unit usually used in specifying the wavelength of light.

Normal Line   A reference line constructed perpendicular to an optical surface.

Object   The source figure in by an optical system.

Object Distance   The distance from the first principal plane of a lens to the object.

Objective Lens   A lens that focuses light coming from an object. The objective lenses are used in microscopes, telescopes, etc.

Optical Axis   A straight line formed by joining the foci of lens.

Optical Coatings   Coatings specifically made for optical components (lenses, prisms, etc.) in light sensitive devices. There are many types of coating materials. One coating helps to protect the optical components from scratches and wear. Some optical components are coated with antireflective (AR) layer(s) to reduce back reflection.

Optical Path   The sum of the optical distances along a specified light ray.

Optical Radiation   Ultraviolet, visible and infrared spectrum (0.35–1.4 μm) that falls in the region of transmittance of the human eye.

Optical Resonator   The mirrors (or reflectors) making up the laser cavity including the laser rod or tube. The mirrors reflect light back and forth to build up amplification.

Optical Surface   The reflecting or refracting surface of an optical element.

Phase   The position of a wave in its oscillation cycle.

Photon   A particle or packet of radiant electromagnetic energy representing a quanta of light.

Photonics   The field of science and engineering encompassing the physical phenomena and associated with the generation, transmission, manipulation, detection and utilization of light.

Polarization   Alignment of the electric and magnetic fields, which comprise an electromagnetic wave. If all light waves from a source have the same alignment, then the light is said to be polarized.

Population Inversion   An excited state of matter in which more atoms (or molecules) are in an upper state than in a lower one. This is a required situation for a laser action.

Prism   An optical element, which is used to change the direction and orientation of a light beam. A prism has polished faces, which are used to transmit and reflect light.

Pulsed Laser   Laser, which delivers energy in the form of a single or sequence of laser pulses.

Ray   Straight lines which represent the path of a light ray.

Rayleigh Scattering   The scattering of light, which results from small impurities in a material or composition.

Rectangular Beamsplitters   Three prisms carefully cemented together along their hypotenuses. Polarization beamsplitters are used in optical devices where the output components are required to exit from the opposite side to the input signal. It also produces a lateral displacement between the two output components.

Reflection   The change in the direction of a light ray when it bounces off of a reflecting surface.

Refraction   The bending of a light ray as it passes from one transparent medium to another of different refractive index.

Refractive Index (n)   The ratio of the speed of light in a vacuum to the speed of light in a specific material.

Right Angle Prism   A prism whose cross-section is a right angle triangle with two 45° interior angles. The prism faces, which are at right angles, are transmitting surfaces, while the hypotenuse face is a reflecting surface.

Scattering   Loss of light due to the presence of atoms in a transparent material.

Secular Reflection   Several rays of a beam of light incident on a smooth, mirror-like, reflecting surface where the reflected rays are all parallel to each other.

Sensor   A device which responds to the presence of energy.

Snell's law   Describes the path that a light ray takes as it goes from one optical medium to another. It is also called Law of Refraction.

Solar Flux   The amount of light from the Sun.

Solid Angle   The ratio of the area on the surface of a sphere to the square of the radius of that sphere. It is expressed in steradians (sr).

Spectra, Spectrum   Spectra is the plural of spectrum, which is a series of energies (like light) arranged according to wavelength or frequency. The electromagnetic spectrum is an array of radiation that is divided into a number of sub-portions, where the boundaries are only vaguely defined. They extend from the shortest cosmic rays, through gamma rays, X-rays, ultraviolet light, visible light, infrared radiation, microwave and all other wavelengths of radio energy.

Spectrograph   A spectroscope that measures wavelengths of light (spectra) and then displays the data as a graph. UVIS is an *imaging* spectrograph, which means it can also display the points of the graph as a picture (see Imaging Spectroscopy).

Spectrometer   A spectroscope equipped with the ability to measure wavelengths.

Spectroscope   A machine (instrument) for producing and observing spectra.

Spectroscopic Measurements   The measurements taken by a spectrograph.

Speed of Light   In vacuum, approximately 3×10^8 m/s.

Spontaneous Emission   Random emission of a photon by decay of an excited state to a lower level. Determined by the lifetime of the excited state.

Stimulated Emission   Coherent emission of radiation, stimulated by a photon absorbed by an atom (or molecule) in its excited state.

Transparent   The adjective used to describe a medium through which light can pass in a percentage.

Total Internal Reflection   Total internal reflection of light occurs when light rays in a high-index medium exceed the critical angle (to the normal to a surface). This is the principal theory for explaining how light travels in the core of a fibre optic cable.

Ultraviolet Light (UV)   (Extreme Ultraviolet and Far Ultraviolet)   A portion of the complete electromagnetic spectrum which has a shorter wavelength than visible light; roughly, with a wavelength interval from 100 to 4000 Å. Ultraviolet radiation from the Sun is responsible for many complex photochemical reactions like the formation of the ozone layer. Extreme and far ultraviolet wavelengths are different portions of the ultraviolet portion of the spectrum, with extreme being between 55.8 and 118 nm and far being between 110 and 190 nm.

Visible Light   Electromagnetic radiation, which is visible to the human eye. It has a wavelength range between 400 and 700 nm.

Wave   One complete cycle of a signal with a fixed period.

Wavelength (λ)   The period of a wave. Distance between successive crests, troughs or identical parts of a wave.

Appendix A: Details of the Devices, Components, Tools, and Parts

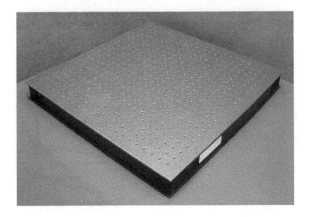

FIGURE A.1 2×2 ft breadboard.

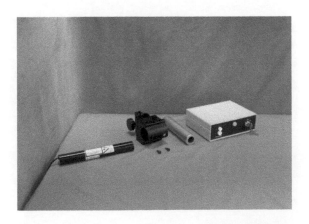

FIGURE A.2 HeNe laser source, laser power supply, and laser mount assembly.

FIGURE A.3 Lens and lens holder/positioner assembly.

FIGURE A.4 Laser sensors.

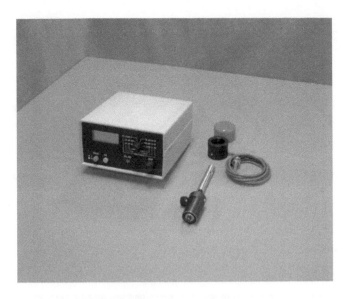

FIGURE A.5 Laser power meter with matching laser power detector.

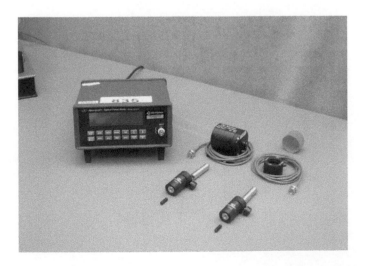

FIGURE A.6 Laser power meter and laser power detectors.

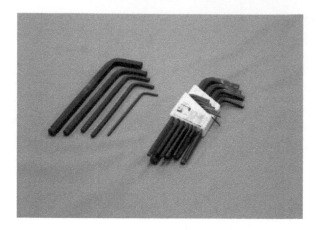

FIGURE A.7 Allen key set.

FIGURE A.8 Black/white card and cardholder.

FIGURE A.9 Rotation stage.

FIGURE A.10 Translation stage.

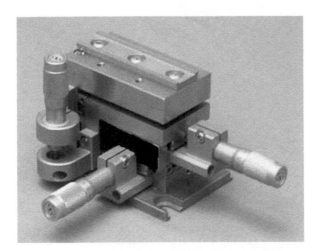

FIGURE A.11 XYZ translation stage.

FIGURE A.12 Multi-translation stage.

FIGURE A.13 Lab jack.

FIGURE A.14 HeNe laser clamp.

FIGURE A.15 Cube prism holder/positioner assembly.

FIGURE A.16 Convex lens and lens holder/positioner assembly.

FIGURE A.17 Prism and prism holder/positioner assembly.

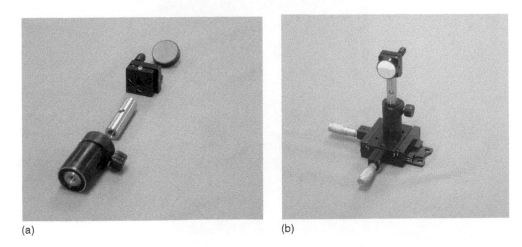

(a)   (b)

FIGURE A.18 (a) Mirror and mirror holder; (b) positioner assembly.

FIGURE A.19 Slide holder/positioner assembly.

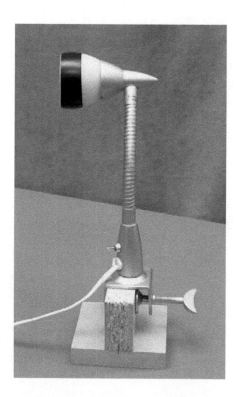

FIGURE A.20 Spot light source.

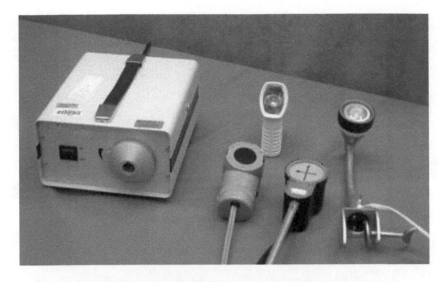

FIGURE A.21 Light sources.

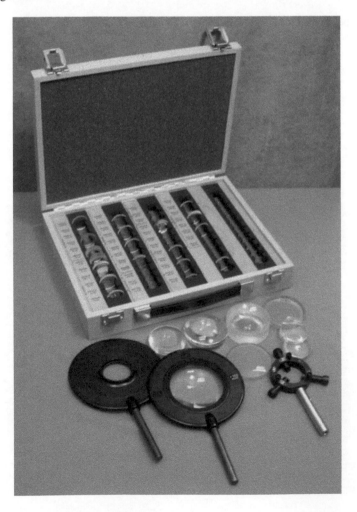

FIGURE A.22 Types of lenses.

FIGURE A.23 Types of prisms.

FIGURE A.24 Types of mirrors.

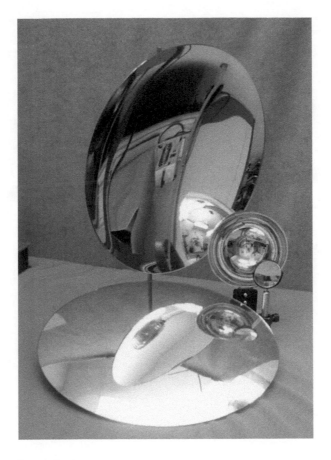

FIGURE A.25 Types of curved mirrors.

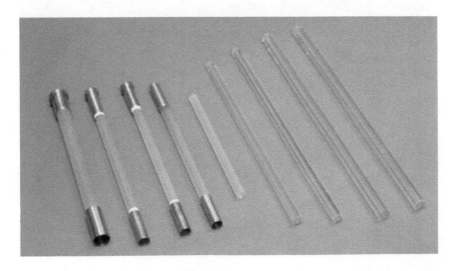

FIGURE A.26 Types of glass rods and tubes.

FIGURE A.27 Water tanks.

FIGURE A.28 Optical spectrum analyser.

Appendix B: International System of Units (SI)

International System of Units (SI) (It is also called metric system). The modern form of the metric system, which has been developed by international standards. The SI is constructed from seven base units for independent physical quantities. The following tables showing these values are included below and are currently used worldwide (Table B.1 through Table B.6).

TABLE B.1
The Common Metric SI Prefixes

Multiplication Factor	Prefix Name	Prefix	Symbol
1 000 000 000 000 000 000 000 000	10^{24}	Yotta	Y
1 000 000 000 000 000 000 000	10^{21}	Zetta	Z
1 000 000 000 000 000 000	10^{18}	Exa	E
1 000 000 000 000 000	10^{15}	Peta	P
1 000 000 000 000	10^{12}	Tera	T
1 000 000 000	10^{9}	Giga	G
1 000 000	10^{6}	Mega	M
1 000	10^{3}	Kilo	k
100	10^{2}	Hecto	h
10	10^{1}	Deka	da
0.1	10^{-1}	Deci	d
0.01	10^{-2}	Centi	c
0.001	10^{-3}	Milli	m
0.000 001	10^{-6}	Micro	μ
0.000 000 001	10^{-9}	Nano	n
0.000 000 000 001	10^{-12}	Pico	p
0.000 000 000 000 001	10^{-15}	Femto	f
0.000 000 000 000 000 001	10^{-18}	Atto	a
0.000 000 000 000 000 000 001	10^{-21}	Zepto	z
0.000 000 000 000 000 000 000 001	10^{-24}	Yecto	y

TABLE B.2
Base Units

Quantity	Unit Name	Unit Symbol
Length	Metre	m
Mass	Kilogram	kg
Time	Second	s
Electric current	Ampere	A
Thermodynamic temperature	Kelvin	K
Amount of substance	Mole	mol
Luminous intensity	Candela	cd

TABLE B.3
SI Derived Units

Quantity	Unit Name	Unit Symbol	Expression in Terms of Other SI Units
Absorbed dose, specific energy imparted, kerma, absorbed dose index	Gray	Gy	J/kg
Activity (of a radionuclide)	Becquerel	Bq	1/s
Celsius temperature	Degree Celsius	°C	K
Dose equivalent	Sievert	Sv	J/kg
Electric capacitance	Farad	F	C/V
Electric charge, quantity of electricity	Coulomb	C	A s
Electric conductance	Siemens	S	A/V
Electric inductance	Henry	H	Wb/A
Electric potential, potential difference, electromotive force	Volt	V	W/A
Electric resistance	Ohm	Ω	V/A
Energy, work, quantity of heat	Joule	J	N m
Force	Newton	N	$Kg\ m/s^2$
Frequency (of a periodic phenomenon)	Hertz	Hz	1/s
Illuminance	Lux	Lx	lm/m^2
Luminous flux	Lumen	Lm	cd sr
Magnetic flux	Weber	Wb	V s
Magnetic flux density	Tesla	T	Wb/m^2
Plane angle	Radian	Rad	m/m
Power, radiant flux	Watt	W	J/s
Pressure, stress	Pascal	Pa	N/m^2
Solid angle	Steradian	Sr	m^2/m^2

Derived units are formed by combining base units and other derived units according to the algebraic relations linking the corresponding quantities. The symbols for derived units are obtained by means of the mathematical signs for multiplication, division, and use of exponents. Some derived SI units were given special names and symbols, as listed in this.

TABLE B.4
Conversion Factors from U.S. Customary Units to Metric Units

To Convert from	Multiply by	To Find
Inches	25.4	Millimetres
	2.54	Centimetres
Feet	30.48	Centimetres
Yards	0.91	Metres
Miles	1.61	Kilometres
Teaspoons	4.93	Millilitres
Tablespoons	14.79	Millilitres
Fluid ounces	29.57	Millilitres
Cups	0.24	Litres
Pints	0.47	Litres
Quarts	0.95	Litres
Gallons	3.79	Litres
Cubic feet	0.028	Cubic metres
Cubic yards	0.76	Cubic metres
Ounces	28.35	Grams
Pounds	0.45	Kilograms
Short tons (2000 lbs)	0.91	Metric tons
Square inches	6.45	Square centimetres
Square feet	0.09	Square metres
Square yards	0.84	Square metres
Square miles	2.6	Square kilometres
Acres	0.4	Hectares

TABLE B.5
Conversion Factors from Metric Units to U.S. Customary Units

To Convert from	Multiply by	To Find
Millimeters	0.04	Inches
Centimeters	0.39	Inches
Meters	3.28	Feet
	1.09	Yards
Kilometers	0.62	Miles
Milliliters	0.2	Teaspoons
Liters	0.06	Tablespoons
	0.03	Fluid ounces
	1.06	Quarts
	0.26	Gallons
	4.23	Cups
	2.12	Pints
Cubic meters	35.32	Cubic feet
	1.35	Cubic yards
Grams	0.035	Ounces

(continued)

Table B.5   *(Continued)*

To Convert from	Multiply by	To Find
Kilograms	2.21	Pounds
Metric ton (1000 kg)	1.1	Short ton
Square centimeters	0.16	Square inches
Square meters	1.2	Square yards
Square kilometers	0.39	Square miles
Hectares	2.47	Acres

Temperature conversion between Celsius and Fahrenheit $°C = (F - 32)/1.8$, $°F = (°C \times 1.8) + 32$

TABLE B.6
The Common Natural Temperatures

Condition	Fahrenheit (°)	Celsius (°)
Boiling point of water	212	100
A very hot day	104	40
Normal body temperature	98.6	37
A warm day	86	30
A mild day	68	20
A cool day	50	10
Freezing point of water	32	0
Lowest temperature Fahrenheit could obtain by mixing salt and ice	0	−17.8

Appendix C: Lighting Lamps

APPENDIX C.1 INCANDESCENT LIGHT LAMPS

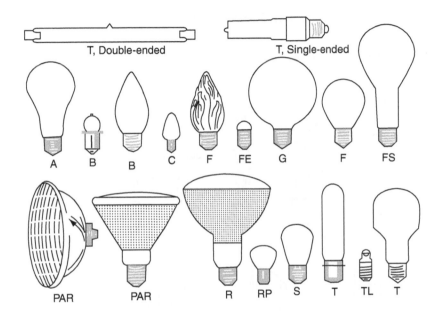

FIGURE C.1 Shape codes.

TABLE C.1
Shape Code and Application

	Shape Code	Application
A	Arbitrary (standard)	Universal use for home lighting
B	Bullet	Decorative
C	One shape	Used mostly for small appliances and indicator lamps
ER	Elliptical reflector	For substitution of incandescent R lamps
F	Flame	Decorative interior lighting
G	Globe	Ornamental lighting and some floodlights
P	Pear	Standard for street-railway and locomotive headlights
PAR	Parabolic aluminized	Used in spotlights and floodlights
S	Straight	Lower wattage lamps-sign and decorative
T	Tubular	Showcase and appliance lighting

APPENDIX C.1.1 LAMP DESIGNATION

For Example: 60A19, it means:
60: Wattage (60 W)
A: Bulb shape
19: Maximum bulb diameter, in eighths of an in.

APPENDIX C.2 TUNGSTEN HALOGEN LAMPS

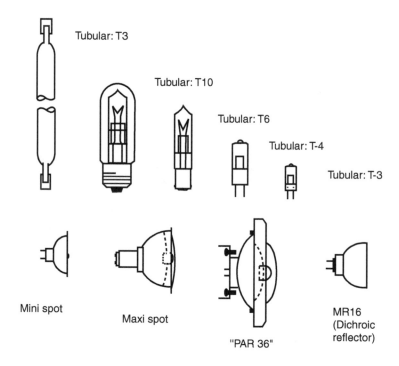

FIGURE C.2 Shape codes.

TABLE C.2
Shape Code and Type

Shape Code	Type
Tubular: T3	Line voltage tungsten halogen lamp-double ended
Tubular: T10	Line voltage tungsten halogen lamp-single ended
Tubular: T6	Line voltage tungsten halogen lamp-single ended
Tubular: T-4	Line voltage tungsten halogen lamp-without reflector
Tubular: T-3	Line voltage tungsten halogen lamp-without reflector
Maxi spot	Line voltage tungsten halogen lamp-with reflector
Mini spot	Line voltage tungsten halogen lamp-with reflector
PAR 36	Line voltage tungsten halogen lamp-PAR 36 reflector
MR 16	Line voltage tungsten halogen lamp-MR 16 reflector

APPENDIX C.3 FLUORESCENT LIGHT LAMPS

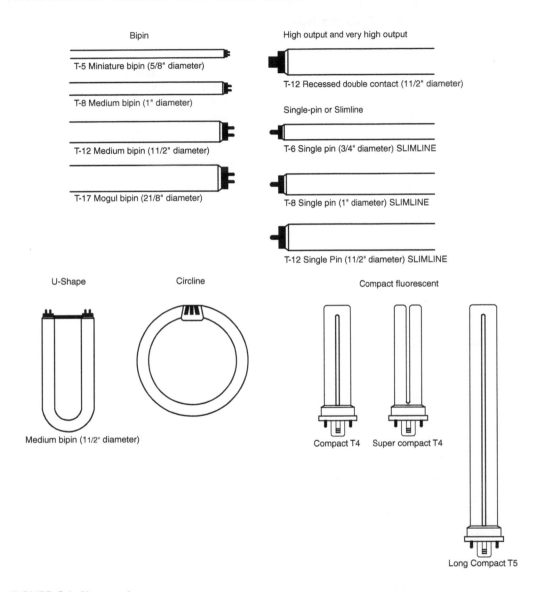

FIGURE C.3 Shape codes.

For Example: F40T12/CW, it means:

F: Fluorescent lamp

40: Wattage (40 W)

T: Tubular bulb shape

12: Maximum tube diameter — in eighths of an in. $(12/8 = 1\frac{1}{2}$ in.)

CW: Cool White Colour

TABLE C.3
Colour Code and Type

Colour Code	Type
C50	Chroma 50 (5000 K, CRI 90+)
C75	Chroma 75 (7500 K, CRI 90+)
CW	Cool white
CWX	Cool white deluxe
D	Daylight
LW	Lite white
LWX	Lite white deluxe
N	Normal
RW	Regal white
SP	Spectrum series
SPX	Spectrum series deluxe
WW	Warm white
WWX	Warm white deluxe
Deluxe	Means better CRI

APPENDIX C.4 MERCURY DISCHARGE LAMPS

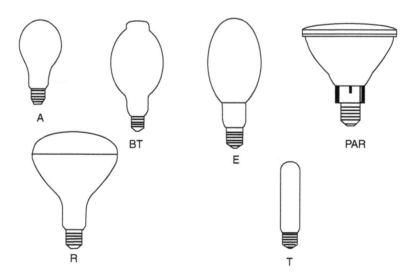

FIGURE C.4 Shape codes.

TABLE C.4
Shape Code and Type

Shape Code	Type
A	Arbitrary
BT	Bulged-tubular
E	Elliptical
PAR	Parabolic aluminized reflector
R	Reflector
T	Tubular

APPENDIX C.5 METAL HALIDE LAMPS

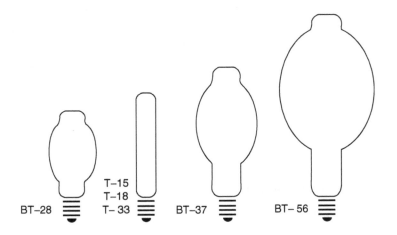

FIGURE C.5 Shape codes.

TABLE C.5
Shape Code and Type

Shape Code	Type
BT	Bulged-tubular
T	Tubular
Numbers	Indicate maximum diameter in eighths of an in.

APPENDIX C.6 HIGH-PRESSURE SODIUM LAMPS

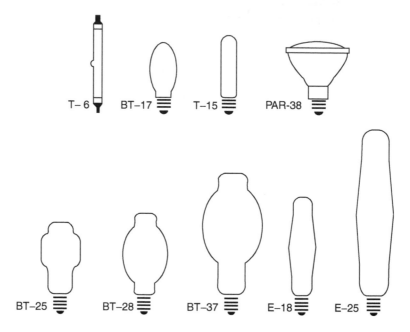

FIGURE C.6 Shape codes.

TABLE C.6
Shape Code and Type

Shape Code	Type
B	Bullet
BT	Bulged-tubular
E	Elliptical
PAR	Parabolic aluminized reflector
T	Tubular
Numbers	Indicate maximum diameter in eighths of an inch

APPENDIX C.7 COMPACT FLUORESCENT LAMPS

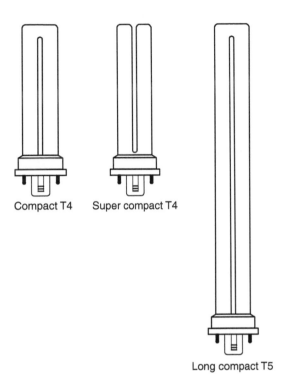

Compact T4   Super compact T4

Long compact T5

FIGURE C.7 Compact fluorescent lamps.

Index